METHODS IN MOLECULAR BIOLOGY™

Series Editor
John M. Walker
School of Life Sciences
University of Hertfordshire
Hatfield, Hertfordshire, AL10 9AB, UK

For other titles published in this series, go to
www.springer.com/series/7651

Clinical Applications of Mass Spectrometry

Methods and Protocols

Edited by

Uttam Garg

Children's Mercy Hospitals and Clinics, Kansas City, MO, USA

Catherine A. Hammett-Stabler

University of North Carolina, Chapel Hill, NC, USA

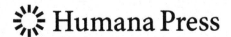

Editors
Uttam Garg
Department of Pathology & Laboratory
  Medicine
Children's Mercy Hospitals and Clinics
2401 Gillham Road
Kansas City, MO 64108
USA
ugarg@cmh.edu

Catherine A. Hammett-Stabler
Department of Pathology & Laboratory
  Medicine
University of North Carolina
Chapel Hill, NC 27599-7525
USA
cstabler@unch.unc.edu

ISSN 1064-3745   e-ISSN 1940-6029
ISBN 978-1-60761-458-6   e-ISBN 978-1-60761-459-3
DOI 10.1007/978-1-60761-459-3

Library of Congress Control Number: 2009939452

Printed on acid-free paper

springer.com

Dedicated to Jyotsna Garg and Thomas Stabler for their patience and understanding throughout this process.

Preface

No longer considered an esoteric technique, mass spectrometry has entered the clinical laboratory where it is being used for a wide range of applications. Methods developed using this technique offer a level of specificity and sensitivity unrealized by spectrophotometric- and immunoassay-based methods. In fact, the technique is considered essential coupled to either gas chromatography or liquid chromatography for the measurement of many clinical analytes.

This volume, *Clinical Applications of Mass Spectrometry*, is intended to provide detailed step-by-step procedures for the analysis of number of analytes of clinical importance. Each chapter provides the brief introduction of the analyte followed by a detailed analytical protocol. The procedures are easy to follow and should be easily reproducible in the laboratory already using mass spectrometry and wanting to introduce new analyses as well as those laboratories considering adding the technology. We do not offer a detailed overview of mass spectrometry since there are many excellent volumes and tutorials on MS theory and principles, but we have included a chapter that discusses some of the basic laboratory practices to remember when introducing the technique into routine use. The analytes for which mass spectrometry methods are described include drugs, hormones, and metabolic compounds spanning the disciplines of toxicology, therapeutic drug monitoring, endocrinology, and pediatric metabolism. Demonstrating the versatility of the technique, we have been able to provide multiple applications for some analytes so that the reader has his/her choice of an LC-MS- or a GC-MS-based method.

We thank all of our colleagues who contributed the methods. We realize that the development of these represent thousands of hours of work. And we hope that those who use this book find it useful.

Uttam Garg
Catherine A. Hammett-Stabler

Contents

Contributors

RYAN ALLENBRAND • *Department of Pathology and Laboratory Medicine, Children's Mercy Hospitals and Clinics, Kansas City, MO, USA*

SUSAN AUCOIN • *Biochemical Genetics Laboratory, Department of Genetics, Children's Hospital of Eastern Ontario, Ottawa, ON, Canada*

MICHAEL J. BENNETT • *Department of Pathology and Laboratory Medicine, University of Pennsylvania and Children's Hospital of Philadelphia, Philadelphia, PA, USA*

MATTHEW W. BJERGUM • *Toxicology and Drug Monitoring Laboratory, Department of Laboratory Medicine and Pathology, Mayo Clinic, Rochester, MN, USA*

BRUNO CASETTA • *Applied Biosystems, Monza, Italy*

JIE CHEN • *Department of Pathology and Laboratory Medicine, The Children's Hospital of Philadelphia, Philadelphia, PA, USA*

CHRIS W. CHRONISTER • *Department of Pathology, Immunology, and Laboratory Medicine, University of Florida College of Medicine, Gainesville, FL, USA*

WEI-LIEN CHUANG • *Genzyme Corporation, Framingham, MA, USA*

KIM COX • *Regional Forensic Science Center, Wichita, KS, USA*

TIM DAHN • *AIT Laboratories, Indianapolis, IN, USA*

AMITAVA DASGUPTA • *Department of Pathology and Laboratory Medicine, University of Texas-Houston Medical School, Houston, TX, USA*

DENNIS J. DIETZEN • *Departments of Pediatrics and Pathology and Immunology, Washington University School of Medicine, St. Louis, MO, USA; St. Louis Children's Hospital, St. Louis, MO, USA*

KERI J. DONALDSON • *Department of Pathology and Laboratory Medicine, University of Pennsylvania, Philadelphia, PA, USA*

CAROLE S. ELBIN • *Genzyme Corporation, Framingham, MA, USA*

CLAUDINE FASCHING • *Pathology and Laboratory Medicine Service, Veterans Affairs Medical Center, Minneapolis, MN, USA*

ROBERT L. FITZGERALD • *Veterans Affairs, San Diego Healthcare System, San Diego, CA, USA; University of California-San Diego, La Jolla, CA, USA*

STEVEN W. FLEMING • *Department of Pathology and Laboratory Medicine, Children's Mercy Hospitals and Clinics, Kansas City, MO, USA*

C. CLINTON FRAZEE III • *Department of Pathology and Laboratory Medicine, Children's Mercy Hospitals and Clinics, Kansas City, MO, USA*

ANGELA M. FERGUSON • *Department of Pathology and Laboratory Medicine, Children's Mercy Hospitals and Clinics, Kansas City, MO, USA*

MEGAN GAMBLE • *Regional Forensic Science Center, Wichita, KS, USA*

UTTAM GARG • *Department of Pathology and Laboratory Medicine, Children's Mercy Hospitals and Clinics, Kansas City, MO, USA*

DIANE CIUFFETTI GEIS • *Core Laboratory, McLendon Clinical Laboratories, UNC Hospitals, Chapel Hill, NC, USA*

MICHAEL GELB • *Departments of Chemistry and Biochemistry, University of Washington, Seattle, WA, USA*

BRUCE A. GOLDBERGER • *Department of Pathology, Immunology, and Laboratory Medicine, University of Florida College of Medicine, Gainesville, FL, USA*

TERRANCE L. GRIFFIN • *Veterans Affairs, San Diego Healthcare System, San Diego, CA, USA*

JOSH GUNN • *AIT Laboratories, Indianapolis, IN, USA*

CATHERINE A. HAMMETT-STABLER • *Department of Pathology and Laboratory Medicine, University of North Carolina at Chapel Hill, Chapel Hill, NC, USA*

LYDIA A. HARRYMAN • *Regional Forensic Science Center, Wichita, KS, USA*

DAVID A. HEROLD • *Veterans Affairs, San Diego Healthcare System, San Diego, CA, USA; University of California-San Diego, La Jolla, CA, USA*

GERRY HUBER • *Department of Pathology and Laboratory Medicine, Children's Mercy Hospitals and Clinics, Kansas City, MO, USA*

LEONARD L. JOHNSON • *Department of Pathology and Laboratory Medicine, Children's Mercy Hospital and Clinics, Kansas City, MO, USA*

PATRICIA M. JONES • *Department of Pathology, University of Texas Southwestern Medical Center and Children's Medical Center, Dallas, TX, USA*

MARK KELLOGG • *Department of Laboratory Medicine, Children's Hospital Boston, Boston, MA, USA*

JOAN M. KEUTZER • *Genzyme Corporation, Framingham, MA, USA*

MICHAEL KISCOAN • *Department of Pathology and Laboratory Medicine, Children's Mercy Hospitals and Clinics, Kansas City, MO, USA*

JEFF M. KNOBLAUCH • *Department of Pathology and Laboratory Medicine, Children's Mercy Hospitals and Clinics, Kansas City, MO, USA*

MAGDALENA KORECKA • *Department of Pathology and Laboratory Medicine, University of Pennsylvania, Philadelphia, PA, USA*

SCOTT KRIGER • *AIT Laboratories, Indianapolis, IN, USA*

VIKRAM S. KUMAR • *Department of Laboratory Medicine, Children's Hospital Boston, Boston, MA, USA*

MARK M. KUSHNIR • *ARUP Institute for Clinical and Experimental Pathology, Salt Lake City, UT, USA*

LORALIE J. LANGMAN • *Toxicology and Drug Monitoring Laboratory, Department of Laboratory Medicine and Pathology, Mayo Clinic, Rochester, MN, USA*

TERENCE LAW • *Department of Laboratory Medicine, Children's Hospital Boston, Boston, MA, USA*

NATHALIE LEPAGE • *Biochemical Genetics Laboratory, Department of Genetics, Children's Hospital of Eastern Ontario, Ottawa, ON, Canada; Department of Pathology and Laboratory Medicine, University of Ottawa, Ottawa, ON, Canada*

STANLEY F. LO • *Department of Pathology, Medical College of Wisconsin, Milwaukee, WI, USA*

STEPHANIE J. MARIN • *ARUP Institute for Clinical and Experimental Pathology, Salt Lake City, UT, USA*

GWENDOLYN A. MCMILLIN • *ARUP Laboratories, Inc., Salt Lake City, UT, USA; Department of Pathology, University of Utah School of Medicine, Salt Lake City, UT, USA*

A. Wayne Meikle • *ARUP Institute for Clinical and Experimental Pathology, Salt Lake City, UT, USA; Departments of Pathology and Medicine, University of Utah, Salt Lake City, UT, USA*

Michele L. Merves • *Department of Pathology, Immunology, and Laboratory Medicine, University of Florida College of Medicine, Gainesville, FL, USA*

Ross J. Molinaro • *Department of Pathology and Laboratory Medicine, Emory University School of Medicine and Emory Healthcare Systems, Atlanta, GA, USA*

Kathleen Neville • *Section of Hematology and Oncology, Children's Mercy Hospitals and Clinics, Kansas City, MO, USA*

Melissa C. Norton • *Regional Forensic Science Center, Wichita, KS, USA*

William E. O'Brien • *Department of Molecular and Human Genetics, Baylor College of Medicine, Houston, TX, USA*

Christine Papadea • *Clinical Neurobiology Laboratory, Institute of Psychiatry, Medical University of South Carolina, Charleston, SC, USA*

Judy Peat • *Department of Pathology and Laboratory Medicine, Children's Mercy Hospitals and Clinics, Kansas City, MO, USA*

William J. Rhead • *Department of Pediatrics, Medical College of Wisconsin and Children's Hospital of Wisconsin, Milwaukee, WI, USA*

James C. Ritchie • *Department of Pathology and Laboratory Medicine, Emory University School of Medicine and Emory Healthcare Systems, Atlanta, GA, USA*

Alan L. Rockwood • *ARUP Institute for Clinical and Experimental Pathology, Salt Lake City, UT, USA; Department of Pathology, University of Utah, Salt Lake City, UT, USA*

Timothy P. Rohrig • *Regional Forensic Science Center, Wichita, KS, USA*

David K. Scott • *Department of Pathology and Laboratory Medicine, Children's Mercy Hospitals and Clinics, Kansas City, MO, USA*

Ronald Scott • *Department of Pediatrics, University of Washington, Seattle, WA, USA*

Leslie M. Shaw • *Department of Pathology and Laboratory Medicine, University of Pennsylvania, Philadelphia, PA, USA*

Jasbir Singh • *Pathology and Laboratory Medicine Service, Veterans Affairs Medical Center, Minneapolis, MN, USA*

Ravinder J. Singh • *Department of Laboratory Medicine and Pathology, Mayo Clinic, Rochester, MN, USA*

Laurie D. Smith • *Section of Clinical Genetics, Dysmorphology, and Metabolism, Children's Mercy Hospitals and Clinics, Kansas City, MO, USA*

Christine L.H. Snozek • *Toxicology and Drug Monitoring Laboratory, Department of Laboratory Medicine and Pathology, Mayo Clinic, Rochester, MN, USA*

Marion L. Snyder • *Department of Pathology and Laboratory Medicine, Emory University School of Medicine and Emory Healthcare Systems, Atlanta, GA, USA*

Douglas F. Stickle • *Department of Pathology, University of Nebraska Medical Center, Omaha, NE, USA*

Judy Stone • *Clinical Laboratories, San Francisco General Hospital, San Francisco, CA, USA*

Qin Sun • *Department of Molecular and Human Genetics, Baylor College of Medicine, Houston, TX, USA*

AMANDA SUTTON • *Department of Pathology and Laboratory Medicine, Children's Mercy Hospitals and Clinics, Kansas City, MO, USA*

ANDREA R. TERRELL • *AIT Laboratories, Indianapolis, IN, USA*

SARAH THOMAS • *Department of Pathology and Laboratory Medicine, Children's Mercy Hospitals and Clinics, Kansas City, MO, USA*

DENISE M. TIMKO • *Department of Pathology, University of Nebraska Medical Center, Omaha, NE, USA*

FRANTISEK TURECEK • *Department of Chemistry, University of Washington, Seattle, WA, USA*

FRANCIS M. URRY • *ARUP Institute for Clinical and Experimental Pathology, Salt Lake City, UT, USA; Department of Pathology, University of Utah, Salt Lake City, UT, USA*

ANNETTE L. WEINDEL • *St. Louis Children's Hospital, St. Louis, MO, USA*

RUTH E. WINECKER • *North Carolina Office of the Chief Medical Examiner and Department of Pathology and Laboratory Medicine University of North Carolina School of Medicine, Chapel Hill, NC, USA*

VELTA YOUNG • *Department of Pathology, Children's Hospital of Wisconsin, Milwaukee, WI, USA*

BINGFANG YUE • *ARUP Institute for Clinical and Experimental Pathology, Salt Lake City, UT, USA*

X. KATE ZHANG • *Genzyme Corporation, Framingham, MA, USA*

Chapter 1

The Evolution of Mass Spectrometry in the Clinical Laboratory

Catherine A. Hammett-Stabler and Uttam Garg

Abstract

Clinical laboratories around the world are recognizing the power of mass spectrometry. This technique, especially when coupled to gas chromatography or liquid chromatography, is revolutionizing the analysis of many analytes. Unlike many other techniques which measure one analyte at a time, these techniques can measure multiple analytes (>40) at one time. In recent years the scope of testing using these techniques has expanded from toxicological purposes to newborn screening to hormones, proteins, and enzymes. It is not uncommon any more to see mass spectrometry being used in the routine clinical laboratories.

Key words: Clinical laboratory testing/mass spectrometry, liquid chromatography/mass spectrometry, gas chromatography/mass spectrometry, tandem mass spectrometry, endocrinology, toxicology, therapeutic drug monitoring, pediatric metabolism, newborn screening.

1. Introduction

In recent years a quiet change has been developing in clinical laboratories throughout the world. It is increasingly common to walk into a clinical laboratory and see technologists using a mass spectrometer coupled to either a liquid chromatograph (LC-MS) or gas chromatograph (GC-MS). Once considered too expensive and cumbersome to use except in forensic and reference settings, such systems are now found alongside automated platforms and used routinely to generate data for patient care.

Although mass spectrometry has long been recognized as an important and powerful analytical tool, there were a number of challenges that had to be overcome to quell the skepticism and doubts that it belonged in the clinical setting for more than a few special applications. GC-MS was introduced into the clinical

U. Garg, C.A. Hammett-Stabler (eds.), *Clinical Applications of Mass Spectrometry*, Methods in Molecular Biology 603, DOI 10.1007/978-1-60761-459-3_1, © Humana Press, a part of Springer Science+Business Media, LLC 2010

laboratory more than two decades before LC-MS. But the advent of relatively small, inexpensive, and user-friendly LC/MS and LC tandem MS (LC/MS-MS) systems along with advances in column chemistries opened the door to many chemical analyses previously unattainable using GC/MS (1). These systems began to remove much of the stigma that mass spectrometry required special personnel beyond the reach of the average clinical laboratory (2). As the versatility of these systems grew, clinical laboratories recognized the potential benefit they offered in terms of facilitating a rapid response to changing clinical needs. Much needed assays could be developed in-house instead of waiting for a manufacturer and the regulatory process. Admittedly this does require some effort, is a new concept for some, and is more than a little intimidating at times. But there is little argument that these techniques provide a higher level of sensitivity and specificity for many analytes compared to other analytical techniques and that patient care has benefited from its use. This is especially true for metabolic analyses, drug and hormone measurements (3–10). Also, these techniques can be cost-effective as they can measure multiple analytes at the same time. Finally, improvements in automation and software have come together with vendor support networks in such a way that clinical laboratories are able to deal with staffing and service issues. A movement that started tentatively is now gaining speed. The fact is that in the first 3 months of 2009, over a thousand papers with mass spectrometry, methods, and clinical laboratories in the MESH headings have been published. Methods are now reported for a range of analytes that were once considered unlikely candidates for MS, but now, thanks to technological improvements, only one's imagination seems to be the limitation.

2. Introducing Mass Spectrometry into the Clinical Laboratory

Mass spectrometry is based on the principle that charged particles moving through a magnetic field can be separated from one another by their mass-to-charge ratios (m/z). As such a mass spectrometer can be divided into four basic elements as seen in **Fig. 1.1.** There must be a means of introducing the sample into the system. Most commonly, this occurs after the sample has undergone an initial separation using either GC or LC, though direct insertion techniques are gaining in popularity. The molecules of the sample must be ionized and the ions must be counted. There are a number of different techniques used to accomplish these tasks. The most common ionization techniques are electron impact, chemical, and electrospray. Finally, these data must be

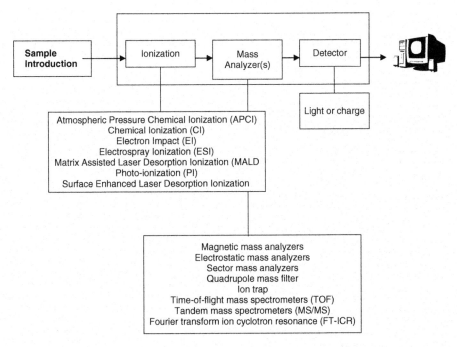

Fig. 1.1. Mass spectrometry options at a glance.

transformed into useful information in the form of a mass spectrum. A detailed description of mass spectrometry, gas chromatography, and liquid chromatography is well beyond the scope of this volume. The interested reader is advised to turn to any of the excellent books, tutorials, and Internet sites devoted to these topics.

Given the choices of instrumentation available, it is easy to see why selection of a GC/MS or LC/MS system can be intimidating. Which mode of ionization is best suited for my analysis? Which mass analyzer is best? Which software package do I need? These are certainly important questions, but they tend to take precedence so that we forget to consider more basic questions: Who will perform the testing? Are there additional quality assurance and regulatory requirements? How will testing need to be scheduled to serve clinical needs?

We have found that the initial fear can be reduced by taking the perspective that all instruments are simply tools to be used by qualified, trained staff to provide accurate laboratory data for patient care. In other words, the GC/MS or LC/MS is just another tool. We have also found the process outlined in **Table 1.1** to be helpful in setting realistic timelines. It also helped to have a strong validation protocol already in place so that everyone was clear that this system and the methods were to be treated like any others we implement. Once we identified the testing and methods we planned

Table 1.1
Steps to implementing and validating MS methods for use in clinical laboratories

- **Method selection**
  - ○ Colleagues, literature, vendors, in-house development
  - ○ Clinical requirements (turnaround time, sample collection, transportation, processing)
  - ○ Chemical features of the analyte (sample preparation, internal standards, reagents)
- **Acquire materials**
  - ○ Obtain the highest quality practical
  - ○ Use stable isotope internal standards where practical
- **Establish quality goals**
  - ○ Method must be able to achieve CLIA requirements
  - ○ Consider biological variation
- **Method validation**
  - ○ Precision, simple and extended
  - ○ Accuracy
  - ○ Analytical measurement range and clinical reportable range
  - ○ Stability
  - ○ Method comparison
  - ○ Reference interval validation or determination
  - ○ Specificity/interference testing
- **Monitoring and statistical analysis**
- **SOP preparation**
- **Additional staff training**
- **Communication of change**

to perform, it became easier to determine the type of system that was needed. Discussions with several vendors and colleagues also helped to achieve more realistic expectations.

The training period is important and it is helpful to have on-hand materials for some of the methods you are planning to initiate. While there are a few commercially available methods, most are home-brew. You will find colleagues more than willing to share methods and offer guidance. Since these are home-brews, the methods fall into the high complexity category under CLIA, and the onus is on the laboratory to prove that the method is reliable. **Table 1.2** provides a listing of resources and documents that may be useful during this process.

Table 1.2
Useful resources and documents during method validations

- Frasier, C.G. (2001) Biological variation: from principles to practice. Washington, DC, AACC Press.
- Westgard, J.O. (2008) Basic method validation, 3rd edition. Madison, WI, Westgard QC.
- CLSI guidelines (current versions)
  - EP10 Preliminary Evaluation of Quantitative Clinical Laboratory Measurement Procedures
  - EP15 User Verification of Performance for Precision and Trueness
  - EP05 Evaluation of Precision Performance of Quantitative Measurement Methods
  - EP06 Evaluation of the Linearity of Quantitative Measurement
  - EP09 Method Comparison and Bias Estimation Using Patient Samples
  - GP31 Laboratory Instrument Implementation, Verification, and Maintenance
  - C43-A2 Gas Chromatography-Mass Spectrometry Confirmation of Drugs
  - EP50 Mass Spectrometry in the Clinical Laboratory
  - C24 Assessment of Laboratory Tests when Proficiency Testing Is Not Available

When acquiring reagents and materials, it pays to use the highest grade of reagents possible – now is not the time to save money. Also we recommend use of stable isotopes as internal standards whenever possible. It is often useful to develop a reagent documentation system so that new lots or vendors are tracked. Such a system will become useful when the method is in routine use as it facilitates troubleshooting. Non-reagent components (phlebotomy tubes, vials, and separation columns) may contain plasticizers and other additives that may also change from time to time. Most of us have encountered at least one episode where a change in one of the assay components has introduced an extraneous peak or some other interference affecting the quality of the analysis and took considerable investigating (11, 12).

3. Method Comparison and Standardization

When performing method comparisons, there are several important points to remember. First, results will likely differ when comparing an MS-based method against an immunoassay. This will relate partially to the specificity of the antibody used in the immunoassay. An equally important and frustrating contributor is the lack of standardized or harmonized reference materials for most analytes. There are efforts within the clinical laboratory community to raise awareness and stimulate development of such materials 13), but until such time as materials are available laboratories must take steps to minimize errors related to preparation of calibrators and controls. You will see throughout the chapters, recommendations to separately prepare stock solutions for calibrators and controls (some recommend these be prepared from different primary stocks on different days by different technologists). Other good practices include recommendations to avoid serial dilutions in the preparation of calibrators and to test new preparations by running aliquots as unknowns against the current batch of materials. It is also useful to prepare a patient pool that can be tracked across changes of calibrator and control lots. Of course, all of these must be monitored for changes and trends.

It is important to notify clients when transitioning from a less specific method to MS when preliminary studies demonstrate a difference between the two methods. The transition may be eased by providing results using both methods for a short time, especially if the results are used for monitoring.

The methods found within this volume are in current use within clinical laboratories – a few have come from colleagues with a forensic focus. There are many more that could have been included but we were faced with limitations of space and time. The authors have taken care to provide sufficient details for the instruments on which they perform the measurements within their laboratories so that one should be able to adapt the methods to another system. It is our hope that this will facilitate the use and acceptance of mass spectrometry methods and will make the challenge of bringing in such a system or a new method, a little less daunting.

References

1. Plumb, R.S. and Balogh, M.P. (2008) The changing face of LC-MS: from experts to users. Current Trends in Mass Spectrometry November:14–19.

2. Balogh, M.P. (2005) A revolution in clinical chemistry. LCGC North America. 23: http://chromatographyonline.findanalytichem.com/

3. Vogeser, M. and Parhofer, K.G. (2007) Liquid chromatography tandem-mass spectrometry (LC-MS/MS) – technique and applications in endocrinology. Exp Clin Endocrinol Diabetes. 115:559–570.

4. Kraemer, T. and Paul, L.D. (2007) Bioanalytical procedures for determination of drugs of abuse in blood. Anal Bioanal Chem. 388:1415–1435.

5. Maure, H.H. (2007) Current role of liquid chromatography-mass spectrometry in clinical and forensic toxicology. Anal Bioanal Chem. 388:1315–1325.

6. van Adel, B.A. and Tarnopolsky, M.A. (2009) Metabolic myopathies: update 2009. Clin Neuromuscl Dis. 10:97–121.

7. Garg, U. and Dasouki, M. (2006) Expanded newborn screening of inherited metabolic disorders by tandem mass spectrometry: Clinical and laboratory aspects. Clin Biochem. 39:315–332.

8. Albrecht, L. and Styne, D. (2007) Laboratory testing of gonadal steroids in children. Pediatr Endocrinol Rev. 5(Suppl 1):599–607.

9. Lakso, H.A., Appelblad, P. and Schneede, J. (2008) Quantification of methylmalonic acid in human plasma with hydrophilic interaction liquid chromatography separation and mass spectrometric detection. Clin Chem. 54:2028–2035.

10. Hutson, J.R., Aleksa, K., Pragst, F. and Koren, G. (2009) Detection and quantification of fatty acid ethyl esters in meconium by headspace-solid-phase microextraction and gas chromatography-mass spectrometry. J Chromatogr B Analyt Technol Biomed Life Sci. 877:8–12.

11. Drake, S.K., Bowen, R.A.R., Bemaley, A.T. and Hortin, G.L. (2004) Potential inter-ferences from blood collection tubes in mass spectrometric analyses of serum polypeptides. Clin Chem. 50:2398–2401.

12. Zhang, X.K., Dutky, R.C. and Fales, H.M. (1996) Rubber stoppers as sources of contaminants in electrospray analysis of peptides and proteins. Anal Chem. 68:3288–3289.

13. Thienpont, L.M. (2008) Accuracy in clinical chemistry – who will kiss Sleeping Beauty awake? Clin Chem Lab Med. 46:1220–1222.

Chapter 2

Measurement of Plasma/Serum Acylcarnitines Using Tandem Mass Spectrometry

Nathalie Lepage and Susan Aucoin

Abstract

Acylcarnitine analysis using tandem mass spectrometry has become a powerful tool in the investigation of pediatric patients presenting with clinical signs and symptoms suggestive of fatty acid oxidation defects. These signs are diverse and include failure to thrive, feeding difficulties, and cardiomyopathy. Because the signs and symptoms are nonspecific, the identification of acylcarnitines characteristic of these inherited diseases is necessary for diagnosis. We describe a method for the analysis of acylcarnitines in plasma or serum samples using electrospray ionization tandem mass spectrometry.

Key words: Fatty acid oxidation disorders, organic acidemia, medium-chain acyl CoA dehydrogenase deficiency, inborn errors of metabolism, tandem mass spectrometry, MSMS.

1. Introduction

Although organic aciduria characterizes a variety of metabolic disorders, many patients who have a fatty acid oxidation defect present with nonspecific organic aciduria or with an uninformative organic acid profile during asymptomatic periods (1). For this reason, the analyses of carnitines and acylcarnitines have become essential for the diagnosis of fatty acid oxidation defects (FAOD). The FAOD defects include at least 11 disease states characterized by deficiencies of specific enzymes, which are associated with decreased concentrations of carnitine, as well as with the identification of specific combinations of abnormal acylcarnitines (2–7).

The development of tandem mass spectrometry (MSMS) methods for the measurement of acylcarnitines is considered one of the most important laboratory advances in the management of metabolic diseases. In the method that follows, a derivatized

U. Garg, C.A. Hammett-Stabler (eds.), *Clinical Applications of Mass Spectrometry*, Methods in Molecular Biology 603, DOI 10.1007/978-1-60761-459-3_2, © Humana Press, a part of Springer Science+Business Media, LLC 2010

plasma or serum extract, containing deuterated acylcarnitine internal standards, is introduced into the MSMS using electrospray ionization (ESI). A combination of the specific primary mass and the unique secondary MS fragment ion is used to selectively monitor the compound to be quantified.

2. Materials

2.1. Specimens

1. Serum or heparinized plasma is used for analysis (*see* **Note 1**). EDTA specimens are stable when stored at −20°C until analysis.

2.2. Reagents

1. Reconstituting solution: Prepare an 80% acetonitrile solution in deionized water. Stable for 1 month when stored at room temperature. Cover with vented aluminum foil and sonicate for 20 min before use to degas.
2. 10% acetonitrile in deionized water. Stable for 1 month when stored at room temperature. Cover with vented aluminum foil and sonicate for 20 min before use to degas.
3. Derivatizing reagent: 3 N HCl in *n*-butanol.
4. L-Carnitine (C0) (Sigma-Aldrich, St. Louis, MO). **Table 2.1** contains a description of L-Carnitine. Protect from light and moisture. Store desiccated at room temperature.

Table 2.1
Description of acylcarnitines–C0

Acylcarnitines: unlabelled	C-nomenclature	Chemical formula	Formula weight	Parent ion mass
L-Carnitine	C0	C7H15NO3	161.20	221.2

2.3. Acylcarnitine Standards (see Note 2)

1. Acylcarnitine standards (*see* **Note 2**): **Table 2.2** contains a listing of the acylcarnitine standards. Protect from light and moisture. Store desiccated at room temperature.
2. Prepare 15 mL of 2.0 mmol/L solutions for short- and medium-chain acylcarnitines (C0–C12) in deionized water. Stable for 1 year when stored at 2–8°C.
3. Prepare 15 mL of 2.0 mmol/L solutions for long-chain acylcarnitines (C14, C16, C18) in methanol. Stable for 1 year when stored at 2–8°C.

Table 2.2
Description of acylcarnitines–deuterated and non-deuterated

Acylcarnitines: Labelled (Deuterated)	C-nomenclature	Formula weight
[methyl-d3]L-Carnitine.HCl	D3-C0	200.686
[d3] Acetyl-L-Carnitine.HCl	D3-C2	242.72
[d3] Propionyl-L-Carnitine.HCl	D3-C3	256.75
[d3] Butyryl-L-Carnitine.HCl	D3-C4	270.78
[d9] Isovaleryl-L-Carnitine.HCl	D9-C5	290.85
[d3] Octanoyl-L-Carnitine.HCl	D3-C8	326.88
[d3] Tetradecanoyl-L-Carnitine.HCl	D3-C14	411.045
[d3] Hexadecanoyl-L-Carnitine.HCl	D3-C16	439.099
Acylcarnitines: Non-deuterated	**C-nomenclature**	**Formula weight**
Acetyl-L-Carnitine.HCl	C2	239.699
Propionyl-L-Carnitine.HCl	C3	253.73
Butyryl-L-Carnitine.HCl	C4	267.75
Isobutyryl-L-Carnitine.HCl	Iso-C4	267.75
Isovaleryl-L-Carnitine.HCl	C5	281.78
Hexanoyl-L-Carnitine.HCl	C6	295.81
Phenylacetyl-L-Carnitine.HCl	C7-Ar*	315.80
Octanoyl-L-Carnitine.HCl	C8	323.86
Decanoyl-L-Carnitine.HCl	C10	351.91
Dodecanoyl-L-Carnitine.HCl	C12	379.97
Tetradecanoyl-L-Carnitine.HCl	C14	408.02
Hexadecanoyl-L-Carnitine.HCl	C16	436.08
Octadecanoyl-L-Carnitine.HCl	C18	464.13

Ar*-Aromatic

4. Working standard solutions: Prepare 2 mL of 100 μmol/L acylcarnitine (grouped) for C0, C2, (C3+C4+C5), (C6+C7+C8+C10), and (C12+C14+C16+C18) in methanol. Stable for 1 year when stored at 2–8°C.

5. Dilute an aliquot of each of the above acylcarnitine solutions to provide 2.0 mL of 1 μmol/L in methanol. Stable for 1 year when stored at 2–8°C.

6. Using 100 µmol/L and 1 µmol/L solutions prepare 200 µL of each calibrator as follows, using methanol as diluent:

   a. (C0) 0.25, 0.5, 1.0, 2.5, 5.0, 10.0, 20.0, 50.0 µmol/L

   b. (C2) 0.25, 0.5, 1.0, 2.5, 5.0, 10.0 µmol/L

   c. (C3+C4+C5) 0.02, 0.05, 0.10, 0.25, 0.5, 1.0 µmol/L

   d. (C6+C8+C10) 0.02, 0.05, 0.10, 0.25, 0.5, 1.0 µmol/L

   e. (C12+C14+C16+C18) 0.02, 0.05, 0.10, 0.25, 0.5, 1.0 µmol/L

2.4. Internal Standard Working Solutions

1. d5-phenylalanine Internal Standard (Cambridge Isotope Laboratories, Andover, MA). Protect from light and moisture. Store desiccated at room temperature. Prepare 2.0 mmol/L d5-phenylalanine in deionized water. Stable for 1 year at 4°C.

2. D-Acylcarnitines (*see* **Note 2** and **Table 2.2**): Prepare 2.0 mmol/L solutions of deuterated acylcarnitine standards in methanol. Stable for 1 year at –20°C. Dilute an aliquot of each acylcarnitine deuterated standard in methanol to 0.2 mmol/L. Stable for 1 year at –20°C.

3. Working deuterated internal standard solution: Prepare by adding the volumes of d5-phenylalanine and acylcarnitines seen in **Table 2.3** to a 50 mL volumetric flask. Bring to volume using methanol. Stable for 1 month at –20°C.

Table 2.3
Preparation of working deuterated internal standard. Final volume is 50 mL, using methanol as diluent

Deuterated amino acid/ acylcarnitine	Concentration of stock solution used (mmol/L)	Final concentration of internal standard in solution (µmol/L)	Volume (µL)
d5-Phe	2.0	200	250
d3-C0	0.2	30.0	375
d3-C2	0.2	15.0	187.5
d3-C3	0.2	5.0	62.5
d3-C4	0.2	5.0	62.5
d3-C5	0.2	5.0	62.5
d3-C8	0.2	3.0	37.5
d3-C14	0.2	6.0	75
d3-C16	0.2	6.0	75

2.5. Quality Control

1. Pool previously analyzed patient plasma found to have acylcarnitine concentrations within the reference range. Centrifuge to remove particulate matter and filter using a 0.22 μm filter. Divide into three portions.

2. Normal control: Aliquot a portion to serve as the normal control.

3. Mid-control: Add acylcarnitine stock standards to concentrations that reflect the upper level of the reference range.

4. High control: Add acylcarnitine stock standards to high concentrations reflecting concentrations observed in fatty acid oxidation defects.

5. The controls are stable up to 1 year when stored at –20°C.

2.6. Instrumentation and Supplies

1. Waters Corporation TQD Tandem Mass Spectrometer
2. Fume hood
3. Nitrogen evaporator and heating block
4. Microfuge centrifuge
5. 1.5 mL disposable polystyrene centrifuge tubes
6. 2 mL plastic transfer pipettes
7. 12 × 75 mm glass culture tubes
8. 0.5 mL HPLC vials

3. Methods

3.1. Sample Preparation

1. Appropriately label a microfuge tube (1.5 mL), 12 × 75 mm glass tube, and 0.5 mL HPLC vial for each specimen.

2. Use distilled water as a blank sample to monitor background/contamination.

3. Mix and centrifuge specimens. Vortex internal standard working solution.

4. Pipet 20 μL of each specimen into the appropriate microfuge tube.

5. Add 400 μL of working deuterated internal standard solution to each.

6. Vortex samples for 10 sec.

7. Allow samples to equilibrate at room temperature for 5 min.

8. Centrifuge samples for 5 min at $1,4000 \times g$.

9. Using a plastic transfer pipet, transfer supernatant to 12 × 75 mm glass tube. Take care not to transfer any precipitate.

10. Evaporate to dryness with gentle stream of nitrogen gas at 37°C.

11. Working in a fume hood, add 100 μL of 3 N HCl in *n*-Butanol to each tube.

12. Cap samples and vortex for 10 sec.

13. Incubate at 55°C for 20 min in heating block.

14. Remove tubes from heating block and evaporate to dryness under nitrogen gas at 37°C in a fume hood.

15. Reconstitute samples with 200 μL of the 80% acetonitrile reconstituting solution.

16. Vortex for 5 sec.

17. Transfer reconstituted samples to HPLC sample vials.

3.2. Tandem Mass Spectrometry Analysis

Purge lines using 100% methanol at a rate of 2.0 mL/min for 20 min. Follow with 80% acetonitrile (also used as the reconstituting solution) at 0.100 mL/min for 20 min to condition before analysis. Operating parameters are given in **Table 2.4**. Examples of three common profiles are shown in **Figs. 2.1**, **2.2**, and **2.3**.

Table 2.4
LC tandem MS operating parameters

Source	
Source: ion mode	Electron spray positive (ES+)
Capillary (kV)	3.50
Cone (V)	40
Extractor (V)	3.00
RF lens (V)	0.0
Collision energy (V)	25
Source temperature (°C)	120
Desolvation temperature (°C)	350
Cone gas flow (L/Hr)	50
Desolvation gas flow (L/Hr)	600
Analyzer	
Time (minutes)	Start 0.08; end 1.8
Scan duration (seconds)	3.8
Mass (m/z) for parents of 85	Start 210; end 580

(continued)

Table 2.4 (continued)

LM 1 resolution	14.2
HM 1 resolution	14.2
Ion energy 1	0.5
Entrance	−2
Collision	25
Exit	1
LM 2 resolution	14.2
HM 2 resolution	14.2
Ion energy 2	0.5
Multiplier (V)	570
Syringe pump flow (µL/min)	10.0
Inject volume (µL)	7.5
Gradient run time (min)	2.5
Solvent	80% acetonitrile
Gradient flow (mL/min) (initial)	0.100
Gradient flow (mL/min) (@0.30 min)	0.015
Gradient flow (mL/min) (@1.20 min)	0.100
Gradient flow (mL/min) (@1.55 min)	0.500
Gradient flow (mL/min) (@1.85 min)	0.100

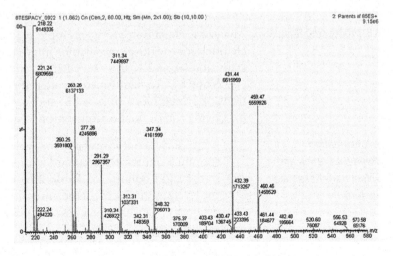

Fig. 2.1. Profile not consistent with inborn errors of metabolism.

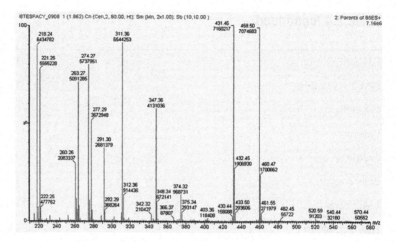

Fig. 2.2. Profile consistent with methylmalonic acidemia.

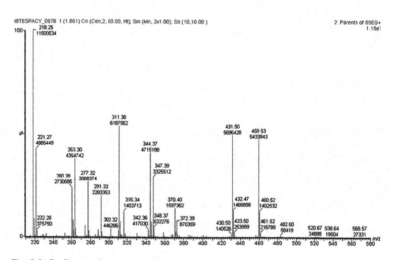

Fig. 2.3. Profile consistent with MCAD.

3.3. Processing Results

Analyze each sample using multiple reaction monitoring of two channels for the d5-phenylalanine internal standard such as the ion transition (m/z 227.2 → m/z 125.2) and primary ions of m/z 85 (*see* **Note 3**). Acylcarnitines are identified by monitoring transitions of primary ions of m/z 85, and acylcarnitine ratios are used for calculation of final concentration (**Table 2.5**).

3.4. Evaluating Standard Curves and Calculating Response Factors

1. Acquire ion count intensity for each acylcarnitine and corresponding deuterated standard in each standard curve as listed in the right column of **Table 2.6**. For each sample, process the spectra and record the intensity of peaks after a background, smoothing process, and peak centering.

2. Concentration of deuterated standard in standard curves (**Table 2.7**): The absolute amount of deuterated standard is derivatized and dried under nitrogen. The sample is then

Table 2.5

Transitions for the acylcarnitines and acylcarnitine ratios used for calculation of final concentration

| | | Transition (*m/z*) for | | |
| | | | | Non-deuterated/ Deuterated ratio used |
Acylcarnitine	Deuterated standard	Acylcarnitine	Deuterated standard	for final concentration
C2 (acetyl)	D3-C2 ([d3]acetyl)	260.2→85	263.2→85	C2/d3-C2 "/">>
C3:1 (propenyl)	D3-C3 ([3,3,3-d3] propionyl)	272.2→85	277.2→85	C3 / d3-C3
C3 (propionyl)	D3-C3 ([3,3,3-d3] propionyl)	274.2→85	277.2→85	C3 / d3-C3
C4 (butyryl)	D3-C4 ([4,4,4-d3] butyryl)	288.2→85	291.2→85	C4 / d3-C4
C5 (isovaleryl)	D9-C5 ([d9] isovaleryl)	302.2→85	311.3→85	C5 / d9-C5
C5:1 (tiglyl)	D9-C5 ([d9] isovaleryl)	300.2→85	311.3→85	C5 / d9-C5
C4-OH (3OH-butyryl)	D3-C4 ([4,4,4-d3] butyryl)	304.2→85	291.2→85	C4 / d3-C4
C6 (hexanoyl)	D9-C5 ([d9] isovaleryl)	316.3→85	311.3→85	C6 / d9-C5
C5-OH/2Me3OH butyryl	D9-C5 ([d9] isovaleryl)	318.2→85	311.3→85	C5 / d9-C5
Benzoyl	D3-C8 ([8,8,8-d3] octanoyl)	322.3→85	347.3→85	C8 / d3-C8
C6-OH (3OH hexanoyl)	D9-C5 ([d9] isovaleryl)	332.3→85	311.3→85	C6 / d9-C5
Phenylacetyl (C7*Ar)	D3-C8 ([8,8,8-d3] octanoyl)	336.2→85	347.3→85	Phenacet/d3-C8
C8:1 (octenoyl)	D3-C8 ([8,8,8-d3] octanoyl)	342.3→85	347.3→85	C8 / d3-C8
C8 (octanoyl)	D3-C8 ([8,8,8-d3] octanoyl)	344.3→85	347.3→85	C8 / d3-C8
C3DC (malonyl)	D3-C8 ([8,8,8-d3] octanoyl)	360.3→85	347.3→85	C8 / d3-C8
C10:3 (decatrienoyl)	D3-C8 ([8,8,8-d3] octanoyl)	366.3→85	347.3→85	C10 / d3-C8

(continued)

Table 2.5 (continued)

Acylcarnitine	Deuterated standard	Transition (*m/z*) for		Non-deuterated/ Deuterated ratio used for final concentration
		Acylcarnitine	Deuterated standard	
C10:2 (decadienoyl)	D3-C8 ([8,8,8-d3] octanoyl)	368.3→85	347.3→85	C10 / d3-C8
C10:1 (decenoyl)	D3-C8 ([8,8,8-d3] octanoyl)	370.3→85	347.3→85	C10 / d3-C8
C10 (decanoyl)	D3-C8 ([8,8,8-d3]octanoyl)	372.3→85	347.3→85	C10 / d3-C8
C4DC (MeMalonyl/ suc-cinyl)	D3-C8 ([8,8,8-d3] octanoyl)	374.3→85	347.3→85	C10 / d3-C8
C5DC (glutaryl)/ 3OHC10	D3-C8 ([8,8,8-d3]octanoyl)	388.3→85	347.3→85	C10 / d3-C8
C12:1 (dodecenoyl)	D3-C14 ([14, 14, 14-d3] tetradecanoyl)	398.4→85	431.4→85	C12 / d3-C14
C12 (dodecanoyl)	D3-C14 ([14, 14, 14-d3] tetradecanoyl)	400.4→85	431.4→85	C12 / d3-C14
C6DC/3MeGlu-taryl	D3-C14 ([14, 14, 14-d3] tetradecanoyl)	402.3→85	431.4→85	C12 / d3-C14
C12-OH (3OHdodecanoyl)	D3-C14 ([14, 14, 14-d3] tetradecanoyl)	416.4→85	431.4→85	C12 / d3-C14
C14:2 (tetradecadienoyl)	D3-C14 ([14, 14, 14-d3] tetradecanoyl)	424.3→85	431.4→85	C14 / d3-C14
C14:1 (tetradecenoyl)	D3-C14 ([14, 14, 14-d3] tetradecanoyl)	426.4→85	431.4→85	C14 / d3-C14
C14 (tetradecanoyl)	D3-C14 ([14, 14, 14-d3] tetradecanoyl)	428.4→85	431.4→85	C14 / d3-C14
C8DC (suberyl)	D3-C14 ([14, 14, 14-d3] tetradecanoyl)	430.4→85	431.4→85	C14 / d3-C14

(continued)

Table 2.5 (continued)

Acylcarnitine	Deuterated standard	Transition (*m/z*) for Acylcarnitine	Deuterated standard	Non-deuterated/ Deuterated ratio used for final concentration
C14:1-OH (3OH tetradecenoyl)	D3-C14 ([14, 14, 14-d3] tetradecanoyl)	442.4→85	431.4→85	C14 / d3-C14
C14-OH (3OH tetradecanoyl)	d3-C14 ([14, 14, 14-d3] tetradecanoyl)	444.4→85	431.4→85	C14 / d3-C14
C16:1 (palmitoleyl)	d3-C16 ([16, 16, 16-d3] hexadecanoyl)	454.4→85	459.4→85	C16 / d3-C16
C16 (palmitoyl)	d3-C16 ([16, 16, 16-d3] hexadecanoyl)	456.4→85	459.4→85	C16 / d3-C16
C10DC (sebacyl)	d3-C16 ([16, 16, 16-d3] hexadecanoyl)	458.4→85	459.4→85	C16 / d3-C16
C16:1-OH (3OH palmitoleyl)	d3-C16 ([16, 16, 16-3] hexadecanoyl)	470.4→85	459.4→85	C16 / d3-C16
C16-OH (3OH palmitoyl)	d3-C16 ([16, 16, 16-d3] hexadecanoyl)	472.4→85	459.4→85	C16 / d3-C16
C18:2 (linoleyl)	d3-C16 ([16, 16, 16-d3] hexadecanoyl)	480.4→85	459.4→85	C18 / d3-C16
C18:1 (oleyl)	D3-C16 ([16, 16, 16-d3] hexadecanoyl)	482.4→85	459.4→85	C18 / d3-C16
C18 (stearoyl)	D3-C16 ([16, 16, 16-d3] hexadecanoyl)	484.4→85	459.4→85	C18 / d3-C16
C18:2-OH (3OH linoleyl)	D3-C16 ([16, 16, 16-d3] hexadecanoyl)	496.4→85	459.4→85	C18 / d3-C16
C18:1-OH (3OH oleyl)	D3-C16 ([16, 16, 16-d3] hexadecanoyl)	498.4→85	459.4→85	C18 / d3-C16

(continued)

Table 2.5 (continued)

| Acylcarnitine | Deuterated standard | Transition (*m/z*) for | | Non-deuterated/ Deuterated ratio used for final concentration |
		Acylcarnitine	Deuterated standard	
C18-OH (3OH stearoyl)	D3-C16 ([16, 16, 16-d3] hexadecanoyl)	500.4→85	459.4→85	C18 / d3-C16
C16DC	D3-C16 ([16, 16, 16-d3] hexadecanoyl)	542.4→85	459.4→85	C18 / d3-C16
C18:1 DC	D3-C16 ([16, 16, 16-d3] hexadecanoyl)	568.5→85	459.4→85	C18 / d3-C16

Table 2.6
Concentration of deuterated standard in standard curves

Deuterated acylcarnitine	Final concentration internal standard in standard curve (μmol/L)	Deuterated acylcarnitine	Final concentration internal standard in standard curve (μmol/L)
d3-C0	3.0	d3-C5	0.5
d3-C2	1.5	d3-C8	0.3
d3-C3	0.5	d3-C14	0.6
d3-C4	0.5	d3-C16	0.6

reconstituted with 200 μL 80% acetonitrile. However, in routine analysis, the internal standard is added into 20 μL of plasma. Therefore, the final concentrations of deuterated standards are ten times lower than for routine samples.

3. For each acylcarnitine standard curve, use the theoretical ratio data (**T** = concentration analyte/concentration internal standard) and the actual ratio (**A** = ion count intensity analyte peak/ion count intensity internal standard peak) data to calculate the response factor.

$$RF = \frac{1}{(A/T)} = \frac{T}{A}$$

Calculate the average response factor for each analyte.

Table 2.7
Mean levels and coefficients of variation (CV) for the three quality control materials used during a 6-month period

	C0	C2	C3	C4	C5	C6	C8	C10	C12	C14	C16	C18
Level 1 (µmol/L)												
Mean	36.7	9.21	0.4	0.2	0.11	0.04	0.09	0.15	0.09	0.03	0.07	0.03
CV (%)	8.5	19.4	11.5	18.2	14.4	20.3	16.4	13.8	18.2	29.4	21.6	30.5
Level 2 (µmol/L)												
Mean	84.1	41.5	2.65	0.97	0.87	0.55	0.61	1.2	0.58	0.6	0.87	0.6
CV (%)	10.0	6.0	11.5	15	9.6	11.3	11.9	11.7	12.3	11.2	12.7	14.0
Level 3 (µmol/L)												
Mean	95.9	57.3	9.3	4.7	5.1	2.5	5.1	2.8	2.9	2.8	2.9	3.0
CV (%)	12.5	9.4	10.6	12.9	10.2	11.5	11.0	11.1	10.6	11.7	9.9	11.1

3.5. Calculation of Concentration for Quality Control and Unknown Samples

Concentrations (µmol/L) of unknown specimens are calculated by the following equation:

(Ion count intensity analyte ÷ Ion count intensity IS)

(IS concentration) (average RF)

1. IS concentrations for routine samples are listed above.
2. Review each peak to check for proper integration and area calculation of the peak.
3. Report acylcarnitine concentrations as ">upper limit of linearity" of standard curve when applicable. Do not extrapolate results.
4. The method is precise and specific and is associated with the following coefficients of variation (CV) for the three quality control materials described above. Examples of expected precision for the various acylcarnitines are seen in **Table 2.7**.

3.6. Reference Intervals

Reference intervals, provided in **Table 2.8**, are based on local data and should be validated by each laboratory.

3.7. Interpretation

Most fatty acid oxidation defects and several organic acidurias will show typical abnormal patterns of acylcarnitines, as analyzed by tandem mass spectrometry. The results should be separated or grouped according to chain length: short-chain (C2, C4), medium-chain (C6, C8, C10, C10:1), long-chain (C12, C14, C14:1, C14:2, C16, C18, C18:1), 3-hydroxy long-chain (C14-

Table 2.8
Reference intervals of acylcarnitines

Acylcarnitine	Reference intervals	Acylcarnitine	Reference intervals
C2 (acetyl)	3.86–23.48	C6DC/3-Meglutaryl	<0.13
C3:1 (propenyl)	<0.03	C12-OH (3OH dodecanoyl)	<0.06
C3 (propionyl)	<0.65	C14:2 (tetradecadienoyl)	<0.19
C4 (butyryl)	<0.52	C14:1 (tetradecenoyl)	<0.18
C5:1 (tiglyl)	<0.06	C14 (tetradecanoyl)	<0.13
C5 (isovaleryl)	<0.38	C8DC (suberyl)	<0.16
C4-OH (3OH-butyryl)	<0.33	C14:1-OH (3OH tetradecenyl)	<0.06
C6 (hexanoyl)	<0.13	C14-OH (3OH tetradecanoyl)	<0.06
C5-OH/2Me3OH butyryl	<0.13	C16:1 (palmitoleyl)	<0.08
C6-OH (3OH-hexanoyl)	<0.03	C16 (palmitoyl)	<0.23
C7*Ar (phenylacetyl)	<0.05	C10DC (sebacyl)	<0.07
C8:1 (octenoyl)	<0.60	C16:1-OH (3OH palmitoleyl)	<0.06
C8 (octanoyl)	<0.33	C16-OH (3OH palmitoyl)	<0.03
C3DC (malonyl)	<0.08	C18:2 (linoleyl)	<0.22
C10:3 (decatrienoyl)	<0.36	C18:1 (oleyl)	<0.21
C10:2 (decadienoyl)	<0.08	C18 (stearoyl)	<0.11
C10:1 (decenoyl)	<0.37	C18:2-OH (3OH linoleyl)	<0.07
C10 (decanoyl)	<0.27	C18:1-OH (3OH oleyl)	<0.06
C4DC (MeMalonyl/ succinyl)	<0.10	C18-OH (3OH stearoyl)	<0.05
C5DC (glutaryl)/ 3OHC10	<0.12	C16DC	<0.10
C12:1 (dodecenoyl)	<0.16	C18:1DC	<0.06
C12 (dodecanoyl)	<0.23		

OH, C16-OH, C18-OH, C18:1-OH), and others indicative of organic acidurias (such as C3, C4DC C5:1, C5-OH, C5DC, etc.). Generally, specific patterns are associated with specific diseases. For example, the expected pattern for medium-chain acyl

CoA dehydrogenase deficiency (MCAD), a common fatty acid oxidation disorder, is an increase in medium-chains C6, C8, C10, C10:1 (8). The reader is referred to recent publications for listing of typical patterns for other diseases (2, 8, 9). While interpreting data, it is important to evaluate data for interferences (*see* **Notes 5–7**).

4. Notes

1. Grossly hemolyzed samples are unsuitable for analysis to avoid inaccurate quantitative results.

2. Acylcarnitine standards were obtained from Dr. Herman J. ten Brink VU Medical Center Metabolic Laboratory, 0A082 P.O. Box 7057 1007 MB Amsterdam, The Netherlands.

3. For each sample, examine the chromatogram and total ion chromatogram (TIC) to determine whether the injection and flow rate are acceptable. An adequate flow is characterized by three main steps: The first is a sharp increase in the intensity (*y*-axis) reflecting proper ionization; the second is a constant intensity for the next 1.4 min; and the last is a return to baseline at the end of the analyzer time. **Figure 2.4** shows an adequate flow as monitored using d5-phenylalanine (MRM of two channels). A plugged and/or dirty electrospray probe tip will affect the chromatogram, resulting in a poor (unacceptable) flow rate such as that seen via **Fig. 2.5**.

4. Response factor information generated from these standards may be stored in a program such as NeoLynx and used to calculate patient concentrations for a period of up to 1 year before re-calibration (**Section 3.4**).

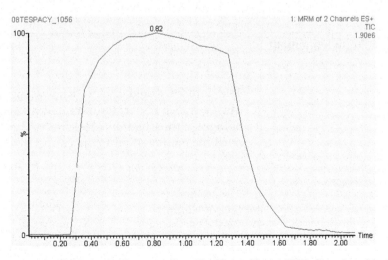

Fig. 2.4. Profile of adequate flow as monitored by d5-phenylalanine (MRM of two channels).

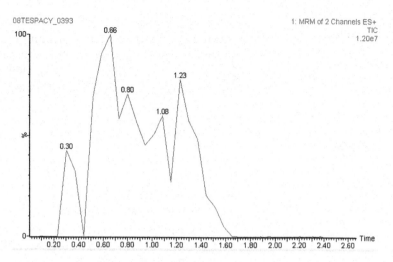

Fig. 2.5. Profile of inadequate flow as monitored by d5-phenylalanine (MRM of two channels).

5. There is potential interference from glutamate for the quantitation of C2 in the parent of 85 scans used to measure acylcarnitines, as both, having an m/z value of 260, produce a fragment ion at 85 m/z.

6. Elevated pivaloylcarnitine concentrations in patients administered the antibiotic pivalic acid may result in falsely elevated isovalerylcarnitine levels, as pivaloylcarnitine is an isomer of isovalerylcarnitine.

7. Incomplete butylation of acylcarnitines in the 260–428 m/z range may interfere with smaller chain acylcarnitines in the same mass range. An excess of 3.0 N HCl n-butanol is used in the assay to minimize this effect.

Acknowledgement

The authors would like to acknowledge the excellent work of Mr. Lawrence Fisher, resource and development laboratory technologist at the Children's Hospital of Eastern Ontario, who developed the method on other mass spectrometry platforms.

References

1. Jellum E. (1977) Profiling of human body fluids in healthy and diseased states using gas chromatography and mass spectrometry, with special reference to organic acids. *J Chromato* 143, 427–472.

2. Valle D, Beaudet AL, Vogelstein B, Kinzler K, Antonarakis S, Ballabio A, eds. (2007) The online metabolic and molecular bases of inherited disease – Chapter 101. McGraw-Hill: New York.

3. Blau N, Duran M, Blaskovics ME, Gibson KM (eds.). (2003) Physician's guide to the laboratory diagnosis of metabolic diseases – 2nd edition. Springer-Verlag: Berlin.

4. Scriver CR, Beaudet AL, Sly WS, Valle D (eds). (2001) The metabolic and molecular bases of inherited disease, 8th edition – Chapter 101. McGraw-Hill medical publishing division: New York.

5. Campoy C, Bayes R, Peinado JM, Rivero M, Lopez C, Molina-Font JA. (1998) Evaluation of carnitine nutritional status in full-term newborn infants. *Early Hum Dev* 53(Suppl), S149–S164.

6. Buchta R, Nyhan WL, Broock R, Schragg P. (1993) Carnitine in adolescents. *J Adolesc Health* 14(6), 440–441.

7. Gloggler A, Bulla M, Puchstein C, Gulotta F, Furst P. (1988) Plasma and muscle carnitine in healthy and hemodialyzed children. *Child Nephrol Urol* 9(5), 277–282.

8. Chace DH, Kalas AT, Naylor EW. (2003) Use of tandem mass spectrometry for multi-analyte screening of dried blood specimens from newborns. *Clin. Chem* 49(11), 1797–1817.

9. Zytkovicz TH, Fitzgerald EF, et al. (2001) Tandem mass spectrometric analysis for amino, organic and fatty acid disorders in newborn dried blood spots: A Two-year summary from the New England newborn screening program. *Clin Chem* 47(11), 1945–1955.

Chapter 3

Comprehensive Determination of Amino Acids for Diagnosis of Inborn Errors of Metabolism

Dennis J. Dietzen and Annette L. Weindel

Abstract

Analysis of clinically relevant amino acids using ion-exchange chromatography coupled to photometric detection has been an indispensable component in the detection of inborn errors of metabolism for six decades. Detection of amino acids using mass spectrometry offers advantages in speed and analytic specificity. Employing methanol extraction and controlled butylation, C8 reversed-phase chromatography, and MS/MS detection, 32 amino acids are quantified in 20 min with clinically appropriate imprecision in plasma, urine, and CSF. Quantitation is linear to 1,000 µmol/L and limits of detection are at least 1.0 µmol/L. Important isobaric amino acids are distinguished by chromatography or by unique patterns of fragmentation following collision-induced dissociation. The technique employs commercially available reagents and may be expanded and customized for specific clinical or research settings.

Key words: Amino acid, mass spectrometry, liquid chromatography, butylation, multiple reaction monitoring, isobar, newborn screening.

1. Introduction

Amino acids are best known for their role in constructing the polypeptide backbones of enzymes and structural proteins. These 20 along with some non-protein amino acids (e.g., citrulline, argininosuccinate) also play critical roles as intermediates in metabolic processes such as the urea cycle. Defects in the catabolism and transport of amino acids (combined incidence of 1:6000) are responsible for a range of clinical syndromes that can cause significant morbidity and mortality if not detected and treated promptly (1). Comprehensive amino acid analysis is indicated in these situations as the signs and symptoms of these disorders are often non-specific. For decades, such amino acid analysis has been

U. Garg, C.A. Hammett-Stabler (eds.), *Clinical Applications of Mass Spectrometry*, Methods in Molecular Biology 603, DOI 10.1007/978-1-60761-459-3_3, © Humana Press, a part of Springer Science+Business Media, LLC 2010

conducted using ion-exchange chromatography with photometric detection of ninhydrin conjugates (2). These techniques are limited by slow throughput (hours) and confounded by co-eluting amines, which cause overestimation of some amino acid concentrations. Homocitrulline/methionine, gabapentin/histidine, and aminoglycosides/phenylalanine commonly co-elute.

Recent adoption of tandem mass spectrometry in newborn screening programs has enabled screening for 29 "core" and many additional "secondary" disorders of amino acid, fatty acid, and organic acid metabolism from a single dried blood spot (3, 4). The mass spectrometry of newborn screening, however, does not distinguish isobaric amino acids. For example, the butyl esters of leucine, isoleucine, allo-isoleucine, and hydroxyproline have identical molecular mass (m/z 188) and similar product ion spectra. Distinction among these compounds is important for the diagnosis and monitoring of children with maple syrup urine disease (MSUD). Elevated concentrations of allo-isoleucine are pathognomonic for MSUD and dietary treatment dictates periodic quantitation of the precursors (leucine, isoleucine, valine) to branched chain keto acids that accumulate in this disorder. The protocol presented is appropriate for screening, confirmation, and follow-up of patients with a broad array of inborn errors of amino acid metabolism and transport. The method employs commercially available reagents and may be customized for specific clinical circumstances (5).

2. Materials

2.1. Samples

1. Serum, plasma (heparin), urine, and spinal fluid are all compatible with this method. The use of other matrices (e.g., tissue homogenates) is likely possible but has not been validated. If analysis must be delayed, samples should be stored at –20°C or colder to prevent in vitro artifacts. Common artifacts include the deamination of glutamine (forming glutamate), decarboxylation of glutamate to γ-aminobutyrate, and the precipitation of cystine.

2.2. Reagents

1. Derivatizing reagent: 3 N HCl/butanol (Regis Technologies, Morton Grove, IL).
2. Mobile phase: 20% acetonitrile/0.1% formic acid.

2.3. Calibration Standards

1. 2.5 mM acidic/neutral amino acid standards (Sigma Chemical Co., St. Louis, MO).
2. 2.5 mM basic amino acid standards (Sigma Chemical Co., St. Louis, MO)

3. Additional 2.5 mM standard stock solution containing glutamine, argininosuccinic acid, allo-isoleucine (all from Sigma Chemical Co., St. Louis, MO), and homocitrulline (Bachem, King of Prussia, PA). Stock is stable for at least 1 year at −20°C.

2.4. Working Internal Standard

50 μM each of glutamine (2,3,3,4,4-D5), glutamate (2,4,4-D3), homocystine (3,3,3',3',4,4,4',4'-D8), cystine (3,3,3',3'-D4), alanine (2,3,3,3-D4), leucine (5,5,5-D3), phenylalanine (Ring-D5,2,3,3-D8), lysine (4,4,5,5-D4), and glycine (2,2-D2); 25 μM tyrosine (3,3-D2), and 10 μM methionine (methyl-D3) (all from Cambridge Isotope Laboratories, Andover, MA). Solubility and performance improved when lower amounts of D3-methionine and D2-tyrosine were used (*see* **Note 1**). Freeze in 1.0 mL aliquots. Stable for a minimum of 1 year at −20°C.

2.5. Instruments and Columns

1. 1100 Series LC System (Agilent Technologies, Santa Clara, CA).
2. API 3000 Tandem Mass Spectrometer (Applied Biosystems, Foster City, CA).
3. Zorbax Eclipse XDB-C8 Column (4.6 mm × 150 mm × 5 μm) (Agilent Technologies, Santa Clara, CA).
4. Zorbax Eclipse XDB-C8 Guard Column (4.6 mm × 12.5 mm × 5 μm) (Agilent Technologies, Santa Clara, CA).

3. Methods

The technique is an adaptation of butylation protocols employed in newborn screening programs with two important modifications. First, butylation time is reduced to avoid significant deamination of glutamine. Second, samples are introduced into the mass spectrometer following liquid chromatography to allow separation and accurate quantitation of isobaric amino acids (e.g., leucine, isoleucine, and allo-isoleucine) that produce identical fragments following collision-induced dissociation (CID). The amino acid menu was developed in the setting of a pediatric tertiary care hospital for three basic populations: (a) patients with defined inborn metabolic disease, (b) neonates referred for follow-up of positive newborn screening results, and (c) patients referred from neurology, gastroenterology, and genetics specialty clinics with signs and symptoms suggestive of inborn errors of metabolism. The choice of internal standards was based on (a) commercial availability, (b) the need to control for the butylation of amino acids with multiple carboxyl groups, (c) the need to space internal standards across the chromatogram, and (d) the imprecision requirements of specific amino acids where tight dietary control is indicated in the prevention of morbidity and mortality associated with some inborn errors of metabolism (e.g., phenylalanine in phenylketonuria or leucine, isoleucine, and valine in MSUD).

3.1. External Calibration

Perform in triplicate using non-extracted standards. Curves are typically stable for at least 1 week.

1. Mix the three calibrator 2.5 mM solutions (basic, acidic/neutral, and supplemental standards of glutamine, allo-isoleucine, argininosuccinate, and homocitrulline), according to **Table 3.1** and dry completely under nitrogen. Discard any remaining calibrator solution after a single use.

2. Add 100 μL of 3 N HCl/butanol to each tube and incubate for 7.5 min at 60°C and dry again under nitrogen (*see* **Note 2**).

3. Redissolve residue in 250 μL of 20% acetonitrile/0.1% formic acid to yield solutions with amino acid concentrations of 0, 50, 100, 500, and 1000 μM.

Table 3.1
Preparation of 5-point calibrator solution for amino acid analysis

Final concentration (μM)	Acidic/neutral standard (μL)	Basic standard (μL)	Supplemental standards (μL)	Internal standard mix (μL)
0	0	0	0	50
50	5	5	5	50
100	10	10	10	50
500	50	50	50	50
1,000	100	100	100	50

3.2. Sample Extraction and Butylation

1. Mix 100 μL patient specimen, 50 μL internal standard mix, and 850 μL methanol into 1.5 mL microcentrifuge tube. Vortex and then centrifuge the mixture at $15,000 \times g$ for 3 min.

2. Transfer 400 μL of supernatant into a 4.0 mL, 15 × 45 mm screw cap borosilicate extraction vial and dry under nitrogen. Add 100 μL 3 N HCl/butanol to the residue and incubate for 7.5 min at 60°C. (*see* **Note 2**).

3. Immediately dry under nitrogen and reconstitute in 250 μL of mobile phase (20% acetonitrile, 0.1% formic acid).

4. Transfer to 2.0 mL, 12 × 32 mm borosilicate autosampler vials with 100 μL conical insert.

3.3. LC/MS/MS Operation and Data Acquisition

1. Isocratic flow of 20% acetonitrile/0.1% formic acid at 1.0 mL/min.

2. Sample injection (10 μL) every 22 min.

3. Flow is split 1:4 prior to entry into the electrospray ionization (ESI) source.

4. Electrospray ionization source set to 325°C and 2000 V (positive mode).

5. Nebulizer gas pressure = 14 mTorr.

6. Curtain gas pressure = 12 mTorr.

7. Collision gas pressure = 10 mTorr.

8. Declustering potential, focusing potential, collision energy, and collision cell exit potential were optimized for each amino acid (5).

9. Data acquired in multiple reaction monitoring mode (Analyst 1.4.2, Applied Biosystems, Foster City, CA). Precursor–product ion pairs (Q1→Q3) were previously determined empirically using aqueous solutions of individual amino acids (*see* **Note 3**).

10. Identification and quantitation are based on a single precursor–product ion pair and retention time. Each amino acid, its respective internal standard, ion pair transition, and retention time are listed in **Table 3.2** (*see* **Notes 4–6**).

11. Chromatography is stable for 2 months or approximately 400 injections.

Table 3.2
Chromatography and MS/MS parameters for each amino acid and deuterated internal standards. Absolute RT = typical C8 retention time; Relative RT = typical retention time relative to retention time of respective internal standard. IS = internal standard

Amino acid	Q1→Q3	Internal standard	Absolute RT	Relative RT
Alanine	146→44	D4-Alanine	2.2	1.0
D4-Alanine	150→48		2.2	
β-Alanine	146→72	D4-Alanine	2.3	1.0
Allo-isoleucine	188→86	D3-Leucine	9.1	0.9
α-Aminoadipic	218→98	D4-Alanine	3.7	1.6
α-Aminobutyric	160→58	D4-Alanine	3.1	1.4
γ-Aminobutyric	160→87	D4-Alanine	2.9	1.3

(continued)

Table 3.2 (continued)

Amino acid	Q1→Q3	Internal standard	Absolute RT	Relative RT
β-Aminoisobutyric	160→86	D4-Alanine	3.1	1.4
Arginine	231→70	D4-Lysine	1.2	1.0
Argininosuccinate	347→70	D4-Alanine	1.6	0.8
Carnosine	283→110	D4-Lysine	1.2	1.0
Citrulline	232→70	D4-Alanine	1.7	0.8
Cystine	297→74	D4-Cystine	1.4	1.0
D4-Cystine	301→154		1.4	
Cystathionine	279→134	D4-Alanine	1.4	0.7
Glutamate	204→84	D3-Glutamate	2.7	1.0
D3-Glutamate	207→87		2.7	
Glutamine	203→84	D5-Glutamine	1.6	1.0
D5-Glutamine	208→89		1.6	
Glycine	132→76	D2-Glycine	1.9	1.0
D2-Glycine	134→78		1.9	
Histidine	212→110	D4-Lysine	1.1	1.0
Homocitrulline	246→127	D4-Alanine	1.9	0.9
Homocystine	325→190	D8-Homocystine	2.0	1.0
D8-Homocystine	333→194		2.0	
Hydroxyproline	188→132	D4-Alanine	2.1	0.9
Isoleucine	188→86	D3-Leucine	9.9	0.9
Leucine	188→86	D3-Leucine	10.8	1.0
D3-Leucine	191→89		10.8	
Lysine	203→84	D4-Lysine	1.2	1.0
D4-Lysine	207→88		1.2	
Methionine	206→104	D3-Methionine	5.8	1.0
D3-Methionine	209→107		5.8	
Ornithine	189→70	D4-Lysine	1.1	1.0
Phenylalanine	222→120	D8-Phenylalanine	15.1	1.0
D8-Phenylalanine	230→128		15.1	
Proline	172→116	D4-Alanine	2.7	1.2

(continued)

Table 3.2 (continued)

Amino acid	Q1→Q3	Internal standard	Absolute RT	Relative RT
Sarcosine	146→90	D4-Alanine	2.0	0.9
Serine	162→60	D4-Alanine	1.7	0.8
Threonine	176→74	D4-Alanine	1.9	0.9
Tyrosine	238→136	D2-Tyrosine	4.9	1.0
D2-Tyrosine	240→138		4.9	
Valine	174→72	D2-Tyrosine	4.9	1.0

3.4. Data Analysis

1. While multiple methods of internal and external calibration and quantitation are feasible, this method employs external calibration because each quantitated analyte is not paired with an identical internal standard. Calibration curves are constructed using the ratio of ion intensities associated with unlabeled amino acid to the appropriate deuterated amino acid. Dose–response curves are fitted to a second order quadratic equation with $1/x$ weighting to minimize imprecision at low concentrations. Two typical standard curves are shown in **Fig. 3.1**. The upper curve is typical for phenylalanine, which is paired with its own deuterated internal standard (D8) and is generally linear. The lower curve is typical for cystathionine, which is paired with a non-identical internal standard (D4-alanine) and is generally not linear.

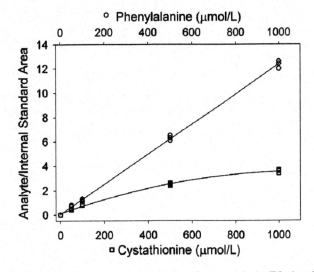

Fig. 3.1. Typical calibration (dose–response) curves for phenylalanine/D8-phenylalanine (O) and cystathionine/D4-alanine (□). Calibrations are performed in triplicate and are stable for at least 1 week.

2. The ratios of specific ion intensities from unlabeled to deuterated amino acids in serum, plasma, urine, or CSF are then compared to the standard curve for quantitation. When 100 μL of patient sample is utilized, amino acid concentrations are multiplied by 2.5, because the final concentration of the external standard is set by final reconstitution in 250 μL, while the amount of unlabeled amino acid in the patient samples is derived from 100 μL.

3. Examples of extracted ion chromatograms for two patient samples are shown in **Fig. 3.2**. Only the peaks with highest abundance are identified. A majority of amino acids elute between 2 and 4 min. The chromatogram on the left displays no abnormalities (A). The chromatogram on the right is from a patient undergoing treatment for MSUD (B). Despite treatment, note the significant elevation of leucine (RT~12 min) and the presence of allo-isoleucine (RT~10 min).

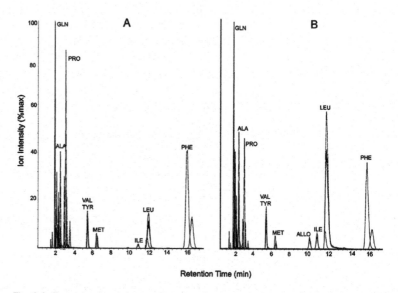

Fig. 3.2. Example chromatogram from unaffected patient (**A**) and patient with MSUD under treatment (**B**). Retention times are shifted slightly from **Table 3.2** as these data were obtained using a different column and lot of mobile phase.

4. Notes

1. Analytic characteristics of this protocol are described in detail in Dietzen et al. (5). Accuracy is compared to cation exchange with post-column ninhydrin. In general, total imprecision ranges from 5 to 20% and recovery ranges from 90 to 110%. The best accuracy, imprecision, and recovery are obtained with clinically critical amino acids that are quantitated with

an identical deuterated internal standard (e.g., phenylalanine, tyrosine, leucine, methionine). Identical internal standards are utilized for 11 of the 32 amino acids in this procedure.

2. The butylation conditions for this protocol were designed to preserve glutamine. Precursor ions of glutamine and glutamate are m/z 203 and 204, respectively, but both form significant product ions at m/z 84 and 130. Longer butylation times lead to significant deamination of glutamine, resulting in underestimation of glutamine and overestimation of glutamate. These inaccuracies may be prevented using two approaches: (A) Deuterated internal standards for glutamine and glutamate that form identical product ions may be used (e.g., 2,3,3,4,4-D5-glutamine and 2,3,3,4,4-D5-glutamate). This combination will allow internal standards to correct for glutamine conversion to glutamate. (B) In this protocol, the time and temperature of butylation are limited to prevent glutamine conversion to glutamate which allows for the use of the cheaper 2,4,4- D3-glutamate as internal standard.

3. Butylated amino acids commonly undergo loss of m/z 102 (butylformate) on CID. This transition is monitored for 12 of 32 amino acids. More specific transitions are monitored for the remaining 20. Quantitation is performed using a single precursor–product ion pair but other product ions may be monitored to insure accurate identification.

4. Four sets of isobars are quantitated in this protocol. Accurate quantitation requires distinction by retention time or by unique product ion formation. The isobars at m/z 188 (hydroxyproline, allo-isoleucine, isoleucine, leucine) are separated chromatographically. Lysine and glutamine (m/z 203) are resolved similarly. Alanine, β-alanine, and sarcosine at m/z 146 are distinguished by both retention time and unique fragmentation. Alanine and sarcosine have similar product ion spectra but are separated by ~15 sec. Alanine and β-alanine have identical retention times, but β-alanine is distinguished on the basis of unique ion at m/z 72 that is not present in the spectra for alanine or sarcosine. Isobars at m/z 160 are resolved in a similar fashion. γ-Aminobutyrate and β-aminoisobutyrate are separated by ~15 sec, but the co-eluting esters of β-aminoisobutyrate and α-aminobutyrate are distinguished using a unique product ion at m/z 58.

5. Six amino acids in this protocol may be butylated on more than one carboxyl group. Five form both mono and dibutyl derivatives (cystine, cystathionine, homocystine, α-aminoadipic acid, and glutamate). In each case, a precursor–product combination derived from the monobutyl esters are monitored because of their reproducible formation and consistent

chromatography. Deuterated cystine, homocystine, and glutamate are commercially available to control for the extent of butylation. Accurate quantitation of cystathionine and α-aminoadipic acid is achieved despite using the monocarboxylic amino acid, D4-alanine, as internal standard.

6. The sixth polybutylated amino acid is argininosuccinate (ASA). ASA is a tricarboxylic acid that may also exist as at least two different dicarboxylic anhydrides. At least a dozen different butylated derivatives of ASA may be formed. In this protocol, the monobutyl and dibutyl derivatives of the tricarboxylate are the principal precursor ions. The monobutyl derivative at m/z 347 and its product ion of m/z 70 is used for quantitation. Although this approach underestimates ASA concentrations compared to ion-exchange separation with ninhydrin detection, it does not limit the detection of abnormally elevated concentrations of ASA.

References

1. Grier RE, Gahl WA, Cowan T, Bernardini I, McDowell GA, Rinaldo P. (2004) American College of Medical Genetics Standards and Guidelines for Clinical Genetics Laboratories, 2003. Genet Med 6, 66–68.

2. Deyl A, Hyanek J, Horakova M. (1986) Profiling of amino acids in body fluids and tissues by means of liquid chromatography. J Chromatogr 379, 177–250.

3. Chace DH, Kalas TA, and Naylor EW. (2003) Use of tandem mass spectrometry for multianalyte screening of dried blood specimens from newborns. Clin Chem 49, 1797–1817.

4. Watson MS, Lloyd-Puryear MA, Mann MY, Rinaldo P, Howell RR (eds). (2006) Newborn screening: toward a uniform screening panel and system [executive summary]. Genet Med 8(5, Suppl), 1S–11S.

5. Dietzen DJ, Weindel AL, Carayannopoulos MO, Landt M, Normansell ET, Reimschisel TE, and Smith CH. (2008) Rapid comprehensive amino acid analysis by liquid chromatography/tandem mass spectrometry: comparison to cation exchange with post-column ninhydrin detection. Rapid Comm Mass Spectrom 22, 3481–3488.

Chapter 4

Identification and Quantitation of Amphetamine, Methamphetamine, MDMA, Pseudoephedrine, and Ephedrine in Blood, Plasma, and Serum Using Gas Chromatography-Mass Spectrometry (GC/MS)

Josh Gunn, Scott Kriger, and Andrea R. Terrell

Abstract

Amphetamine, methamphetamine, MDMA, pseudoephedrine, and ephedrine are measured in blood, serum, and plasma using gas chromatography coupled to mass spectrometry (GC/MS). Following a simple liquid–liquid extraction, analytes are derivatized with heptafluorobutyric anhydride (HFBA) and 1 μL injected onto a HP-5MS 15-meter capillary column. Quantitation of each analyte is accomplished using a multi-point calibration curve and deuterated internal standards. The method provides a simple, robust, and reliable means to identify and measure these analytes.

Key words: Amphetamine, methamphetamine, MDMA, pseudoephedrine, ephedrine, gas chromatography, mass spectrometry, Heptafluorobutyric Anhydride (HFBA), liquid–liquid extraction.

1. Introduction

The amphetamines are similar to cocaine in their ability to modify the actions and levels of catecholamines. Amphetamines act to stimulate the sympathomimetic nervous system both centrally and peripherally. This again is achieved through increasing levels of dopamine and norepinephrine; however, the mechanism by which the amphetamines achieve this differs slightly from other stimulants such as cocaine. Protein molecules that would normally transport dopamine and norepinephrine back into the nerve terminal from the synaptic cleft fail to distinguish between them. Thus, the structural similarity between the amphetamines, dopamine,

U. Garg, C.A. Hammett-Stabler (eds.), *Clinical Applications of Mass Spectrometry*, Methods in Molecular Biology 603, DOI 10.1007/978-1-60761-459-3_4, © Humana Press, a part of Springer Science+Business Media, LLC 2010

and norepinephrine facilitates the entry of amphetamines into the presynaptic terminal. Once in the presynaptic terminal, amphetamines act to release dopamine and norepinephrine from vesicles resulting in increased levels of free catecholamines in the nerve ending. Amphetamines also inhibit monoamine oxidase (MAO), an enzyme responsible for the deactivation of free catecholamines in the presynaptic terminal. As a result, excess levels of dopamine and norepinephrine are transported out of the presynaptic terminal and into the synapse where they produce feelings of pleasure and euphoria (1–4).

Due to the increasing abuse and synthesis of amphetamine-like stimulants, there is a need for sensitive methodologies capable of detecting low levels of these drugs. Methamphetamine is the most widely abused amphetamine and reliable detection of low concentrations is extremely important to the medical and law enforcement communities alike.

2. Materials

2.1. Specimen Collection and Storage

1. Whole Blood or plasma (NaF/Potassium Oxalate or NaF/EDTA), serum (see **Note 1**). Samples are stable 6 weeks when refrigerated and for at least 1 year when frozen at –20°C or colder.

2.2. Reagents and Solutions

1. Adult bovine serum (Sigma Chemical Co) or drug free serum
2. Heptafluorobutyric Anhydride (HFBA) (Campbell Science)
3. 1% (v/v) HCl in methanol
4. 2.0 M Sodium hydroxide

2.3. Standards and Calibrators

1. Stock Standards, (1 mg/mL each in methanol): Amphetamine, Pseudoephedrine, Methamphetamine, MDMA and Ephedrine (Cerilliant Inc., Round Rock, TX).
2. Deuterated Stock Standards (1 mg/mL each in methanol): Amphetamine d6, Pseudoephedrine d3, Methamphetamine d9, MDMA d5, Ephedrine d3 (Cerilliant Inc).
3. Amphetamine Working Standard (Amphetamine-100 ng/mL, Methamphetamine-100 ng/mL, MDMA-100 ng/mL, Pseudoephedrine-100 ng/mL, Ephedrine-50 ng/mL): Prepare by adding 200 μL of amphetamine, methamphetamine, MDMA, and pseudoephedrine stock standards (1 mg/mL) to 100 μL of ephedrine stock standard (1 mg/mL) and bring to 1.1 mL with acetonitrile. Stable for 1 year when stored at –20°C or below.

4. Amphetamine Internal Standard, 10 µg/mL amphetamine, pseudoephedrine, methamphetamine, MDMA and ephedrine: Prepare by diluting 1 mL of each deuterated stock standard to 100 mL with methanol. Stable for 1 year when stored at −20°C or below.

5. Unextracted Standard, 12.5 µg/mL for amphetamine, methamphetamine, pseudoephedrine, and MDMA and 6.25 µg/mL for ephedrine: Prepare by diluting 250 µL of the amphetamine, methamphetamine, MDMA, and pseudoephedrine 1 mg/mL stock standard and 125 µL of the ephedrine 1 mg/mL stock standard to 20 mL with methanol. Stable for 1 year when stored at −20°C or below.

6. Amphetamine Calibrators: To 2970 µL negative serum add 30 µL of the 100 µg/mL working standard to get a concentration of 1,000 ng/mL of amphetamine, methamphetamine, MDMA, pseudoephedrine and 500 ng/mL of ephedrine. Prepare 1,000, 500, 250, 100, 50 ng/mL calibrators for amphetamine, methamphetamine, pseudoephedrine, and MDMA, and 500, 250, 125, 50, and 25 ng/mL for ephedrine according to **Table 4.1**. Calibrators are prepared fresh on each day of use.

Table 4.1
Preparation of the working standard calibration curve

Desired concentration* (ng/mL)	Volume of standard (µL)	Volume of negative serum (µL)
1,000/500	1,000	0
500/250	500	500
250/125	250	750
100/50	100	900
50/25	50	950
Neg	0	1,000

*The lower concentrations are for ephedrine.

7. Controls:
   A. High Control (Amphetamine-800 ng/mL, Methamphetamine-800 ng/mL, MDMA-800 ng/mL, Pseudoephedrine-800 ng/mL, Ephedrine-400 ng/mL): Prepare by diluting 160 µL of amphetamine, methamphetamine, MDMA, and pseudoephedrine stock standard (1 mg/mL) and 80 µL of ephedrine stock standard (1 mg/mL) to

200 mL with acetonitrile. Controls are divided into single use aliquots and are stable for 1 year when stored at –20°C or below.

B. Low Control (Amphetamine-200 ng/mL, Methamphetamine-200 ng/mL, MDMA-200 ng/mL, Pseudoephedrine-200 ng/mL, Ephedrine-100 ng/mL): Prepare by diluting 40 μL of amphetamine, methamphetamine, MDMA, and pseudoephedrine 1 mg/mL stock standard and 20 μL of ephedrine 1 mg/mL stock standard to 200 mL with acetonitrile. Controls are divided into single use aliquots and are stable for 1 year when stored at –20°C or below.

C. Negative bovine serum is used as the negative control.

2.4. Supplies

1. 12 × 75 mm glass test tubes

2. 13 × 100 mm glass tubes with Teflon lined screw caps

3. Glass GC autosampler vials

4. 11 mm crimp caps

5. Column: HP-5MS Length: 15 m, Internal Diameter: 0.25 mm, Film Thickness: 0.25 μm (J&W Scientific)

---

3. Methods

3.1. Preparation of the Samples, Controls, and Calibrators

1. To each 1 mL sample, control or calibrator add 250 μL of 2.0 M NaOH.

2. Add 50 μL of 10 μg/mL internal standard to each sample.

3. Vortex all tubes for 10–15 sec.

4. Add 5 mL of ethyl acetate to each tube.

5. Cap all tubes with large Teflon disposable caps and place on block vortex for 10 min.

6. Centrifuge at $3000 \times g$ for 5 min.

7. Transfer the organic layer to a small test tube using a glass transfer pipette.

8. Add 100 μL of 1% (v/v) HCl in methanol to each tube.

9. Dry samples down under nitrogen.

10. Add 25 μL of ethyl acetate and 25 μL HFBA to each sample tube.

11. Cap with non-vented caps and place in the 75°C heat block for 30 min.

12. Dry down under nitrogen.

13. Reconstitute with 40 µL of ethyl acetate.

14. Transfer sample to autosampler vial for analysis.

3.2. Unextracted Sample

Transfer 40 µL of the 12.5 µg/mL standard and 50 µL of the 10 µg/mL internal standard into a 12 × 75 mm test tube. Process the sample as above starting with step 8 by addition of 100 µL of 1% (v/v) HCl in methanol.

3.3. Instrument Operating Conditions

1. **Gas Chromatography (GC):** Chromatographic separations were performed on an Agilent 6890 GC-MSD instrument equipped with a 5973 MSD, 6890 GC, and 6890 series autosampler. Separations were achieved on an HP-5MS 15 m capillary column having an internal diameter of 0.25 mm and film thickness of 0.25 µm (J&W Scientific). The carrier gas was helium (UHP purity). The GC conditions are provided in **Tables 4.2** and **4.3**.

Table 4.2
GC conditions

Inlet/Column/Detector Conditions

Inlet temperature	175°C	Inlet pressure	5.15 psi
Pulse pressure	20 psi	Pulse time:	0.20 min
Purge flow	20 mL/min	Purge time:	0.20 min
Total flow	24.8 mL/min	Gas type:	Helium
Detector temp	280°C		
Injection volume	1 µL		

Table 4.3
GC conditions continued

Oven Conditions

Time	Rate	Temperature
Initial	N/A	60°C
1 min	N/A	60°C
	20°C/min	220°C
	Total run time 9.0 min	
	Re-equilibration time 0.50 min	

2. **Mass Spectrometry:** Mass spectrometric detection was performed using an Agilent 5973 MSD equipped with a 70 eV electron ionization (EI) source operating in positive ion mode. The MSD was operated in the selected ion monitoring (SIM) mode with a solvent delay of 4 min. The quadrupole temperature was maintained at 150°C and the source temperature was maintained at 230°C. **Table 4.4** outlines the ions monitored for the analysis, and the **Fig. 4.1** shows the total ion chromatogram of various drugs.

Table 4.4
Ions monitored for the data analysis

Analyte	Ion (*m/z*)	Retention time (min)	Purpose	Dwell time
Amphetamine	240	4.66	Quantitation	20
	118		Qualifier	20
	91		Qualifier	20
Methamphetamine	254	5.30	Quantitation	20
	210		Qualifier	20
	118		Qualifier	20
Ephedrine	254	5.54	Quantitation	20
	210		Qualifier	20
	344		Qualifier	20
Pseudoephedrine	254	5.82	Quantitation	20
	210		Qualifier	20
	344		Qualifier	20
MDMA	254	7.14	Quantitation	20
	210		Qualifier	20
	162		Qualifier	20

3.4. Data Analysis

The data was processed using the ChemStation® software. The results for the batch are considered accurate, reliable and acceptable if the quantitation of the quality control samples agree within ± 20% of the expected values. If either of the QC values is outside of the acceptance criteria, the data is considered to not be reliable. The calibration curve is considered acceptable if $r^2 \geq 0.98$ and each of the individual data points agree to within ±20% of the expected value. Analyte quantitation was accomplished using the standard calibration curve and the ratio of the peak area of the analyte quantifying ion to the peak area of the internal standard quantifying ion.

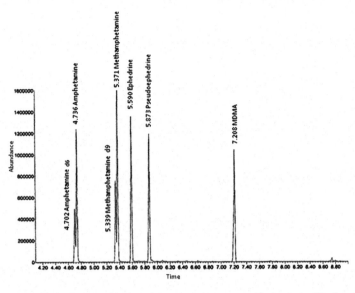

Fig. 4.1. Total Ion Chromatogram (TIC) illustrating the separation and retention time for each component.

4. Notes

1. Avoid the use of gel-barrier tubes for serum or plasma samples. For serum, use tubes without barriers such as a "plain" red top or royal blue. Barrier tubes have been shown to bind some drugs and/or to release chemicals, which can interfere with chromatographic analyses.

References

1. Drummer, O.H., *The Forensic Pharmacology of Drugs of Abuse*. 2001, London: Arnold.
2. Di Maio, T.G. and V.J.M. Di Maio, *Excited Delirium Syndrome Cause of Death and Prevention*. 2006, Boca Raton: Taylor & Francis Group.
3. McCann, U. and G. Ricaurte, *Amphetamine neurotoxicity: Accomplishments and remaining challenges*. Neurosci Biobehav Rev, 2004, 27(8): 821–826.
4. Guyton, A.C., *Structure and Function of the Nervous System*. 1972, Philadelphia, PA: W.B Saunders Company.

Chapter 5

Quantification of Antidepressants Using Gas Chromatography-Mass Spectrometry

Ruth E. Winecker

Abstract

Antidepressants are of great interest to clinical and forensic toxicologists as they are frequently used in suicidal gestures; they can be the source of drug interactions and some have narrow therapeutic indices making the potential for toxicity more likely. There are five categories of antidepressants based on function and/or structure. These are monoamine oxidase inhibitors (MAOI), cyclic antidepressants including tricyclic and tetracyclic compounds (TCA), selective serotonin reuptake inhibitors (SSRI), serotonin-norepinephrine reuptake inhibitors (SNRI), and atypical compounds. This method is designed to detect the presence of antidepressant drugs in blood/serum, urine, and tissue specimens using gas chromatography/mass spectrometry (GC/MS) following liquid–liquid extraction (LLE) and identified by relative retention times and mass spectra.

Key words: Antidepressants, Liquid–Liquid Extraction (LLE), Gas Chromatography-Mass Spectrometry (GCMS).

1. Introduction

Antidepressants (AD) are among the most frequently prescribed drugs with 172.3 million prescriptions written annually in the United States (1). The most common rationale for the administration of AD is major depression, although they are used for the treatment of chronic pain, bipolar disorder, various anxiety disorders, and attention deficit hyperactivity disorder as well as others (2). It is estimated that 8–12% of the population will suffer from depression at least once during their lifetime. Antidepressants are of great interest to clinical and forensic toxicologists as they are frequently used in suicidal gestures; they can be the source of drug interactions and

U. Garg, C.A. Hammett-Stabler (eds.), *Clinical Applications of Mass Spectrometry*, Methods in Molecular Biology 603,
DOI 10.1007/978-1-60761-459-3_5, © Humana Press, a part of Springer Science+Business Media, LLC 2010

some have narrow therapeutic indices making the potential for toxicity more likely (2). In general, AD can be divided into groups based on structure or function: monoamine oxidase inhibitors (MAOI), cyclic antidepressants including tricyclic and tetracyclic compounds (TCA), selective serotonin reuptake inhibitors (SSRI), serotonin-norepinephrine reuptake inhibitors (SNRI), and atypical compounds.

Quantification is important to monitor treatment compliance, for therapeutic drug monitoring in AD with narrow therapeutic indices, and to assess drug interactions caused by metabolic interferences via inspection of parent metabolite ratios (2). Fortunately, all of these compounds are weakly basic in nature and are easily separated from biological matrices using a modification of a previously published liquid–liquid extraction (LLE) procedure (3).

2. Materials

2.1. Reagents

1. n-Butyl Chloride:Ethyl Ether: In a 1l graduated cylinder, place 750 mL of n-butyl chloride. Add 250 mL of ethyl ether. Mix well. Store with cap tightly closed. Stable at room temperature for 6 months.

2. 2 N Sulfuric Acid: In a 500 mL volumetric flask, place approximately 300 mL of deionized water. Slowly add 29 mL of concentrated sulfuric acid and gently vortex. Dilute to volume with de-ionized water. Mix well. Stable at room temperature for 3 months.

3. 20 µg/mL Alphaprodine Internal Standard: In a 25 mL volumetric flask, add 0.5 mL of 1.0 mg/mL Alphaprodine. Dilute to volume with methanol. Mix well. Store at 4°C. Stable for 1 year.

4. Mixed Antidepressant Calibration Standards. *See* **Table 5.1** for preparation details. Store at 4°C. Stable for 1 year.

5. Mixed Quality Controls Standards. *See* **Table 5.2** for preparation details. Store at 4°C. Stable for 1 year

6. Hexane, HPLC grade

7. n-Butyl Acetate, HPLC grade

8. Concentrated Ammonium Hydroxide. Stable for 1 month stored tightly closed at room temperature.

9. Drug-free blood. Store at 4°C. Stable for 3 months.

Table 5.1
Preparation of working standards

	Stock source	Stock concentration (mg/mL)	Volume of stock used	Final standard concentration (ug/mL)*
Standard 1 drugs				
Bupropion	Alltech	1	0.5	20
Threoamino-bupropion	Toronto Research Chemicals**	1	0.5	20
Venlafaxine	Cerilliant	1	0.5	20
Desmethylvenlafaxine	Wyeth**	1	0.5	20
Amitriptyline	Cerilliant	1	0.5	20
Nortriptyline	Cerilliant	1	0.5	20
Mirtazapine	Cerilliant	1	0.25	10
Trazodone	Alltech	1	0.5	20
Standard 2 drugs				
Fluoxetine	Cerilliant**	1	0.5	20
Norfluoxetine	Cerilliant	1	0.5	20
Fluvoxamine	Cerilliant	1	0.5	20
Doxepin	Cerilliant	1	0.5	20
Nordoxepin	Cerilliant	1	0.5	20
Sertraline	Cerilliant	1	0.5	20
Desmethylsertraline	Cerilliant	1	0.5	20
Citalopram	Alltech**	1	0.5	20
Paroxetine	Cerilliant	1	0.5	20
Standard 3 drugs				
Imipramine	Cerilliant	1	0.5	20
Desipramine	Cerilliant	1	0.5	20
Clomipramine	Cerilliant	1	0.5	20
Amoxapine	Alltech	1	0.5	20

* Prepared by transferring the denoted volume of stock solution of each drug to a 25 mL volumetric flask and diluting with acetonitrile.
**Alltech (State College, PA, USA) Cerilliant (Round Rock, TX, USA) Toronto Research Chemicals (North York, Ontario, Canada) Wyeth (Philadelphia, PA, USA).

Table 5.2
Low and high control preparation

	Stock source	Stock concentration (mg/mL)	Volume of stock	Final standard concentration (ug/mL)
Low control*				
Fluoxetine	Cerilliant	1	0.5	10
Norfluoxetine	Cerilliant	1	0.5	10
Sertraline	Cerilliant	1	0.5	10
Desmethylsertraline	Cerilliant	1	0.5	10
Citalopram	Alltech	1	0.5	10
Paroxetine	Cerilliant	1	0.5	10
Bupropion	Alltech	1	0.5	10
Threoamino-bupropion	Toronto Research Chemicals	1	0.5	10
Venlafaxine	Cerilliant	1	0.5	10
Desmethylvenlafaxine	Wyeth	1	0.5	10
Amitriptyline	Cerilliant	1	0.5	10
Nortriptyline	Cerilliant	1	0.5	10
Mirtazapine	Cerilliant	1	0.25	5
Trazodone	Alltech	1	1.0	20
High control**				
Fluoxetine	Cerilliant	1	1.0	40
Norfluoxetine	Cerilliant	1	1.0	40
Sertraline	Cerilliant	1	1.0	40
Desmethylsertraline	Cerilliant	1	1.0	40
Citalopram	Alltech	1	1.0	40
Paroxetine	Cerilliant	1	1.0	40
Bupropion	Alltech	1	1.0	40
Threoamino-bupropion	Toronto Research Chemicals	1	1.0	40
Venlafaxine	Cerilliant	1	1.0	40

(continued)

Table 5.2 (continued)

	Stock source	Stock concentration (mg/mL)	Volume of stock	Final standard concentration (ug/mL)
Desmethylvenlafaxine	Wyeth	1	1.0	40
Amitriptyline	Cerilliant	1	1.0	40
Nortriptyline	Cerilliant	1	1.0	40
Mirtazapine	Cerilliant	1	0.5	20
Trazodone	Alltech	1	1.0	40

*Prepared by transferring the denoted volume of stock solution of each drug to a 50 mL volumetric flask and diluting with acetonitrile.
**Prepared by transferring the denoted volume of stock solution of each drug to a 25 mL volumetric flask and diluting with acetonitrile.

2.2. Specimens

Typically, blood specimens are analyzed for antidepressants; however, when not available for testing, other specimens such as serum, urine, vitreous, and tissue may be substituted. Specimen volumes used for testing vary and depend on specimen availability; the standard volume used when screening for antidepressants is 2 mL of blood or serum, 1 g of tissue homogenate (1:4 dilution), or 1 mL of urine (*see* **Note 1**). Other volumes are used as needed for specimens exceeding the upper limit of quantification.

2.3. Equipment and Supplies

1. Gas Chromatograph Mass Spectrometer (GCMS): Agilent 6890 GC and 5973 MS (Wilmington, DE) or Thermo-Fisher DSQ II (Austin, TX).

2. Capillary analytical column: 15 m, 5% phenyl substituted di-methyl polysiloxane with 0.25 mm id and 0.25 μm film thickness.

3. Methods

3.1. Agilent Instrument Conditions

a. GC: A 2-μL aliquot is injected into a split/splitless injection port connected to a mass selective detector utilizing helium as the carrier gas at a flow rate of 1.2 mL/min. The split ratio is 6:1. The injection port and transfer line temperatures are set to 275 and 280°C, respectively. The GC oven is programmed to an initial temperature of 70°C, held for 2 min followed by a 15°C/min ramp to a final temperature of 300°C held for 7.67 min for a total run time of 25 min.

b. MS: Ions are generated by electron impact with the emission current at 70 eV. The quadrupole and source temperatures are at 150 and 230°C, respectively. The solvent delay is 2.5 min. The detector is operated in scan mode with low and high masses of 40 and 550 m/z, respectively.

3.2. Thermo-Fisher Instrument Conditions

a. GC: A 2-µL aliquot is injected into a split/splitless injection port connected to a mass selective detector utilizing helium as the carrier gas at a flow rate of 2.0 mL/min. The split flow is 10 mL/min. The injection port and transfer line temperatures are set to 260 and 290°C, respectively. The GC oven is programmed to an initial temperature of 70°C, held for 1 min followed by a 15°C/min ramp to a final temperature of 310°C held for 7.00 min for a total run time of 24 min.

b. MS: Ions are generated by electron impact with the emission current at 70 eV. The source temperature is 250°C. The solvent delay is 2.5 min. The detector is operated in scan mode with low and high masses of 40 and 650 m/z, respectively. The detector gain is 4×10^5 and the scan rate is 5634.30.

3.3. Stepwise Extraction

1. Prepare six controls by labeling 16 × 125 mm glass culture tubes with screw tops "Standard 1," "Standard 2," "Standard 3," "Blank," "QC Low," and "QC High."

2. To all standards and patient samples add 100 µL of 20 µg/mL Alphaprodine internal standard solution.

3. Add 100 µL of Standard 1 mix to the blood control labeled "Standard 1," Add 100 µL of Standard 2 mix to the blood control labeled "Standard 2," Add 100 µL of Standard 3 mix to the blood control labeled "Standard 3," Add 50 µL of QC Low mix to the blood control labeled "QC Low," Add 100 µL of QC High mix to the blood control labeled "QC High."

4. Pipette 2 mL of negative (drug-free) control blood into all standard, blank, and QC samples (*see* **Note 1**).

5. Pipette 2 ml of specimen (unless otherwise directed upon assignment) into 16 × 125 mm glass culture tubes labeled with the case number.

6. Add *0.5 mL concentrated ammonium hydroxide* to each sample and vortex for 5 sec (*see* **Note 2**).

7. Add *7 mL n-butyl chloride/ethyl ether mixture* to each tube. Cap each tube and shake vigorously for 2 min, or place on rotator for 10 min and then centrifuge for 10 min at 3,000 × g (*see* **Note 3**).

8. Transfer the top organic layer to a new 16 × 125 mm glass culture tube.

9. Add *2.5 mL of 2 N sulfuric acid*. Cap each tube and shake vigorously for 2 min, or place on rotator for 10 min and then centrifuge for 10 min at 3,000 × g.

10. Aspirate the top n-butyl chloride/ethyl ether layer to waste.

11. Add *2 mL of hexane* to the remaining aqueous layer. Cap each tube and shake vigorously for 2 min, or place on rotator for 10 min and then centrifuge for 5 min at 3,000 × g.

12. Aspirate the top hexane layer of hexane to waste (*see* **Note 4**).

13. Add *1 mL concentrated ammonium hydroxide* to the remaining aqueous layer, vortex for 5 sec and transfer each sample to a 5 mL screw cap conical tube (*see* **Note 5**).

14. Add *100 µL butyl acetate* to each sample with a micropipette, vortex for 5 sec and centrifuge for 5 min at 3,000 × g.

15. With a Pasteur pipette, withdraw all but about 100 µL of the bottom aqueous layer to waste.

16. Centrifuge at ~3,000 × g for 5 min.

17. Transfer ~50 µL of the top organic layer to an appropriately labeled autosampler vial with a micropipette.

3.4. Screening and Quantification

1. Drug presence for screening purposes is conducted by searching available mass spectral libraries (*see* **Note 6**).

2. Quantification is based on a one-point calibration for each drug using quantifying ion area ratios of target drug versus internal standard. The analysis is considered acceptable if calculated concentrations of drugs in the controls are within 20% of target values. Peak identity is confirmed if retention time (RT) is within 0.02 min of expected RT and confirming ions are within 20% of target ratios. **Table 5.3** details brand names, elution order, RT, and relative retention times (RRT) of target antidepressants. **Table 5.4** lists quantification and qualifier ions

Table 5.3

Brand names, elution order, retention times (RT) and relative retention times (RRT) of common antidepressants

Drug	Brand name	Elution order	RT	RRT*	Notes
Alphaprodine	Nisentil	4	9.91	1.0	Internal standard
Amitriptyline	Elavil	10	12.50	1.27	
Amoxapine	Moxadil	22	14.98	1.52	
Bupropion	Wellbutrin	1	8.35	0.84	
Citalopram	Celexa, Lexapro	19	13.75	1.39	
Clomipramine	Anafranil	20	13.76	1.39	
Desipramine	Norpramin	14	12.79	1.29	Imipramine metabolite
Desmethylsertraline	n/a	17	13.41	1.36	Sertraline metabolite
Doxepin	Sinequan	13	12.71	1.29	
Fluoxetine	Prozac	6	10.35	1.05	
Fluvoxamine	Luvox	7	10.50	1.07	
Imipramine	Tofranil	12	12.67	1.28	
m-CPP	n/a	3	9.39	0.95	Trazodone metabolite
Mirtazapine	Remeron	16	12.82	1.30	
Nordoxepin	n/a	15	12.80	1.30	Doxepin metabolite
Norfluoxetine	n/a	5	10.20	1.03	Fluoxetine metabolite
Nortriptyline	Pamelor	11	12.61	1.28	Amitriptyline metabolite
Desmethylvenlafaxine	Pristiq	9	12.21	1.24	Venlafaxine metabolite
Paroxetine	Paxil	21	14.68	1.49	
Sertraline	Zoloft	18	13.49	1.37	
Threoamino-bupropion	n/a	2	9.18	0.93	Bupropion metabolite
Trazodone	Desyrel	23	18.36	1.86	
Venlafaxine	Effexor	8	11.84	1.20	

*RRT = RT of antidepressant/RT of alphaprodine.

Table 5.4
Quantification (Q), Qualifier ions (Q1, Q2), Limit of Detection (LOD), Limit of Quantification (LOQ), and co-elutions

Drug	Q	Q1	Q2	LOD/ LOQ*	Co-elutions
Alphaprodine	172	187	261		
Amitriptyline	58	202	275	50/100	
Amoxapine	245	193	257	50/100	
Bupropion	44	100	224	50/100	
Citalopram	58	238	324	50/100	clomipramine, norpropoxyphene
Clomipramine	269	314	228	50/100	citalopram, norpropoxyphene
Desipramine	208	195	266	50/100	mirtazapine, nordoxepin, cyclobenzaprine
Desmethylsertraline	119	246	290	50/100	
Doxepin	58	220	219	50/100	
Fluoxetine	44	104	309	50/100	
Fluvoxamine	276	187	172	100/200	
Imipramine	234	193	280	50/100	
m-CPP	154	196	139	25/50	
Mirtazapine	195	167	265	25/50	cyclobenzaprine, desipramine, nordoxepin
Nordoxepin	44	178	165	50/100	cyclobenzaprine, mirtazapine, desipramine
Norfluoxetine	134	191	104	50/100	
Nortriptyline	44	220	263	50/100	
Desmethylvenlafaxine	58	120	165	50/100	
Paroxetine	329	192	138	100/200	
Sertraline	274	159	239	50/100	
Threoamino- bupropion	44	100	208	50/100	
Trazodone	205	176	278	200/400	
Venlafaxine	58	134	179	50/100	

*ng/mL.

as well as limit of detection (LOD) and limit of quantification (LOQ) of target antidepressants. The upper limit of quantification is equivalent to the concentration in the high QC. Smaller sample volumes may be used if dilutions are needed. **Figure 5.1** illustrates the total ion chromatograms of the standards as well as other drugs detected by this procedure (*see* **Note 7**).

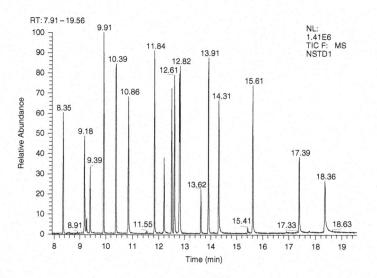

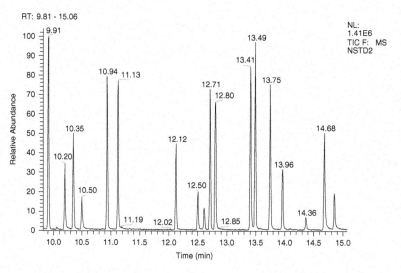

Fig. 5.1. Total ion chromatogram for standards 1, 2, and 3 (**A–C**, respectively). Peaks may be identified by RTs listed in **Table 5.3**. Note: This assay is applicable for drugs other than antidepressants and therefore there are more peaks present than those discussed in this chapter.

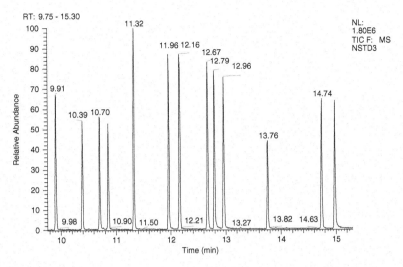

Fig. 5.1. (continued)

Table 5.5
Additional antidepressants detected by this method

Drug	Brand name	Class
Maprotiline	Ludiomil	TCA
Mianserin	Norval	TCA
Nefazodone	Serzone	Atypical
Olanzapine	Symbyax*	Atypical
Protriptyline	Vivactil	TCA
Quetiapine	Seroquel	Atypical
Trimipramine	Stangyl	TCA

*in combination with fluoxetine.

4. Notes

1. The method is validated using whole blood calibrators and controls for whole blood, serum, plasma, and urine. Matrix matched calibrators and controls are necessary for tissue specimens. Alternatively, the method of standard addition may be employed for quantification of these drugs in tissue matrices.

2. The ammonium hydroxide should be fresh for best results. Repeated opening and closing of the stock bottle will result in a decrease of pH and poor extraction recovery.

3. Specimens with excess lipid content may form an emulsion at this step. The emulsion is eliminated by the addition of 2 mL of ethyl ether, shaking for 60 sec, and re-centrifugation.

4. Remove hexane completely for best results. Vacuum aspiration is the most efficient method.

5. To avoid contamination of the organic layer with aqueous use a conical tube with a sharp point rather than a round bottom.

6. **Table 5.5** lists other antidepressants and adjunct agents, also detected by this procedure.

7. This procedure is applicable to many other weakly basic drugs including most sympathomimetic amines, narcotic analgesics, antihistamines, benzodiazepines, antipsychotics, and some cardiac agents.

References

1. "Top 200 generic drugs by units in 2007." *Drug Topics, Feb 18, 2008.* URL: http://drugtopics.modernmedicine.com/drugtopics/data/articlestandard/drugtopics/072008/491207/article.pdf%20%20%20. Accessed 12/01/08.

2. Baldessarini, R. L. (2001) Drugs and the treatment of psychiatric disorders: Depression and anxiety disorders. In: Hardeman, J.G., and Limbard, L.E. (eds), Goodman and Gilman's The Pharmacological Basis of Therapeutics, pp. 447–484.

3. Foerster, E.H., Hatchett, D., and Garriott, J.C. (1978) A rapid, comprehensive screening procedure for basic drugs in blood or tissue by gas chromatography. J. Anal. Toxicol. 2: 50–55.

Chapter 6

Quantitation of Argatroban in Plasma Using Liquid Chromatography Electrospray Tandem Mass Spectrometry (UPLC-ESI-MS/MS)

Ross J. Molinaro

Abstract

The following chapter describes a method to measure argatroban in plasma samples using ultra-performance liquid chromatography combined with electrospray positive ionization tandem mass spectrometry (UPLC-ESI-MS/MS). Samples are pre-treated using methanol containing the internal standard (IS) diclofenac. The plasma extracts are dried under a stream of nitrogen. The residue is reconstituted in ammonium acetate–formic acid–water. The reconstituted plasma is injected into the UPLC-ESI-MS/MS. Argatroban and diclofenac show similar retention times in plasma. Quantification of argatroban in the samples is made by multiple reaction monitoring using the hydrogen adduct mass transitions, from a seven-point calibration curve.

Key words: Argatroban, diclofenac, mass spectrometry, liquid chromatography, anticoagulant, heparin-induced thrombocytopenia.

1. Introduction

Argatroban was recently approved for prophylaxis of thrombosis in patients with heparin-induced thrombocytopenia (HIT) and as an anticoagulant in patients undergoing percutaneous coronary intervention who have or are at risk for HIT (1, 2). Argatroban is derived from L-arginine, which directly and reversibly binds to the catalytic site of either free or clot-bound thrombin, making it a direct thrombin inhibitor (3).

Pharmacokinetic and pharmacodynamic studies have revealed that renal function, age, and sex do not have a clinical effect on metabolism, distribution, elimination, or anticoagulation of

U. Garg, C.A. Hammett-Stabler (eds.), *Clinical Applications of Mass Spectrometry*, Methods in Molecular Biology 603, DOI 10.1007/978-1-60761-459-3_6, © Humana Press, a part of Springer Science+Business Media, LLC 2010

argatroban (4, 5). Hepatic metabolism of argatroban occurs primarily via cytochrome P-450 3A4 (CYP3A4), so dosage modification is necessary in patients with hepatic insufficiency. Steady-state blood levels and anticoagulant effect of argatroban are usually obtained 1–3 h after initiation of therapy. Both steady-state concentrations and anticoagulant effects are well correlated and predictable. No significant interactions between argatroban and aspirin, erythromycin, acetaminophen, digoxin, lidocaine, or warfarin have been demonstrated or reported (4, 6, 7). The use of argatroban and warfarin concomitantly, however, can increase the risk of bleeding, so an assay that can specifically measure argatroban, such as mass spectrometry, would be beneficial for drug monitoring (7).

2. Materials

2.1. Samples

1. Plasma is obtained from centrifuged (2,500 × g for 15 min) whole blood specimens collected in 3.2% sodium citrate blood collection tubes. Serum samples have not been validated for this assay.

2. Plasma should be tested within 24 h if stored at room temperature or at 2–4°C. Plasma should be tested within 2 weeks at –20°C and within 6 months if stored at –80°C. If immediate testing is to be done, the plasma may remain on the packed cells or separated.

2.2. Reagents and Buffers

1. Formic acid and methanol (Fisher Scientific, Fair Lawn, NJ or Sigma-Aldrich, St. Louis, MO) and were of analytical or chromatography grade.

2. 1 M Ammonium acetate (77 g anhydrous ammonium acetate in 1 L water). Store at room temperature 18–24°C. Stable for 1 year.

3. Mobile Phase A: (2 mM ammonium acetate-1 mL/L formic acid in water). Store at room temperature 18–24°C. Stable for 1 year.

4. Mobile Phase B: (2 mM ammonium acetate-1 mL/L formic acid in methanol). Store at room temperature 18–24°C. Stable for 1 year.

5. Human drug-free pooled plasma.

2.3. Standards and Calibrators

1. Primary argatroban standard: GlaxoSmithKline Argatroban Injection U.S.P. 100 mg/mL (*see* **Note 1**).

2. Secondary argatroban standard (1 mg/mL): Prepare by transferring 100 µL of argatroban primary standard into a 10 mL

Table 6.1
Preparation of plasma calibrators

Calibrator	Drug-free plasma (mL)	Tertiary plasma standard (mL)	Concentration (ng/mL)
1	8.00	2.00	2,000
2	9.00	1.00	1,000
3	9.40	0.60	600
4	9.70	0.30	300
5	9.90	0.10	100
6	9.97	0.03	30
7	10.00	0	0

volumetric flask and diluting to volume with 50% analytical grade water/50% methanol. Secondary standard is stable for 1 year at $-80°$ C.

3. Tertiary standard (10 μg/mL): Prepare by transferring 100 μL to a 10 mL volumetric flask and filling with drug-free plasma. Tertiary standard is stable for 1 year at $-80°$ C.

4. Calibrators are made according to **Table 6.1** using 10 mL volumetric flasks. The calibrators are stable for 6 months when stored at $-20°$ C.

2.4. Internal Standard and Quality Controls

1. Diclofenac primary internal standard (IS), diclofenac sodium salt (Sigma-Aldrich Co.): Prepare by dissolving 100 mg of diclofenac sodium salt into 10 ml 50% analytical grade water/50% methanol using a 10 ml volumetric flask.

2. Working IS solution (500 ng/mL): Prepare by transferring 5.0 μL of diclofenac I (10 mg/ml) to a 100 mL volumetric flask and diluting with analytical-grade methanol. This Working IS Solution is stable for 6 months at -20°C.

3. Quality control samples: Prepare according to **Table 6.3** using 10 mL volumetric flasks. Independent stock solutions of argatroban in plasma are used to prepare the controls listed in **Table 6.2**. The controls are stable for 6 months when stored at $-80°$ C.

2.5. Supplies

1. Target DP^{TM} Vials C4000-1 W autosampler vials, 0.300 mL limited volume Target PP Polyspring inserts, DP^{TM} Blue Cap (T/RR Septa) (National Scientific, Rockwood, TN, USA)

2. Seal-Rite® 1.5 mL tubes (USA Scientific, Inc., Ocala, FL, USA)

3. 15 × 75 mm glass tubes (Fisher Scientific, Fair Lawn, NJ, USA)

Table 6.2

Preparation of in-house plasma controls

Control	Negative PLASMA (mL)	10 ug/mL argatroban plasma standard (mL)	Concentration (ng/mL)
Low	9.95	0.05	50
Medium	9.50	0.50	500
High	8.50	1.50	1,500

2.6. Equipment

1. Waters ACQUITY UPLC® TQD and Masslink software
2. Zymark TurboVap® IV Evaporator
3. ACQUITY BEH C18 column, 1.7 μm, 2.1 × 50 mm.

3. Methods

3.1. Stepwise Procedure

1. Label 1.5 mL microcentrifuge tubes for each standard, control, and sample to be analyzed.
2. Pipette 500 μL of working IS solution to each microcentrifuge tube.
3. Add 100 μL of plasma standard, plasma control, or plasma sample into appropriate tubes.
4. Vortex for 2 min (*see* **Note 2**).
5. Centrifuge samples at $11,356 \times g$ for 10 min.
6. Label a 15 × 75 mm glass tube for each standard, control, and sample.
7. Accurately transfer 300 μL supernatant from step 5 to labeled 15 × 75 mm glass tubes (*see* **Note 3**).
8. Concentrate eluate to dryness under N_2 in a water bath at room temperature.
9. Add 300 μL mobile phase A to residue in glass tubes.
10. Vortex for 15 sec to ensure adequate reconstitution.
11. Label autosampler vials and add glass inserts into labeled vials.
12. Transfer entire amount of reconstituted sample into appropriately labeled vial inserts.

3.2. Instrument Operating Conditions

1. Inject 4 μLs plasma onto UPLC-ESI-MS/MS.
2. The instrument's operating conditions are given in **Table 6.3**.

Table 6.3
UPLC-ESI-MS/MS operating conditions

A. UPLC[a]

Column temp. (°C)	30	
Flow (mL/min)	0.25 mL/min	
Gradient	Time (min)	% Mobile Phase A
	0.0	70 (Start)
	0.5	70
	1.0	10
	3.0	10
	3.5	70 (End)

B. MS/MS Tune Settings[b]

Capillary (kV)	1.00
Cone (V)	50
Source temp. (°C)	150
Desolvation temp (°C)	300
Cone gas flow (L/Hr)	5
Desolvation gas flow (L/Hr)	550
LM 1 resolution	5.00
LM 2 resolution	5.00
HM 1 resolution	5.00
HM 1 resolution	5.00

[a]Mobile phase A, 2 mM ammonium acetate-1 mL/L formic acid in water; mobile phase B, 2 mM ammonium acetate-1 mL/L formic acid in methanol.
[b]Tune settings may vary slightly between instruments.

3.3. Data Analysis

1. Argatroban and diclofenac precursor and product ions used for quantification are described in **Table 6.4**. These ions were optimized based on tuning parameters also listed in **Table 6.4** and can change based on tuning parameters.

2. The data are analyzed using TargetLynx Software (Waters Corp., Milford, MA, USA) or similar software. Multiple reaction monitoring and peak area ratios of analyte vs. IS of quantifying ions is used to quantify the concentration of argatroban in samples. Ratio limits of precursor and primary product ions were set at 10% to confirm the presence of the argatroban and diclofenac. The run is considered acceptable if calculated concentrations of argatroban in the controls are within 2 standard deviations of the target values. Liquid

Table 6.4
Precursor and primary productions for argatroban and diclofenac

	Precursor ion (H+ Transitions)	Product ions[a,b]	Dwell (sec)	Cone (V)	Collision (eV)
Argatroban	509.4	384.4	0.1	20	20
Diclofenac	296.1	250.2	0.1	50	40

[a]Optimized *m/z* may change based on tuning parameters.
[b]Quantitation trace for primary ion.

chromatography retention time window limits for argatroban and were set at 0.3 and 0.2 min, respectively (all compounds with transitions specific for argatroban and diclofenac within the stated window limit will be labeled argatroban and diclofenac).

3. The linearity/limits of quantitation of the plasma method is 0.04–2,000 µg/mL. Samples in which the argatroban concentrations exceed the upper limit of quantitation can be diluted with appropriate pooled plasma.

4. Typical coefficient of correlation of the standard curve is >0.99.

5. Typical intra- and inter-assay variations are <5%.

4. Notes

1. Argatroban is light sensitive and should be protected by wrapping all standard, control, and IS solutions with aluminum foil or another material and secure from light exposure.

2. Ensure the microcentrifuge tube caps are secured properly.

3. Be careful not to transfer any of the pellets from the bottom of the tubes after microcentrifugation.

References

1. Warkentin T.E., Greinacher A., Koster A., Lincoff A.M., American College of Chest Physicians. (2008) Treatment and prevention of heparin-induced thrombocytopenia: American College of Chest Physicians Evidence-Based Clinical Practice Guidelines (8th Edition). *Chest*, Jun;133(6 Suppl):340S–380S.

2. Messmore H.L., Jeske W.P., Wehrmacher W.H., Walenga J.M. (2003) Benefit-risk assessment of treatments for heparin-induced thrombocytopenia. *Drug Saf*, 26:625–641.

3. McKeage K., Plosker G.L. (2001) Argatroban. *Drugs*, 61:515–522.

4. GlaxoSmith-Kline. (2003) Prescribing Information. Argatroban. Research Triangle Park, NC.

5. Swan S.K., Hursting M.J. (2000) The pharmacokinetics and pharmacodynamics of argatroban: effects of age, gender, and hepatic or renal dysfunction. *Pharmacotherapy*, 20:318–329.

6. Brown P.M., Hursting M.J. (2002) Lack of pharma-cokinetic interactions between argatroban and warfarin. *Am J Health Syst Pharm*, 59:2078–2083.

7. Inglis A.L., Sheth S.B., Hursting M.J., Tenero D.M., Graham A.M., DiCicco R.A. (2002) Investigation of the interaction between argatroban and acetaminophen, lidocaine, or digoxin. *Am J Health Syst Pharm*, 59:1258–1266.

Chapter 7

Quantitation of Amobarbital, Butalbital, Pentobarbital, Phenobarbital, and Secobarbital in Urine, Serum, and Plasma Using Gas Chromatography-Mass Spectrometry (GC-MS)

Leonard L. Johnson and Uttam Garg

Abstract

Barbiturates are central nervous system depressants with sedative and hypnotic properties. Some barbiturates, with longer half-lives, are used as anticonvulsants. Their mechanism of action includes activation of γ-aminobutyric acid (GABA) mediated neuronal transmission inhibition. Clinically used barbiturates include amobarbital, butalbital, pentobarbital, phenobarbital, secobarbital, and thiopental. Besides their therapeutic use, barbiturates are commonly abused. Their analysis is useful for both clinical and forensic proposes. Gas chromatography mass spectrometry is a commonly used method for the analysis of barbiturates. In the method described here, barbiturates from serum, plasma, or urine are extracted using an acidic phosphate buffer and methylene chloride. Barbital is used as an internal standard. The organic extract is dried and reconstituted with mixture of trimethylanilinium hydroxide (TMAH) and ethylacetate. The extract is injected into a gas chromatogram mass spectrometer where it undergoes "flash methylation" in the hot injection port. Selective ion monitoring and relative retention times are used for the identification and quantitation of barbiturates.

Key words: Gas chromatography, mass spectrometry, barbiturates, amobarbital, butalbital, pentobarbital, Phenobarbital, secobarbital.

1. Introduction

Barbiturates are CNS suppressants which are medically used as sedative-hypnotics, anticonvulsants, and for reduction of cerebral edema secondary to head injury (1, 2). Barbiturates used in clinical practice include amobarbital, butalbital, pentobarbital, phenobarbital, secobarbital, and thiopental. Depending on their half-life barbiturates are classified as ultrashort, short, intermediate, and long-acting. Short-acting barbiturates such as secobarbital and

U. Garg, C.A. Hammett-Stabler (eds.), *Clinical Applications of Mass Spectrometry*, Methods in Molecular Biology 603, DOI 10.1007/978-1-60761-459-3_7, © Humana Press, a part of Springer Science+Business Media, LLC 2010

pentobarbital are used to induce anesthesia and long-acting barbiturates such as phenobarbital and methobarbital are used as anticonvulsants. The major mechanism of action of barbiturates is believed to take place by potentiating the action of γ-aminobutyric acid (GABA) at its receptor. GABA is the principal inhibitory neurotransmitter in the human CNS. The use of barbiturates has declined over time because of their low therapeutic index and high potential of abuse and they have been replaced by benzodiazepines for many purposes.

Laboratory analysis of barbiturates is needed due to their narrow therapeutic index and potential for abuse (3, 4). The methods of analysis include immunoassays and colorimetric and chromatographic methods. Colorimetric methods are nonspecific and are thus not normally used. Immunoassays are available for the detection of barbiturates as a class or for specific barbiturates (5–7). Immunoassays are the most commonly used method for screening. However, chromatographic methods are the reference methods and are preferred over immunoassays and are generally needed to confirm immunoassay positive results for forensic purposes. Gas chromatography mass spectrometry (GC-MS) is a commonly used technique for analysis and confirmation of barbiturates (8, 9). Although GC-MS analysis can be performed without derivatization of barbiturates, derivatization is preferred as it improves chromatography. Methylation is a commonly used derivatization technique (10, 11). In the methylation technique known as flash methylation, the sample extract containing the drugs is mixed with a methylation agent such as trimethylanilinium hydroxide and injected into the hot injection port. The high temperature of the injection port causes the methylation reaction to take place within the injection port. This GC-MS method is described in this chapter.

2. Materials

2.1. Samples

Serum, plasma (collected using heparin or EDTA anticoagulants), or urine are acceptable samples for this procedure. Samples are stable for 1 week when refrigerated and 3 months when frozen at −20°C.

2.2. Reagents and Buffers

1. 0.4 M Phosphate Solution A: Dissolve 11.0 g monobasic sodium phosphate monohydrate into 200 mL deionized water (stable for 1 year at room temperature).
2. 0.4 M Phosphate Solution B: Dissolve 10.7 g dibasic sodium phosphate heptahydrate into 100 mL deionized water. Stable for 1 year at room temperature.

3. 0.4 M Phosphate buffer: Add phosphate solution A to a 500 mL beaker. Adjust the pH to 6.0 ± 0.1 by slowly adding phosphate solution B (*see* **Note 1**). Stable for 1 year at room temperature.

4. Extraction Tubes: 3 mL of methylene chloride and 1 mL of 0.4 M phosphate buffer in 13 × 100 mm extraction tubes.

5. Trimethylamilimium hydroxide (TMAH) 0.2 M (Pierce, Rockford, IL).

2.3. Standards and Calibrators

1. Primary standards: Amobarbital, butalbital, secobarbital, pentobarbital, and phenobarbital at concentrations of 1 mg/mL in methanol (Cerilliant Corporation, Round Rock, TX).

2. Secondary combo standards (100 μg/mL): Prepare by transferring 1.0 mL of each primary standard (amobarbital, butalbital, secobarbital, pentobarbital, phenobarbital) into a 10 mL volumetric flask and diluting with methanol. Secondary combo standard is stable for 1 year at –20°C.

3. Tertiary combo standards (10 μg/mL): Prepare by transferring 1.0 mL of secondary combo standard to a 10 mL volumetric flask and diluting with methanol. Tertiary combo standard is stable for 1 year at –20°C.

4. Working calibrators are made according to **Table 7.1** using 10 mL volumetric flasks. The calibrators are stable for six months when stored at –20°C.

Table 7.1
Preparation of calibrators

Secondary combo standard (mL)	Tertiary combo standard (mL)	Drug-free serum or urine (mL)	Final concentration (ng/mL)
0	0.05	9.95	50
0	0.50	9.5	500
0.50	0	9.5	5,000

2.4. Internal Standard and Quality Controls

1. The primary internal standards: Barbital at concentrations of 1 mg/mL (Cerilliant Corporation, Round Rock, TX).

2. Working internal standards (20 μg/mL): Prepare by transferring 1.0 mL of the primary internal standard to a 50 mL volumetric flask and diluting with methanol. This is stable for 1 year at –20°C.

3. Quality control samples: Commercial controls (Biorad Laboratories, Irvine, CA).

2.5. Supplies

1. 13 × 100 mm screw-cap glass tubes. These tubes are used for extraction and concentrating drug extracts.
2. Transfer pipettes.
3. Auto sampler vials (12 × 32 mm with crimp caps) with 0.3 mL limited volume inserts.
4. GC column: Zebron ZB-1 with dimensions of 15 m × 0.25 mm × 0.25 μm. (Phenomenex, Torrance, California).

2.6. Equipment

1. A gas chromatograph/mass spectrometer system (GC/MS; 6890/5975 or 5890/5972) with autosampler and operated in electron impact mode (Agilent Technologies, Wilmington, DE).
2. TurboVap® IV Evaporator (Zymark Corporation, Hopkinton, MA, USA).

3. Methods

3.1. Stepwise Procedures

1. Add 300 μL urine, serum, or plasma to the extraction tubes.
2. Add 30 μL working barbital internal standard to each tube.
3. Cap the tubes and rock for 5 min.
4. Centrifuge for 5 min at ~1,600 × g.
5. Remove and discard the top aqueous layer (*see* **Note 2**).
6. Transfer organic phase to the concentration tubes (*see* **Note 3**).
7. Prepare an unextracted standard by combining 60 μL tertiary barbiturate combo standard and 30 μL working internal standard in a concentration tube.
8. Evaporate all tubes to dryness under nitrogen in a water bath at 45°C (*see* **Note 4**).
9. Add 10 μL TMAH in each tube.
10. Add 300 μL of ethylacetate to each tube and vortex gently to mix.
11. Transfer the contents to autosampler vials containing glass inserts and inject the samples to GC/MS for analysis.

3.2. Instrument Operating Conditions

See **Table 7.2** for instrument's operating conditions.

3.3. Data Analysis

1. Representative GC-MS chromatogram of amobarbital, butalbital, secobarbital, pentobarbital, phenobarbital, and barbital (internal standard) is shown in **Fig. 7.1**. GC-MS

Table 7.2
GC operating conditions

Column pressure	5 psi
Purge time	0.5 min
Injector temperature	250°C
Detector temperature	280°C
Initial temperature	90°C
Initial time	1 min
Ramp 1	32°C/min
Temperature 2	170°C
Time 2	2 min
Ramp 2	20°C/min
Final temperature	270°C
Final time	1.5 min
MS mode	Electron impact at 70 eV, selected ion monitoring
MS tune	Auto-tune

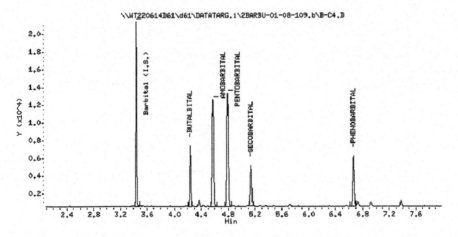

Fig. 7.1. GC-MS chromatogram of dimethyl derivatives of barbital, butalbital, amobarbital, pentobarbital, secobarbital, and phenobarbital (1250 ng/mL each).

selected ion chromatograms are shown in **Fig. 7.2**. Electron impact ionization mass spectra of these compounds are shown in **Figs. 7.3, 7.4, 7.5, 7.6, 7.7**, and **7.8** respectively (*see* **Note 5**). Ions used for identification and quantification are listed in **Table 7.3**.

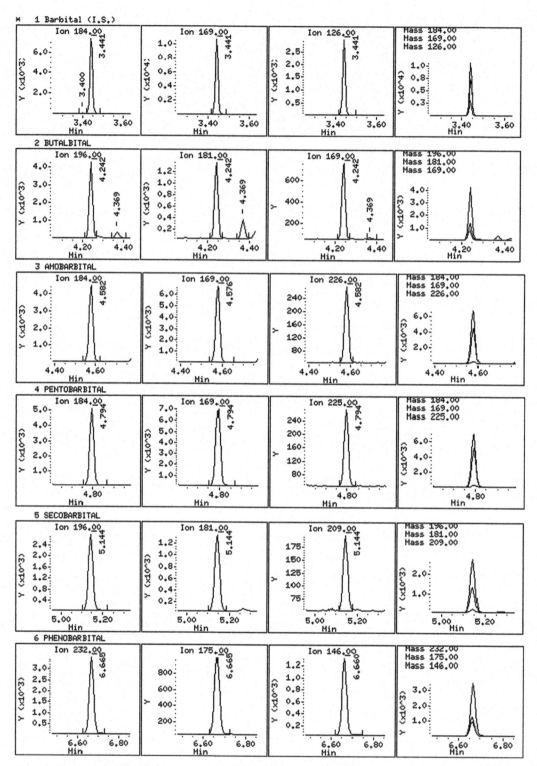

Fig. 7.2. Selected ion chromatograms of dimethyl derivatives of barbital, butalbital, amobarbital, pentobarbital, secobarbital, and phenobarbital.

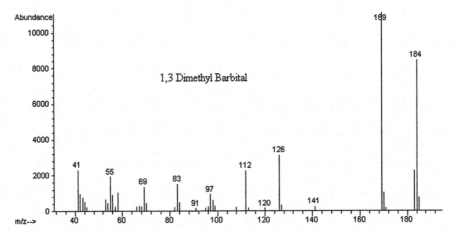

Fig. 7.3. Electron impact ionization mass spectra of dimethyl derivatives of barbital.

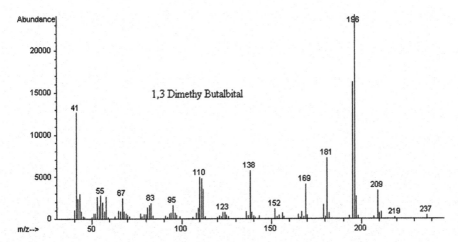

Fig. 7.4. Electron impact ionization mass spectra of dimethyl derivatives of butalbital.

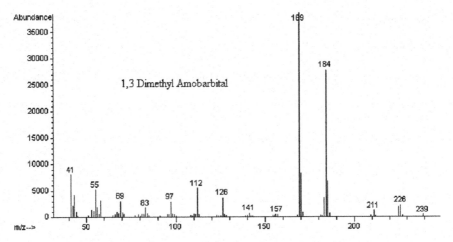

Fig. 7.5. Electron impact ionization mass spectra of dimethyl derivatives of amobarbital.

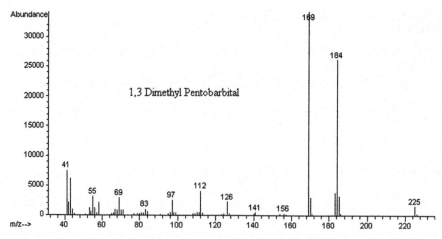

Fig. 7.6. Electron impact ionization mass spectra of dimethyl derivatives of pentobarbital.

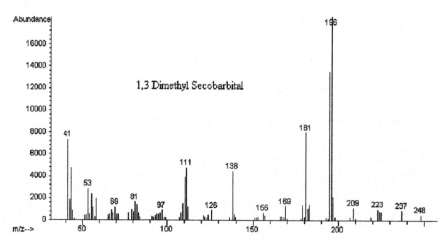

Fig. 7.7. Electron impact ionization mass spectra of dimethyl derivatives of secobarbital.

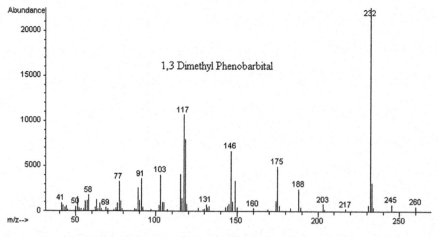

Fig. 7.8. Electron impact ionization mass spectra of dimethyl derivatives of phenobarbital.

Table 7.3
Quantitation and qualifying ions for barbiturates

Analyte	Quantitation ion	Qualifier ion(s)
Barbital	184	126, 169
Butalbital	196	169, 181
Amobarbital	184	226, 169
Pentobarbital	184	225, 169
Secobarbital	196	209, 181
Phenobarbital	232	175, 146

2. Analyze data using Target Software (Thru-Put Systems, Orlando, FL) or similar software. The quantifying ions (**Table 7.3**) are used to construct standard curves of the peak area ratios (calibrator/internal standard pair) vs. concentration. These curves are then used to determine the concentrations of the controls and unknown samples.

3. The linearity/limit of quantitation of the method is 5–200 µg/mL. Samples in which the drug concentrations exceed the upper limit of quantitation should be diluted with negative serum or plasma and retested.

4. Standard curves should have a correlation coefficient (r^2) >0.99.

5. Typical intra- and inter-assay imprecision is <10%.

6. Quality control: The run is considered acceptable if calculated concentrations of drugs in the controls are within +/−20% of target values. Quantifying ion in the sample is considered acceptable if the ratios of qualifier ions to quantifying ion are within +/−20% of the ion ratios for the calibrators.

4. Notes

1. It takes approximately 30 mL of solution B to adjust the pH to 6.0.

2. It is important to avoid disturbing the organic phase when removing the aqueous phase. It is okay to remove interface between the aqueous and organic layers.

3. Be careful not to transfer any remaining aqueous phase into the concentration tubes. This will result in failed assay, as the drugs will not derivatize.

4. Do not over-dry the extract. This will result in poor recovery and failed run.

5. Electron impact ionization spectra are needed in the initial stages of method set up to establish retention times and later on if there is a need for change in quantifying or qualifying ions. They are not needed for routine quantitaion.

References

1. Lopez-Munoz, F., Ucha-Udabe, R. and Alamo, C. (2005) The history of barbiturates a century after their clinical introduction. *Neuropsychiatr Dis Treat*, 1, 329–43.

2. Smith, M.C. and Riskin, B.J. (1991) The clinical use of barbiturates in neurological disorders. *Drugs*, 42, 365–78.

3. Coupey, S.M. (1997) Barbiturates. *Pediatr Rev*, 18, 260–4; quiz 265.

4. Morgan, W.W. (1990) Abuse liability of barbiturates and other sedative-hypnotics. *Adv Alcohol Subst Abuse*, 9, 67–82.

5. Charlier, C.J. and Plomteux, G.J. (2000) Evaluation of Emit tox benzodiazepine and barbiturate assays on the Vitalab Viva analyser and FPIA on the Abbott ADx analyser. *Clin Chem Lab Med*, 38, 615–8.

6. Cannon, R.D., Wong, S.H., Gock, S.B. and Jentzen, J.J. (1999) Comparison of the serum barbiturate fluorescence polarization immunoassay by the COBAS INTEGRA to a GC/MS method. *Ther Drug Monit*, 21, 553–8.

7. Adamczyk, M., Douglas, J., Grote, J. and Harrington, C.A. (1998) A barbiturate screening assay for the Abbott AxSYM analyzer. *J Anal Toxicol*, 22, 105–11.

8. Hall, B.J. and Brodbelt, J.S. (1997) Determination of barbiturates by solid-phase microextraction (SPME) and ion trap gas chromatography-mass spectrometry. *J Chromatogr A*, 777, 275–82.

9. Meatherall, R. (1997) GC/MS confirmation of barbiturates in blood and urine. *J Forensic Sci*, 42, 1160–70.

10. Liu, R.H., McKeehan, A.M., Edwards, C., Foster, G., Bensley, W.D., Langner, J.G. and Walia, A.S. (1994) Improved gas chromatography/mass spectrometry analysis of barbiturates in urine using centrifuge-based solid-phase extraction, methylation, with d5-pentobarbital as internal standard. *J Forensic Sci*, 39, 1504–14.

11. DeGraeve, J. and Vanroy, J. (1976) Simultaneous determination of methylated barbiturates and other anticonvulsant drugs by high-resolution gas chromatography. *J Chromatogr*, 129, 171–9.

Chapter 8

Quantitation of Benzodiazepines in Blood and Urine Using Gas Chromatography-Mass Spectrometry (GC-MS)

Bruce A. Goldberger, Chris W. Chronister, and Michele L. Merves

Abstract

The benzodiazepine assay utilizes gas chromatography-mass spectrometry (GC-MS) for the analysis of diazepam, nordiazepam, oxazepam, temazepam, lorazepam, α-hydroxyalprazolam, and α-hydroxytriazolam in blood and urine. A separate assay is employed for the analysis of alprazolam. Prior to solid phase extraction, urine specimens are subjected to enzyme hydrolysis. The specimens are fortified with deuterated internal standard and a five-point calibration curve is constructed. Specimens are extracted by mixed-mode solid phase extraction. The benzodiazepine extracts are derivatized with N-methyl-N-(tert-butyldimethylsilyl)trifluoroacetamide (MTBSFTA) producing tert-butyldimethyl silyl derivatives; the alprazolam extracts are reconstituted in methanol without derivatization. The final extracts are then analyzed using selected ion monitoring GC-MS.

Key words: Benzodiazepine, diazepam, nordiazepam, oxazepam, temazepam, lorazepam, α-hydroxyalprazolam, α-hydroxytriazolam, alprazolam, gas chromatography, mass spectrometry, solid phase extraction.

1. Introduction

Benzodiazepines are a class of central nervous system depressant drugs utilized for their anxiolytic, sedative-hypnotic, anticonvulsant, muscle relaxant, and amnesic properties in the treatment of an equally wide range of disorders. Benzodiazepines are also used as preanesthetic and intraoperative medications. Chronic use of many of the drugs within this class may produce physical dependence with withdrawal symptoms of insomnia, agitation, irritability, muscle tension, and, in more severe cases, hallucinations, psychosis, and seizures (1, 2, 3).

U. Garg, C.A. Hammett-Stabler (eds.), *Clinical Applications of Mass Spectrometry*, Methods in Molecular Biology 603,
DOI 10.1007/978-1-60761-459-3_8, © Humana Press, a part of Springer Science+Business Media, LLC 2010

Most benzodiazepines are well absorbed from the gastrointestinal tract and distributed widely. The duration of action of benzodiazepines is classified as ultra-short, short-, and long-acting. Because of their lipid solubility, they readily penetrate the blood-brain barrier and are stored in slow-releasing lipophilic tissues. Hepatic metabolism is quite extensive with both pharmacologically active and inactive metabolites produced. The half-lives of benzodiazepines and their metabolites range between 5 and 250 h (1, 2, 3).

The widespread use and the potential for abuse of benzodiazepines are responsible for the interest in clinical and forensic drug analysis. Testing for benzodiazepines consists of an initial immunoassay screen, followed by confirmation and quantitation using gas chromatography-mass spectrometry (GC-MS). Analysis of urine specimens typically includes enzymatic hydrolysis of conjugates prior to the isolation of the benzodiazepine. Solid phase extraction methods have been reported for the isolation of drug from blood and urine matrices, and derivatization is often essential in order to obtain satisfactory chromatographic performance (4, 5, 6, 7). Since the chemical structures vary greatly, a limitation to both immunoassay and chromatographic analyses is that not all benzodiazepines (and all metabolites) are detected by any given assay. The methods described below are validated for the analysis of alprazolam, diazepam, nordiazepam, oxazepam, temazepam, lorazepam, α-hydroxyalprazolam, and α-hydroxytriazolam in blood and urine (*see* **Note 1**).

2. Materials

2.1. Chemicals, Reagents, and Buffers

1. Blood, drug-free prepared from human whole blood purchased from a blood bank and pretested to confirm the absence of analyte or interfering substance.

2. N-methyl-N-(tert-butyldimethylsilyl)trifluoroacetamide, MTBSFTA (United Chemical Technologies)

3. β-Glucuronidase Enzyme Solution (5,000 units/mL): Dissolve 0.0892 g of β-glucuronidase (Sigma-Aldrich, Inc.) in 40 mL of water in a 50 mL volumetric flask. Mix well and Q.S. to 50 mL with water. Stable for 1 month at 0–8°C.

4. 0.1 M Phosphate Buffer, pH 6: Dissolve 13.61 g of potassium phosphate monobasic in 900 mL of water. Adjust the pH to 6.0 with 5.0 M potassium hydroxide. Q.S. to 1,000 mL with water. Stable for 1 year at 0–8°C.

5. 1.0 M Potassium Hydroxide: Dissolve 5.6 g of potassium hydroxide in 50 mL of water in a 100 mL volumetric flask. Mix well and Q.S. to 100 mL with water. Stable for 1 year at room temperature.

6. 5.0 M Potassium Hydroxide: Dissolve 70.13 g of potassium hydroxide in 100 mL of water in a 250 mL volumetric flask. Mix well and Q.S. to 250 mL with water. Stable for 1 year at room temperature.

7. 10% Methanol Solution: Mix 10 mL of methanol with 90 mL of water. Prepare fresh.

8. 5% Ammonium Hydroxide Solution: Mix 5 mL of ammonium hydroxide with 95 mL of water. Prepare fresh.

9. 3% Ammonium Hydroxide:Ethyl Acetate Solution: Mix 3 mL of ammonium hydroxide with 97 mL of ethyl acetate. Prepare fresh. Note: Sonicate solution for 20–30 min prior to use.

10. 1.0 M Acetic Acid: Add 28.6 mL acetic acid to 400 mL deionized water. Dilute to 500 mL with deionized water. Stable for 1 year at room temperature.

11. 0.1 M Acetate Buffer, pH 4: Add 570 µL of glacial acetic acid into a 100 mL volumetric flask filled with 80 mL of water. Add 1.6 mL of 1.0 M potassium hydroxide. Mix well, check pH and adjust if necessary to pH 4.0. Q.S. to 100 mL with water. Stable for 1 year at room temperature.

12. Methylene Chloride:Isopropanol:Ammonium Hydroxide Solution (78:20:2; v:v:v): To 20 mL of isopropanol, add 2 mL of ammonium hydroxide. Add 78 mL of methylene chloride. Mix well. Prepare fresh.

2.2. Preparation of Calibrators (see Note 2)

2.2.1. Benzodiazepine Assay

1. Benzodiazepine Internal Standard Solution (10 µg/mL): Add the contents of 100 µg/mL D5-diazepam, D5-nordiazepam, D5-oxazepam, D5-temazepam, D4-lorazepam, D5-α-hydroxyalprazolam, and D4-α-hydroxytriazolam vials (Cerilliant Corporation) to a 10 mL volumetric flask and bring to volume with methanol. Store at ≤–20°C.

2. Benzodiazepine Standard Stock Solution (100 µg/mL): Add the contents of 1.0 mg/mL diazepam, nordiazepam, oxazepam, temazepam, and lorazepam vials (Cerilliant Corporation) to a 10 mL volumetric flask and bring to volume with methanol. Store at ≤–20°C.

3. Benzodiazepine Standard Solution (10 µg/mL): Dilute 1.0 mL of the 100 µg/mL benzodiazepine standard stock solution in addition to the contents of the 100 µg/mL α-hydroxyalprazolam and α-hydroxyalprazolam vials (Cerilliant Corporation) to a 10 mL volumetric flask and bring to volume with methanol. Store at ≤–20°C.

Table 8.1
Preparation of the benzodiazepine calibration curve

Calibrator	Calibrator concentration (ng/mL)	Volume of 1.0 µg/mL standard solution	Volume of 10 µg/mL standard solution
1	50	50 µL	–
2	100	100 µL	–
3	250	–	25 µL
4	500	–	50 µL
5	1,000	–	100 µL

4. Benzodiazepine Standard Solution (1.0 µg/mL): Dilute 1.0 mL of the 10 µg/mL benzodiazepine standard solution to 10 mL with methanol in a 10 mL volumetric flask. Store at ≤–20°C.

5. Aqueous Benzodiazepine Calibrators: Add standard solutions to 1.0 mL of drug-free blood according to **Table 8.1**.

2.2.2. Alprazolam Assay

1. Alprazolam Internal Standard Stock Solution (10 µg/mL): Add the contents of 100 µg/mL D5-alprazolam vial (Cerilliant Corporation) to a 10 mL volumetric flask and bring to volume with methanol. Store at ≤–20°C.

2. Alprazolam Internal Standard Solution (1.0 µg/mL): Dilute 1.0 mL of the 10 µg/mL D5-alprazolam standard stock solution to 10 mL with methanol in a 10 mL volumetric flask. Store at ≤–20°C.

3. Alprazolam Standard Stock Solution (100 µg/mL): Add the contents of 1.0 mg/mL alprazolam vial (Cerilliant Corporation) to a 10 mL volumetric flask and bring to volume with methanol. Store at ≤–20°C.

4. Alprazolam Standard Solution (10 µg/mL): Dilute 1.0 mL of the 100 µg/mL alprazolam standard stock solution to 10 mL with methanol in a 10 mL volumetric flask. Store at ≤–20°C.

5. Alprazolam Standard Solution (1.0 µg/mL): Dilute 1.0 mL of the 10 µg/mL alprazolam standard solution to 10 mL with methanol in a 10 mL volumetric flask. Store at ≤–20°C.

6. Alprazolam Standard Solution (0.1 µg/mL): Dilute 1.0 mL of the 1.0 µg/mL alprazolam standard solution to 10 mL with methanol in a 10 mL volumetric flask. Store at ≤–20°C.

7. Aqueous Alprazolam Calibrators: Add standard solutions to 1.0 mL of drug-free blood according to **Table 8.2**.

Table 8.2
Preparation of the alprazolam calibration curve

Calibrator	Calibrator concentration (ng/mL)	Volume of 0.1 µg/mL standard solution	Volume of 1.0 µg/mL standard solution
1	5	50 µL	–
2	10	100 µL	–
3	25	–	25 µL
4	50	–	50 µL
5	100	–	100 µL

2.3. Preparation of Control Samples (see Note 2)

2.3.1. Benzodiazepine Assay

1. Negative (drug-free) Control: Prepare using human whole blood. Blood is pretested to confirm the absence of benzodiazepine or interfering substance.

2. Benzodiazepine Control Stock Solution (100 µg/mL): Add the contents of 1.0 mg/mL diazepam, nordiazepam, oxazepam, temazepam, and lorazepam vials (Cerilliant Corporation) to a 10 mL volumetric flask and bring to volume with methanol. Store at ≤–20°C.

3. Benzodiazepine Control Solution (10 µg/mL): Dilute 1.0 mL of the 100 µg/mL benzodiazepine control stock solution in addition to the contents of the 100 µg/mL α-hydroxyalprazolam and α-hydroxyalprazolam vials (Cerilliant Corporation) to 10 mL with methanol in a 10 mL volumetric flask. Store at ≤–20°C.

4. Aqueous Benzodiazepine Controls: Add benzodiazepine control solution to 1.0 mL of drug-free blood according to **Table 8.3**.

2.3.2. Alprazolam Assay

1. Alprazolam Control Stock Solution (100 µg/mL): Add the contents of 1.0 mg/mL alprazolam vial (Cerilliant Corporation) to a 10 mL volumetric flask and bring to volume with methanol. Store at ≤–20°C.

Table 8.3
Preparation of benzodiazepine controls

Control	Quality control concentration (ng/mL)	Volume of 10 µg/mL control solution
Low	250	25 µL
High	750	75 µL

2. Alprazolam Control Solution (10 μg/mL): Dilute 1.0 mL of the 100 μg/mL alprazolam control stock solution to 10 mL with methanol in a 10 mL volumetric flask. Store at ≤−20°C.

3. Alprazolam Control Solution (1.0 μg/mL): Dilute 1.0 mL of the 10 μg/mL alprazolam control solution to 10 mL with methanol in a 10 mL volumetric flask. Store at ≤−20°C.

4. Aqueous Alprazolam Controls: Add alprazolam control solution to 1.0 mL of drug-free blood according to **Table 8.4**.

Table 8.4
Preparation of alprazolam controls

Control	Quality control concentration (ng/mL)	Volume of 1.0 μg/mL control solution
Low	25	25 μL
High	75	75 μL

2.4. Supplies

1. Clean Screen Extraction Columns (United Chemical Technologies).

2. Autosampler vials.

3. Disposable glass culture tubes.

4. Volumetric pipet with disposable tips.

2.5. Equipment

1. 6890 Gas Chromatograph (Agilent Technologies, Inc.) or equivalent.

2. 5973 Mass Selective Detector (Agilent Technologies, Inc.) or equivalent.

3. 7673 Automatic Liquid Sampler (Agilent Technologies, Inc.) or equivalent.

4. ChemStation with DrugQuant Software (Agilent Technologies, Inc.) or equivalent.

5. Extraction manifold.

6. Heat Block.

7. Vortex Mixer.

8. Centrifuge.

9. Caliper Life Sciences TurboVap connected to nitrogen gas.

3. Methods

3.1. Stepwise Procedure (Excluding Alprazolam)

1. Label culture tubes for each calibrator, control and specimen and add 1.0 mL of the appropriate specimen to each corresponding tube.

2. Add 25 μL of the 10 μg/mL benzodiazepine internal standard solution to all culture tubes except the negative control.

3. Add 1.0 mL of 0.1 M phosphate buffer (pH 6).

4. Add 2.0 mL of the β-glucuronidase solution to urine specimens.

5. Cap and vortex all specimens.

6. Incubate for at least 2 h at 37° ± 2°C.

7. Add 1.0 mL of 0.1 M phosphate buffer (pH 6) and vortex.

8. Centrifuge at ∼1,500 × g for 5 min.

9. Place columns into the extraction manifold.

10. Prewash the columns with 3.0 mL of the methylene chloride:isopropanol:ammonium hydroxide solution.

11. Pass 2.0 mL of methanol through each column. Do not permit the columns to dry.

12. Pass 2.0 mL of 0.1 M acetate buffer (pH 4) through each column. Do not permit the columns to dry.

13. Pour specimen into column. Slowly draw specimen through column (at least 2 min) under low vacuum.

14. Pass 2 × 5 mL of 10% methanol through each column.

15. Pass 1 mL of 5% ammonium hydroxide solution through each column.

16. Dry column with full vacuum.

17. Turn off vacuum. Dry tips. Place labeled disposable culture tubes into column reservoir.

18. Add 3.0 mL of 3% ammonium hydroxide:ethyl acetate solution to each column and collect in disposable culture tubes. Note: Sonicate the elution solvent for 20–30 min before use.

19. Evaporate to dryness at 60°C ± 5°C with a stream of nitrogen.

20. Add 50 μL of MTBSTFA. Cap tightly and vortex.

21. Place tubes in heating block for 45 min at 70°C ± 5°C.

22. Transfer extract to autosampler vial and submit for GC-MS analysis.

3.2. Stepwise Procedure (Alprazolam Only)

1. Label culture tubes for each calibrator, control and specimen and add 1.0 mL of the appropriate specimen to each corresponding tube.

2. Add 25 µL of the 1.0 µg/mL alprazolam internal standard solution to all culture tubes except the negative control.

3. Add 2.0 mL of 0.1 M phosphate buffer (pH 6) and vortex well.

4. Centrifuge at ~1,500 × *g* for 5 min.

5. Place columns into the extraction manifold.

6. Prewash the columns with 3.0 mL of the methylene chloride: isopropanol:ammonium hydroxide solution.

7. Pass 3.0 mL of methanol through each column. Do not permit the column to dry.

8. Pass 3.0 mL of water through each column. Do not permit the column to dry.

9. Pass 1.0 mL of 0.1 M phosphate buffer through each column. Do not permit the column to dry.

10. Pour specimen into column. Slowly draw specimen through column (at least 2 min) under low vacuum.

11. Pass 3.0 mL of deionized water through each column.

12. Pass 1.0 mL of 1.0 M acetic acid through each column.

13. Dry column with full vacuum.

14. Pass 2.0 mL of hexane though each column.

15. Pass 3.0 mL of methanol through each column.

16. Dry column with full vacuum.

17. Turn off vacuum. Dry tips. Place labeled disposable culture tubes into column reservoir.

18. Add 3.0 mL of the methylene chloride:isopropanol:ammonium hydroxide solution to each column and collect in disposable culture tubes.

19. Evaporate to dryness at 40°C ± 5°C with a stream of nitrogen.

20. Add 50 µL of methanol to each tube and vortex well.

21. Transfer extract to autosampler vial and submit for GC-MS analysis.

3.3. Instrument Operating Conditions

1. The GC-MS operating parameters are presented in **Tables 8.5, 8.6, 8.7,** and **8.8**.

2. The capillary column phase used for the benzodiazepine and alprazolam procedures is either 100% methylsiloxane or 5% phenyl-95% methylsiloxane.

Table 8.5
GC operating conditions for the analysis of benzodiazepines

Initial oven temp.	140°C
Initial time	0.5 min
Ramp 1	25°C/min
Final temp.	320°C
Final time	5.10 min
Total run time	12.80 min
Injector temp.	250°C
Detector temp.	290°C
Purge time	0.5 min
Column flow	1.0 mL/min

Table 8.6
GC operating conditions for the analysis of alprazolam

Initial oven temp.	140°C
Initial time	0.5 min
Ramp 1	30°C/min
Final temp.	320°C
Final time	5.00 min
Total run time	11.50 min
Injector temp.	250°C
Detector temp.	300°C
Purge time	0.5 min
Column flow	1.0 mL/min

3. Setup of the autosampler should include the exchange of solvent in both autosampler wash bottles with fresh solvent (bottle 1 – acetonitrile for benzodiazepines, methanol for alprazolam; bottle 2 – ethyl acetate).

4. A daily autotune must be performed with perfluorotributylamine (PFTBA) as the tuning compound prior to each GC-MS run.

Table 8.7
Quantitative and qualifier ions for the benzodiazepine assay

Analyte	Quantitative ion (m/z)	Qualifier ions (m/z)
Diazepam	283	284, 256
D5-Diazepam	261	289
Nordiazepam	327	328, 329
D5-Nordiazepam	332	333
Oxazepam	457	458, 459
D5-Oxazepam	462	463
Temazepam	357	358, 359
D5-Temazepam	362	363
Lorazepam	491	492, 493
D4-Lorazepam	495	496
α-Hydroxyalprazolam	381	382, 383
D5-α-Hydroxyalprazolam	386	387
α-Hydroxytriazolam	415	416, 417
D5-α- Hydroxytriazolam	419	420

Table 8.8
Quantitative and qualifier ions for alprazolam assay

Analyte	Quantitative ion (m/z)	Qualifier ions (m/z)
Alprazolam	279	204, 308
D5-Alprazolam	284	313

3.4. Data Analysis

1. The review of the data requires the following information: retention times obtained from the selected ion chromatograms; ion abundance and ion peak ratios obtained from the selected ion chromatograms; and quantitative ion ratios between the native drug and its corresponding deuterated internal standard.

2. Typical Agilent ChemStation reports for benzodiazepines and alprazolam are illustrated in **Figs. 8.1** and **8.2**.

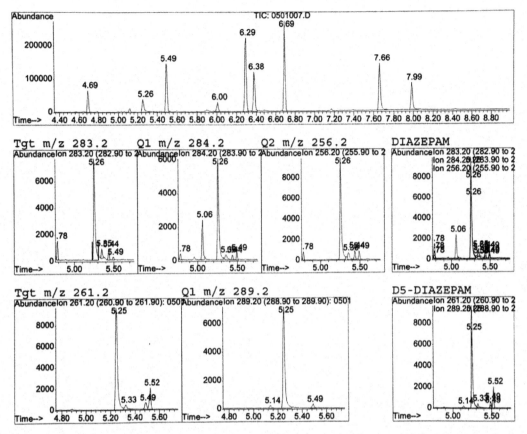

Fig. 8.1. Typical Agilent ChemStation report for benzodiazepines. **Top panel** is the total ion chromatogram; **middle and bottom panels** are selected ion chromatograms for diazepam (retention time 5.26 min) and deuterated diazepam (retention time 5.25 min), respectively. Nordiazepam (retention time 5.49 min), oxazepam (retention time 6.29 min), temazepam (retention time 6.38 min), lorazepam (retention time 6.69 min), α-hydroxyalprazolam (retention time 7.66 min), and α-hydroxytriazolam (retention time 7.99 min) are also illustrated in the top panel. The retention time for each deuterated internal standard is similar to its native drug.

3. In order for a result to be reported as "Positive", the following criteria must be satisfied: (1) the retention times of the ion peaks must be within ± 1% of the corresponding ions of the intermediate calibrator; (2) the ion peak ratio for the specimen must be within ± 20% of corresponding ion peak ratio of the intermediate calibrator; and (3) the correlation coefficient for the calibration curve must be 0.99 or greater if linear regression is used. Failure to meet one of the above criteria requires that the specimen be reported as "None Detected."

4. The limit of detection for the benzodiazepine assay (excluding alprazolam) is 12.5 ng/mL and the range of linearity is 50–1,000 ng/mL. The limit of detection for the alprazolam assay is 1.25 ng/mL and the range of linearity is 5–100 ng/mL. The reporting criteria for blood are presented in

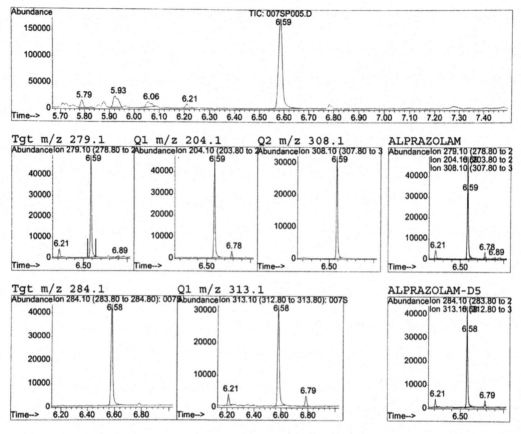

Fig. 8.2. Typical Agilent ChemStation report for alprazolam. **Top panel** is the total ion chromatogram; **middle and bottom panels** are selected ion chromatograms for alprazolam (retention time 6.59 min) and deuterated alprazolam (retention time 6.58 min), respectively.

Tables 8.9 and **8.10**. Urine results are qualitative only and are reported as "Positive" when the concentration is greater than the limit of detection.

5. The intra- and inter-assay variability (%CV) for all benzodiazepines is less than <10% for all analytes.

Table 8.9
Reporting criteria for benzodiazepines in blood

Concentration (ng/mL)	Reported result
12.5–<25	trace
25–<50	<50 ng/mL
50–1,200	report concentration
>1,200 ng/mL	reanalyze "diluted" specimen

Table 8.10
Reporting criteria for alprazolam in blood

Concentration (ng/mL)	Reported result
1.25–<2.5	trace
2.5–<5.0	<5.0 ng/mL
5.0–120	report concentration
>120	reanalyze "diluted" specimen

4. Notes

1. The benzodiazepine assay and alprazolam assay can be applied to other specimens including those obtained at autopsy.

2. Separate sources of benzodiazepine analyte must be used when preparing standard and control solutions.

References

1. Drummer, O.H. (2002) Benzodiazepines: Effects on human performance and behavior. *Forensic Sci Rev*, 14: 1–14. Review.

2. Charney, D.S., Mihic, S.J. and Harris, R.A. (2006) Hypnotics and sedatives. In Brunton L.L., Lazo J.S. and Parker K.L. (eds), *Goodman & Gilman's the pharmacological basis of therapeutics*, 11th edition, McGraw-Hill, New York, pp. 401–427.

3. Lee, D.C. (2006) Sedative-hypnotics. In Flomenbaum N.E., Goldfrank L.R., Hoffman R.S., Howland M.A., Lewin N.A. and Nelson L.S. (eds), *Goldfrank's toxicologic emergencies*, 8th edition. McGraw-Hill, New York, pp. 1098–1111.

4. Clean Screen Extraction Column Application Manual, United Chemical Technologies (2009).

5. Drummer, O.H. (1998) Methods for the measurement of benzodiazepines in biological samples. *J Chromatogr B Biomed Sci Appl*, 713, 201–225.

6. Gunnar, T., Ariniemi, K. and Lillsunde, P. (2005) Determination of 14 benzodiazepines and hydroxy metabolites, zaleplon and zolpidem as tert-butyldimethylsilyl derivatives compared with other common silylating reagents in whole blood by gas chromatography-mass spectrometry. *J Chromatogr B*, 818, 175–189.

7. Tiscione, N.B., Shan, X., Alford, I. and Yeatman, D.T. (2008) Quantitation of benzodiazepines in whole blood by electron impact-gas chromatography-mass spectrometry. *J Anal Toxicol*, 32, 644–652.

Chapter 9

LC-MS/MS Analysis of 13 Benzodiazepines and Metabolites in Urine, Serum, Plasma, and Meconium

Stephanie J. Marin and Gwendolyn A. McMillin

Abstract

We describe a single method for the detection and quantitation of 13 commonly prescribed benzodiazepines and metabolites: α-hydroxyalprazolam, α-hydroxyethylflurazepam, α-hydroxytriazolam, alprazolam, desalkylflurazepam, diazepam, lorazepam, midazolam, nordiazepam, oxazepam, temazepam, clonazepam and 7-aminoclonazepam in urine, serum, plasma, and meconium. The urine and meconium specimens undergo enzyme hydrolysis to convert the compounds of interest to their free form. All specimens are prepared for analysis using solid-phase extraction (SPE), analyzed using liquid chromatography coupled to tandem mass spectrometry (LC-MS/MS), and quantified using a three-point calibration curve. Deuterated analogs of all 13 analytes are included as internal standards. The instrument is operated in multiple reaction-monitoring (MRM) mode with an electrospray ionization (ESI) source in positive ionization mode. Urine and meconium specimens have matrix-matched calibrators and controls. The serum and plasma specimens are quantified using the urine calibrators but employing plasma-based controls. Oxazepam glucuronide is used as a hydrolysis control.

Key words: Benzodiazepines, urine, serum, plasma, meconium, solid phase extraction (SPE), liquid chromatography tandem mass spectrometry (LC-MS/MS).

1. Introduction

Benzodiazepines are a schedule IV class of psychotropic drugs prescribed for their sedative, anxiolytic, and anticonvulsant properties. Indeed, of the top 200 generic drugs ranked by number of prescriptions written in 2006 (1) five were benzodiazepines: alprazolam was ranked number 7, lorazepam 17, clonazepam 23, diazepam 41, and temazepam 60. Though of great therapeutic benefit to many patients, physiological and psychological dependence can lead to misuse and abuse. Benzodiazepines are widely

U. Garg, C.A. Hammett-Stabler (eds.), *Clinical Applications of Mass Spectrometry*, Methods in Molecular Biology 603, DOI 10.1007/978-1-60761-459-3_9, © Humana Press, a part of Springer Science+Business Media, LLC 2010

distributed and extensively metabolized. Many have common metabolites, and many of the metabolites are also prescribed drugs (2). Most are eliminated to a considerable extent as a glucuronide conjugate.

Detection of benzodiazepines in urine is clinically important for compliance, monitoring, and identification of abuse. Measurement of the serum or plasma concentration may help optimize chronic dosing, verify compliance, and identify changes in pharmacokinetics. Detection in meconium can identify neonates exposed to drugs during the prenatal period to guide treatment and improve outcomes for children exposed to drugs in utero.

Many methods are reported describing the quantification of benzodiazepines in urine or blood using LC-MS/MS (4–15), but these methods fail to include analysis of the drugs in meconium. These three specimen types have unique compositions that can lead to different analytical interferences. Described here is a single method for the identification and quantitation of 13 commonly prescribed benzodiazepines and metabolites (α-hydroxyalprazolam, α -hydroxyethylflurazepam, α -hydroxytriazolam, alprazolam, desalkylflurazepam, diazepam, lorazepam, midazolam, nordiazepam, oxazepam, temazepam, clonazepam, and 7-aminoclonazepam) in urine, serum, plasma, and meconium using deuterated internal standards, solid-phase extraction (SPE) and liquid chromatography tandem mass spectrometry (LC-MS/MS) (16). This method also includes the use of oxazepam glucuronide as a control to assure adequate enzyme hydrolysis of each batch of urine and meconium specimens analyzed. The use of a single method offers distinct advantages of efficiency and throughput.

2. Materials

2.1. Sample Preparation and Solid-Phase Extraction

1. Drug-free urine: Obtain commercially or from volunteers. Confirm each pool is free of benzodiazepines and metabolites prior to use.

2. Drug-free meconium: Residual negative specimens may be pooled. Confirm each pool is free of benzodiazepines and metabolites prior to use.

3. Drug-free plasma: Human plasma with potassium oxalate or sodium fluoride preservative.

4. Beta-glucuronidase enzyme from abalone, 5,000 units/mL (Campbell Scientific, Logan, UT). Prepare 1 L of 5,000 units/mL solution, taking into consideration purity of the reagent. Mix

with 0.1 M sodium acetate buffer, pH 5.0, and dilute to 1 L. Transfer the solution to a labeled, silanized amber bottle. Stable at 2–8°C for 3 months.

5. 0.1 M sodium acetate buffer pH 5.0: Add 54.4 g of sodium acetate to 3.2 L of water while stirring. Allow solution to stir until the chemical is completely dissolved. Adjust the solution pH to 5.0 using acetic acid. Dilute the solution to a final volume of 4 L with water. Stable for 1 year at room temperature.

6. 0.1 M sodium bicarbonate buffer, pH 9.0: Add 33.6 g of sodium bicarbonate to 3.2 L of water while stirring. Allow solution to stir until the chemical is completely dissolved. Adjust the solution pH to 9.0 using 1 M HCl or 1 M NaOH. Dilute to a final volume of 4 L with water. Stable for 1 year at room temperature.

7. Elution solvent, ethyl acetate:ammonium hydroxide 98:2 (v/v): The elution solvent must be prepared fresh just prior to use. Prepare 100 mL by mixing 98 mL of ethyl acetate and 2 mL of ammonium hydroxide. Mix well to completely dissolve the ammonium hydroxide.

8. Mobile phase buffer, 100 mM ammonium formate, pH 3.0: Dissolve 1.287 g ammonium formate (Sigma-Aldrich, St Louis, MO) into 200 mL of water. Add 4.5 mL 99% formic acid. Replace every 3 days.

2.2. Calibrators and Controls

1. All calibration standards and internal standards are of 99% purity and purchased from Cerilliant (Austin, TX) in sealed ampules containing 1.0 mg of analyte in 1 mL of methanol or acetonitrile: α-hydroxyalprazolam, α-hydroxyethylflurazepam, α-hydroxytriazolam, alprazolam, desalkylflurazepam, diazepam, lorazepam, midazolam, nordiazepam, oxazepam, temazepam, clonazepam and 7-aminoclonazepam, α-hydroxyalprazolam-d5, α a-hydroxyethylflurazepam-d4, α -hydroxytriazolam-d4, alprazolam-d5, desalkylflurazepam-d4, diazepam-d5, lorazepam-d4, midazolam-d4, nordiazepam-d5, oxazepam-d5, temazepam-d5, clonazepam-d4 and 7-aminoclonazepam-d4 (*see* **Note 1**).

2. Oxazepam glucuronide Cerilliant (Austin, TX) in sealed ampules containing 100 μg in 1 mL of methanol.

3. Calibration standard solution, 1 ng/μL of each analyte: Add 50 mL of methanol to a 100 mL class A volumetric flask. Add 100 μL of each of the thirteen 1.0 mg/mL reference standards and fill to volume with methanol. Mix thoroughly and equilibrate 1 h at room temperature. Stable for 3 months at 2–8°C and for 1 year at -65–75°C.

4. Internal standard solution, 1 ng/μL each deuterated analyte: Add approximately 50 mL of methanol to a 100 mL class A volumetric flask. Add 100 μL of each of the thirteen

deuterated benzodiazepine 1.0 mg/mL reference standards and fill to volume with methanol. Mix thoroughly and equilibrate 1 h at room temperature. Stable for 3 months at 2–8°C and for 1 year at –65 to –75°C.

5. Urine positive control, 50 ng/mL of each analyte in urine: Transfer 50 mL of drug-free urine into a class A 100 mL volumetric flask. Add 5 μL of each of the thirteen 1.0 mg/mL control reference solutions. Fill to the mark with drug-free urine. Mix thoroughly and equilibrate 1 h at room temperature. Stable at 2–8°C for 3 months or –65 to –75°C for 1 year.

6. Plasma positive control, 50 ng/mL of each analyte in drug-free plasma: Transfer 50 mL of drug free plasma into a 100 mL class A volumetric flask, and transfer 5 μL of each of the thirteen 1.0 mg/mL control reference solutions. Fill to the mark with drug-free plasma. Mix thoroughly and equilibrate 1 h at room temperature. Stable at 2–8°C for 3 months or –65 to –75°C for 1 year.

7. Meconium positive control spike solution, 1 ng/μL of each analyte: Transfer 25 mL of methanol to a 50 mL volumetric flask. Add 50 μL of each of the thirteen 1.0 mg/mL reference standards. Fill to volume with methanol. Mix thoroughly and equilibrate 1 h at room temperature. Stable at 2–8°C for 3 months or –65 to –75°C for 1 year.

8. Urine hydrolysis control, 81 ng/mL oxazepam glucuronide (50 ng/mL free oxazepam after hydrolysis): Transfer 50 mL of drug-free urine into a 100 mL volumetric flask and add 81 μL of the oxazepam glucuronide reference standard. Fill to the mark with drug-free urine. Mix thoroughly and equilibrate 1 h at room temperature. Stable at 2–8°C for 3 months or –65 to –75°C for 1 year.

9. Meconium hydrolysis control spike solution, 1 ng/μL oxazepam glucuronide: Transfer 25 mL of methanol to a 50 mL volumetric flask and add 807 μL of the oxazepam glucuronide reference standard. Fill to the mark with methanol. Mix thoroughly and equilibrate 1 h at room temperature. Stable at 2–8°C for 3 months or –65 to –75°C for 1 year.

10. Prepare an unextracted control by placing 20 μL (20 ng) calibration standard solution and 100 μL of internal standard solution into a 2 mL max recovery autosampler vial. Set aside (this will be added to steps after solid-phase extraction before evaporation of elution solvent and reconstitution of samples) (*see* **Note 2**).

2.3. Supplies

1. 16 × 50 mm, 5 mL polypropylene transport tubes (VWR, West Chester, PA).

2. 16 × 100 test tubes (VWR, West Chester, PA).

3. Trace-B (35 mg/3 mL), solid-phase separation columns (SPEware Inc., San Pedro, CA).

4. 2 mL max recovery autosampler vials (VWR).

5. Waters XTerra® MS C18 Analytical Column, 3.5 μm particle size, 2.1 × 150 mm (Waters Corporation, Milford, MA).

6. Waters XTerra® MS C18 Guard column, 3.5 μm particle size, 2.1 × 10 mm (Waters Corporation, Milford, MA).

2.4. Equipment

1. Omni Tissuemiser Homogenizer (Thermo-Fisher Scientific).

2. Jouan centrifugal vacuum evaporator (CVE) system, model RC10.10 (Thermo-Fisher Scientific).

3. CEREX 48 place positive-pressure manifold (SPEWare).

4. CEREX 48 place sample concentrator station (SPEWare).

5. LC-MS/MS analysis was performed on a Waters/Micromass Quattro Micro LC-MS/MS system equipped with a Waters Alliance® HT HPLC system. The HPLC system included a solvent delivery/separation module, sample management system (autosampler), and column oven.

3. Methods

Organization of a batch of samples is summarized in **Table 9.1**.

3.1. Sample Preparation Meconium Specimens

1. Weigh six 1.00±0.02 g aliquots of drug-free meconium into individual 16 × 50 mm 5 mL transport tubes. (Cal 1, Cal 2, Cal 3, positive control, hydrolysis control, and negative control).

2. Spike three drug-free meconium samples (Cal 1, Cal 2 and Cal 3) with 20, 50, and 200 μL, respectively, of calibration standard solution.

3. Label the fourth tube containing drug-free meconium as the meconium-positive control and add 50 μL of meconium positive control spike solution.

4. Label the fifth tube containing drug free meconium as the meconium hydrolysis control and add 50 μL of meconium hydrolysis control spike solution.

5. The sixth tube should contain only drug-free meconium (this is the negative control and will contain internal standards only).

6. Weigh 1.00 ± 0.02 g of each meconium specimen into individually labeled 16 × 50 mm 5 mL polypropylene transport tubes.

Table 9.1
Batch organization for benzodiazepine analysis

Sample ID	All specimen types		Urine, Serum, Plasma			Meconium		
	Calibration standard solution, μL	Internal standard solution, μL	Drug-free urine, mL	Pre-made urine or plasma positive control, mL	Pre-made urine hydrolysis control, mL	Drug free meconium, g	Meconium positive control solution, μL	Meconium hydrolysis control solution, μL
Unextracted Control	20	100						
Cal 1 (20 ng/mL or ng/g)	20	100	1			1		
Cal 2 (50 ng/mL or ng/g)	50	100	1			1		
Cal 3 (200 ng/ mL or ng/g)	200	100	1			1		
Negative control		100	1			1		
Positive control		100		1		1	50	
Hydrolysis control		100			1	1		50
Patient specimens*		100						

*1 mL urine or plasma, 1 g meconium.

7. Add 100 µL of internal standard solution to all tubes (calibrators, controls, and specimens).

8. Add 3 mL of methanol to each tube.

9. Homogenize each meconium specimen using the Omni Tissuemiser until uniform.

10. Centrifuge each sample at $150 \times g$ (radian/sec meter) or $1.47 \times gs$ (radian/sec^2) and 0°C for 15 min and transfer the supernatant to individual 16×100 mm test tubes.

11. Place the tubes in the CVE and evaporate to less than 1 mL at 60°C.

3.2. Sample Preparation Urine Specimens

1. Pipet four 1 mL aliquots of drug-free urine into individual 16×100 mm test tubes (Cal 1, Cal 2, Cal 3, and the negative control).

2. Add 20, 50, and 200 µL, respectively, of calibration standard solution to the three calibration samples.

3. One tube should contain only drug-free urine (this is the negative control and will contain internal standards only).

4. Label a tube for the urine-positive control and pipet 1 mL of urine-positive control into the tube.

5. Label a tube for the urine-hydrolysis control and pipet 1 mL of urine-hydrolysis control into one tube.

6. Pipet 1 mL of each urine specimen into individually labeled tubes.

7. Add 100 µL internal standard solution to all tubes (calibrators, controls, and specimens).

3.3. Sample Preparation Serum and Plasma Specimens

1. Label a tube for the plasma-positive control and pipet 1 mL of plasma-positive control into the tube.

2. Pipet 1 mL of each serum or plasma specimen into individually labeled tubes.

3. Add 100 µL internal standard solution to each tube.

4. Add 2 mL of 0.1 M sodium acetate buffered to pH 5.0 to each tube.

Serum and plasma specimens are quantitated using the urine calibrators.

3.4. Hydrolysis

1. Add 2 mL of beta-glucuronidase enzyme from abalone (5,000 units/mL in 0.1 M sodium acetate buffer, pH 5.0) to each urine and meconium specimen and briefly vortex.

2. Incubate at 60°C for 2 h.

3.5. Solid-Phase Extraction

1. Centrifuge all samples (including the serum and plasma samples) at 0°C, $37.4 \times g$ ($3.67 \times gs$) for 5 min.

2. Load the samples onto the Trace-B columns at 1 drop per 4 seconds.

3. Wash each sample at 1 drop per second with 1 mL pH 9.0 sodium bicarbonate buffer and 1 mL water.

4. Dry the samples on the columns using nitrogen at 25 psi for 10 min.

5. Elute the samples into 2 mL max recovery autosampler vials with 1 mL of elution solvent.

6. Dry the samples (include the unextracted control samples) in the autosampler vials using the CEREX 48 place sample concentrator with nitrogen at 40°C for approximately 15 min.

7. Reconstitute the dried urine and serum/plasma samples in 200 µL of 1:1 acetonitrile:water.

8. Reconstitute the dried meconium samples in 150 µL of 1:1 acetonitrile:water.

3.6. LC-MS/MS analysis

The instrument is operated in multiple reaction monitoring (MRM) mode with an ESI probe in positive electrospray ionization mode and controlled using Micromass MassLynx software. For each analyte and its internal standard, a quantitative and a qualitative mass transition is selected (**Table 9.2**). The quantitative transition is listed first, followed by the qualitative transition for all analytes and internal standards. Retention time, cone voltage, and collision voltage are also listed (*see* **Note 5**).

1. LC Column. Conditions:
   a. XTerra® MS C18 Analytical Column, 3.5 µm particle size, 2.1 × 150 mm, Waters part number: 186000408.
   b. XTerra® MS C18 Guard column, 3.5 µm particle size, 2.1 × 10 mm, Waters part number: 186000632.

2. Mobile Phase Solvents: A: Acetonitrile, HPLC grade; B: Water; C: 100 mM ammonium formate in HPLC grade water, pH 3.0.

3. LC Conditions:
   a. Column Temp: 30°C
   b. Isocratic: 55% A, 40% B, 5% C
   c. Flow rate: 0.125 mL/min
   d. Injection volume: 10-20 µL
   e. LC column effluent split flow: no split

4. Injection wash solvents:
   a. Seal wash solvent: methanol: water (50:50)
   b. Needle wash solvent: methanol: water:sodium hydroxide (80:20:0.1)
   c. Purge solvent: acetonitrile

Table 9.2
Retention Times (RT, min), MRM Transitions Monitored (*m/z*), cone and collision voltages for benzodiazepines

Analyte	RT	MRM	Cone	Collision	Internal Std	RT	MRM	Cone	Collision
α-OH-alprazolam	4.48	325.00 > 297.00	35	25	α-hydroxy-alprazolam-d5	4.48	330.00 > 302.05	35	25
		325.00 > 205.10	35	45			330.00 > 210.15	35	45
α-hydroxyethylflurazepam	5.12	333.00 > 109.00	30	25	α-hydroxyethylflurazepam-d4	5.12	337.00 > 113.05	30	25
		333.00 > 211.10	30	35			337.00 > 215.15	30	35
α-hydroxytriazolam	4.41	358.95 > 176.05	35	25	α-hydroxytriazolam-d4	4.41	362.95 > 176.05	35	25
		358.95 > 277.00	35	35			362.95 > 281.00	35	35
alprazolam	4.93	309.00 > 205.10	35	40	alprazolam-d5	4.93	314.00 > 210.15	35	40
		309.00 > 274.25	35	25			314.00 > 279.25	35	25
desalkylflurazepam	5.64	289.00 > 140.05	35	30	desalkylflurazepam-d4	5.64	293.00 > 140.05	35	30
		289.00 > 226.10	35	25			293.00 > 230.15	35	30
diazepam	7.75	285.00 > 154.05	35	25	diazepam-d5	7.59	290.05 > 154.05	35	25
		285.00 > 193.10	35	30			290.05 > 198.15	35	30
lorazepam	5.03	320.90 > 229.10	25	30	lorazepam-d4	5.03	324.95 > 233.10	25	30
		320.90 > 194.10	25	40			324.95 > 198.15	25	40
midazolam	4.03	326.00 > 291.05	35	25	midazolam-d4	4.03	330.00 > 295.20	35	25
		326.00 > 249.15	35	35			330.10 > 227.10	35	35
nordiazepam	6.02	271.00 > 140.00	35	25	nordiazepam-d5	5.98	276.00 > 140.05	35	25
		271.00 > 208.10	35	25			276.00 > 213.15	35	30
oxazepam	4.93	286.95 > 241.05	25	25	oxazepam-d5	4.93	295.00 > 246.10	25	25
		286.95 > 104.00	25	35			295.00 > 109.05	25	35

(continued)

Table 9.2 (continued)

Analyte	RT	MRM	Cone	Collision	Internal Std	RT	MRM	Cone	Collision
temazepam	5.95	300.95 > 255.05 300.95 > 193.15	20 20	20 35	temazepam-d5	5.95	306.00 > 260.10 306.00 > 198.15	20 20	20 35
clonazepam	5.23	316.20 > 214.20 241.20 > 241.20	35 35	35 35	clonazepam-d4	5.23	320.20 > 274.10 320.00 > 245.10	35 35	25 35
7-aminoclonazepam	3.93	286.40 > 222.10 286.40 > 250.10	35 35	25	7-aminoclonazepam-d4	3.93 20	290.40 > 226.10 290.40 > 254.10	35 35	25 20

Reproduced from the *Journal of Analytical Toxicology* by permission of Preston Publications, A Division of Preston Industries, Inc.

5. Autosampler conditions:
   a. Injection type: sequential
   b. Fill mode: Partial loop
   c. Sample temperature: 15°C
   d. Flush time (s): 6
   e. Wash time (s): 15
   f. Wash cycles: 1
   g. Secondary wash volume (μL): 600
   h. Wash sequence: Wash-Purge

6. MS Source parameters:
   a. Capillary (kV): 1.00
   b. Extractor (V): 2.00
   c. RF lens (V): 0.2
   d. Source temperature (°C): 120
   e. Desolvation temperature (°C): 400
   f. Cone gas flow (L/hr): 0
   g. Desolvation gas flow (L/hr): 800

7. Analyzer parameters:
   a. LM1 resolution: 13.0
   b. HM1 resolution: 13.0
   c. Ion energy 1: 0.1
   d. Entrance: −5
   e. Exit: 1
   f. LM2 resolution: 13.0
   g. LM2 resolution: 13.0
   h. Ion energy 2: 1.5
   i. Multiplier (V): 650
   j. Peak width at half-height: 0.70

3.7. Data Analysis

1. Quantitation parameters are summarized in **Table 9.3**.

2. Urine, serum, and plasma samples are quantified using the urine calibrators. Meconium samples are quantified using the meconium calibrators.

3. The LOD is 10 ng/mL or ng/g for all analytes. The LLOQ for all analytes is 20 ng/mL or ng/g. The ULOQ is 5,000 ng/mL of urine, 2,500 ng/mL for serum/plasma, and 5,000 ng/g for meconium for all analytes (*see* **Note 3**).

4. Qualitative Criteria:
   a. Retention time ±2% of the retention time for the Cal 2 standard; all monitored ions present with ion ratios ±25% for all specimen types.

Table 9.3
Quantitation parameters for benzodiazepines

	urine (ng/mL)	serum/plasma (ng/mL)	meconium (ng/g)
LOD	10	10	10
LLOQ	20	20	20
ULOQ	5,000	2,500	5,000
CRR	10–250,000	10–50,000	10–5,000

(*see* **Note 4**)

    b. Relative retention time (RRT). The retention time of the analyte with respect to the retention time of its internal standard must be within 2%.

    c. Peaks must be at least 85% baseline resolved.

5. Quantitative Criteria: In addition to the qualitative criteria, the calibrators and controls must be within ±15% of the target concentration for urine and serum/plasma and ±20% for meconium. The three-point linear calibration curve must have a correlation coefficient ≥0.995 and a y-intercept ≤LOD. The negative control must demonstrate the presence of all 13 deuterated internal standards and have a concentration ≤LOD for all non-deuterated analytes.

4. Notes

1. When preparing control solutions, it is best to use material from a different vendor (or at least a different lot of material from the same vendor) than that used to prepare the calibration standard solution.

2. The unextracted control is used to check instrument performance. If the batch fails because of poor recovery of benzodiazepines from the extracted controls and samples but good recovery from the unextracted control, it shows the problem was probably with the extraction and not the LC-MS/MS instrumentation.

3. For many of the analytes, the LOD is <10 ng/mL or ng/g. We choose to use an LOD of 10 ng/mL or ng/g across all analytes and all specimen types for continuity and ease of reporting. The LLOQ for all analytes and specimen types was selected to be equal to two times the LOD (20 ng/mL or ng/g). The LOD met all qualitative criteria (retention time

±2%, all monitored ions present a signal to noise ratio ≥3:1). The LLOQ for all analytes met all quantitative criteria and had a signal to noise ratio of ≥5:1.

4. Serum/plasma may be diluted up to 25-fold, urine up to 50-fold. The clinically reportable range (CRR) for each specimen type is listed in **Table 9.3**.

5. Ion suppression should be evaluated for each matrix by extracting a blank urine, plasma, and meconium sample and injecting the sample while infusing a sample containing all 13 analytes and internal standards. **Figure 9.1** shows the total ion chromatograms (TIC) for an unextracted control (A), and a drug free urine (B), plasma (C), and meconium (D) injected during infusion. All three specimen types showed very minimal ion suppression. None of the observed ion suppression caused a loss in sensitivity significant enough to interfere with

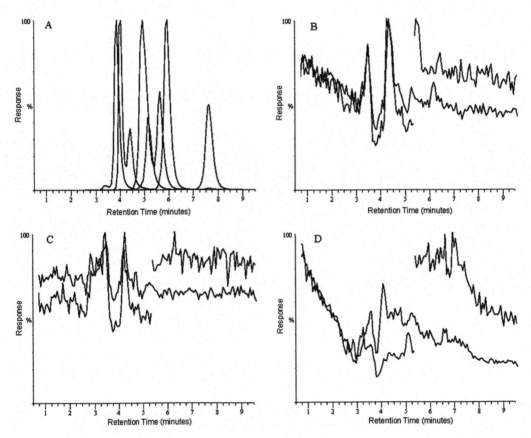

Fig. 9.1. Ion suppression study showing an unextracted control spiked at 100 ng/mL or ng/g of each analyte and 500 ng/mL or ng/g of each internal standard (**A**), and a drug-free urine (**B**), plasma (**C**), and meconium (**D**) sample injected while infusing a sample containing all of the analytes and internal standards. Reproduced from the *Journal of Analytical Toxicology* by permission of Preston Publications, A Division of Preston Industries, Inc.

accurate quantitation. The use of a deuterated internal standard for all 13 analytes compensates for the minimal ion suppression that occurs.

6. **Tables 9.4**, **9.5**, and **9.6** show the linear correlation for 15 spiked samples, total imprecision, and average recovery determined during validation of this method (15).

Table 9.4
Linear correlation data for spiked samples ($n=15$)

	Urine		plasma		meconium	
	equation	R^2	equation	R^2	equation	R^2
nordiazepam	y = 0.938x + 12.310	0.9992	y = 0.983x + 10.780	0.9982	y = 0.957x − 18.868	0.9967
oxazepam	y = 1.004x + 11.050	0.9995	y = 0.968x + 11.330	0.9980	y = 0.966x − 27.317	0.9943
temazepam	y = 0.922x + 16.090	0.9995	y = 0.945x + 16.795	0.9970	y = 0.845x + 5.167	0.9990
diazepam	y = 0.888x + 8.727	0.9995	y = 0.969x + 0.060	0.9919	y = 0.869x + 2.263	0.9981
lorazepam	y = 1.011x − 2.602	0.9988	y = 0.962x − 4.070	0.9913	y = 1.006x − 4.251	0.9966
alprazolam	y = 1.061x − 11.280	0.9989	y = 0.980x + 10.018	0.9985	y = 0.958x − 2.588	0.9992
α-OH-alprazolam	y = 1.042x + 10.399	0.9994	y = 1.018x + 3.552	0.9992	y = 0.992x − 27.292	0.9959
α-OH-triazolam	y = 0.968x + 9.431	0.9970	y = 0.988x + 10.713	0.9995	y = 0.947x − 6.016	0.9988
α-OH-ethylflurazepam	y = 1.029x − 3.245	0.9998	y = 0.963x + 16.052	0.9962	y = 0.982x − 27.107	0.9959
desalkylflurazepam	y = 1.053x − 14.844	0.9995	y = 1.007x + 5.828	0.9998	y = 0.929x − 2.911	0.9990
midazolam	y = 1.146x − 31.166	0.9989	y = 1.001x + 10.230	0.9996	y = 0.978x − 26.091	0.9951
clonazepam	y = 0.993x + 16.124	0.9981	y = 0.979x + 11.322	0.9990	y = 0.950x + 8.937	0.9987
7-aminoclonazepam	y = 1.125x − 25.450	0.9977	y = 1.097x − 15.383	0.9986	y = 1.102x − 41.644	0.9971

All linear correlation data had slopes between 0.85 and 1.15, y-intercepts less then 20 ng/mL or 20 ng/g (LLOQ), and a correlation coefficient greater than 0.99. 15 samples spiked with all 13 analytes between the LOD and the ULOQ were used

Table 9.5
Total imprecision (%CV)

Analyte	Urine			Serum/Plasma			Meconium		
	20 ng/mL	200 ng/mL	5,000 ng/mL	20 ng/mL	200 ng/mL	2,500 ng/mL	20 ng/g	200 ng/g	5,000 ng/g
nordiazepam	11.2	6.9	4.7	4.5	9.6	7.1	10.0	5.1	5.3
oxazepam	8.5	8.4	3.4	14.7	9.8	3.3	12.1	5.4	7.5
temazepam	9.7	10.8	0.6	8.2	12.4	3.3	8.6	5.3	4.0
diazepam	9.7	7.9	2.5	8.5	7.3	4.8	9.5	4.7	4.0
lorazepam	8.0	7.9	4.9	14.0	8.3	3.4	15.1	5.7	16.4
alprazolam	7.0	6.7	3.3	9.1	6.8	3.1	11.4	7.1	6.5
α-OH-alprazolam	5.8	3.8	3.1	7.5	5.3	4.5	9.5	6.1	8.9
α-OH-triazolam	7.5	9.9	5.0	14.0	12.1	4.3	11.2	9.1	6.0
α-OH-ethylflurazepam	9.1	6.2	5.2	7.5	9.3	3.1	13.9	5.8	9.0
desalkylflurazepam	7.8	8.3	5.6	9.9	11.6	4.2	11.2	7.1	5.5
midazolam	3.7	4.4	8.3	2.5	4.2	8.6	9.7	5.9	6.7
clonazepam	6.3	3.1	4.4	2.0	5.5	3.6	9.9	6.0	4.4
7-aminoclonazepam	9.5	7.8	6.4	9.1	8.9	9.0	16.8	8.2	13.6

Total imprecision was the square root of the sum of the squares of the CVs for within-run and between-run imprecision at each target concentration. Samples prepared and analyzed in triplicate on three different days (nine results for each concentration for each specimen type) spiked at the LLOQ, high calibrator and the ULOQ were used. Total imprecision also required CVs within ±15% (±20% for meconium) of the target concentration.
Reproduced from the *Journal of Analytical Toxicology* by permission of Preston Publications, A Division of Preston Industries, Inc.

Table 9.6
Average recovery (%)

Analyte	Urine			Serum/Plasma			Meconium		
	20 ng/mL	200 ng/mL	5,000 ng/mL	20 ng/mL	200 ng/mL	2,500 ng/mL	20 ng/g	200 ng/g	5,000 ng/g
nordiazepam	99.3	107.8	90.0	110.1	102.9	92.9	102.5	102.6	92.1
oxazepam	98.7	99.6	86.9	96.1	95.8	89.5	97.1	104.2	91.6
temazepam	105.1	105.2	85.5	107.8	98.4	88.4	101.8	101.5	82.1
diazepam	105.1	106.7	87.3	106.1	96.2	87.6	102.3	102.4	83.4
lorazepam	97.1	98.7	94.9	95.4	94.9	88.9	97.4	103.9	96.1
alprazolam	101.6	102.1	89.7	104.2	98.8	93.7	101.6	103.3	92.5
α-OH-alprazolam	109.7	110.5	98.3	109.3	107.7	103.4	98.7	106.6	97.2
α-OH-triazolam	97.7	97.2	86.5	100.3	97.6	91.5	99.2	100.7	87.3
α-OH-ethylflurazepam	106.0	103.1	89.5	100.3	96.6	91.1	100.5	103.7	96.4
desalkylflurazepam	100.4	105.2	91.9	102.8	100.6	95.1	101.1	101.2	89.0
midazolam	109.4	110.6	101.0	110.0	110.5	101.6	104.0	107.7	95.4
clonazepam	111.0	107.5	87.9	112.7	105.2	89.0	113.0	103.5	87.2
7-aminoclonazepam	104.4	104.1	109.4	103.2	102.3	107.0	103.9	104.2	99.4

Average recovery was the mean of all samples at a specific target concentration for each specimen type. Samples were prepared and analyzed in triplicate on three different days (nine results for each concentration for each specimen type) spiked at the LLOQ; high calibrator and the ULOQ were used. Average recovery also required CV's within ±15% (±20% for meconium) of the target concentration.

Reproduced from the *Journal of Analytical Toxicology* by permission of Preston Publications, A Division of Preston Industries, Inc.

References

1. Chi, J. editor-in-chief. (2004–2008) Top 200 Generic Drugs by Units. In: *Drug Topics Magazine*.

2. Baselt, R.C. (2008) *Disposition of toxic drugs and chemicals in man*, 8th ed. Foster City, CA: Biomedical Publications.

3. Gareri, J., Klein, J., Koren, G. (2006) Drugs of abuse testing in meconium *Clin Chim Acta* 366, 101–11.

4. Feng, J., Wang, L., Dai, I. Harmon, T., Bernert, J.T. (2007) Simultaneous determination of multiple drugs of abuse and relevant metabolites in urine by LC-MS-MS *J Anal Toxicol* 31, 359–68.

5. Hegstad, S., Oiestad, E.L., Johansen, U., Christophersen, A.S. (2006) Determination of benzodiazepines in human urine using solid-phase extraction and high-performance liquid chromatography-electrospray ionization tandem mass spectrometry *J Anal Toxicol* 30, 31–7.

6. Laloup, M., Fernandez Mdel, M., Wood, M., et al. (2007) Detection of diazepam in urine, hair and preserved oral fluid samples with LC-MS-MS after single and repeated administration of Myolastan and Valium *Anal Bioanal Chem* 388, 1545–56.

7. Laloup, M., Ramirez Fernandez Mdel, M., De Boeck, G., Wood, M., Maes, V., Samyn, N. (2005) Validation of a liquid chromatography-tandem mass spectrometry method for the simultaneous determination of 26 benzodiazepines and metabolites, zolpidem and zopiclone, in blood, urine, and hair *J Anal Toxicol* 29, 616–26.

8. Maurer, H.H. (2005) Multi-analyte procedures for screening for and quantification of drugs in blood, plasma, or serum by liquid chromatography-single stage or tandem mass spectrometry (LC-MS or LC-MS/MS) relevant to clinical and forensic toxicology *Clin Biochem* 38, 310–8.

9. Maurer, H.H. (2007) Current role of liquid chromatography-mass spectrometry in clinical and forensic toxicology *Anal Bioanal Chem* 388, 1315–25.

10. Quintela, O., Sauvage, F.L., Charvier, F., Gaulier, J.M., Lachatre, G., Marquet, P. (2006) Liquid chromatography-tandem mass spectrometry for detection of low concentrations of 21 benzodiazepines, metabolites, and analogs in urine: method with forensic applications *Clin Chem* 52, 1346–55.

11. Rivera, H.M., Walker, G.S., Sims, D.N., Stockham, P.C. (2003) Application of liquid chromatography-tandem mass spectrometry to the analysis of benzodiazepines in blood *Eur J Mass Spectrom* (Chichester, Eng) 9, 599–607.

12. Smink, B.E., Brandsma, J.E., Dijkhuizen, A., et al. (2004) Quantitative analysis of 33 benzodiazepines, metabolites and benzodiazepine-like substances in whole blood by liquid chromatography-(tandem) mass spectrometry *J Chromatogr B Analyt Technol Biomed Life Sci* 811, 13–20.

13. Smink, B.E., Mathijssen, M.P., Lusthof, K.J., de Gier, J.J., Egberts, A.C., Uges, D.R. (2006) Comparison of urine and oral fluid as matrices for screening of thirty-three benzodiazepines and benzodiazepine-like substances using immunoassay and LC-MS(-MS) *J Anal Toxicol* 30, 478–85.

14. Drummer, O.H. (1998) Methods for the measurement of benzodiazepines in biological samples *J Chromatogr B Biomed Sci Appl* 713, 201–25.

15. Nakamura, M., Ohmori, T., Itoh, Y., Terashita, M., Hirano, K. (2008) Simultaneous determination of benzodiazepines and their metabolites in human serum by liquid chromatography-tandem mass spectrometry using a high-resolution octadecyl silica column compatible with aqueous compounds *Biomed Chromatogr* Oct 20 (epub ahead of print)

16. Marin, S.J., Coles, R., Merrell, M., McMillin, G.A. (2008) Quantitation of Benzodiazepines in Urine, Serum, Plasma, and Meconium by LC-MS-MS *J Anal Toxicol* 32, 491–8.

Chapter 10

Simultaneous Determination and Quantification of 12 Benzodiazepines in Serum or Whole Blood Using UPLC/MS/MS

Josh Gunn, Scott Kriger, and Andrea R. Terrell

Abstract

The benzodiazepines are a large, commonly prescribed family of psychoactive drugs. We describe a method permitting the simultaneous detection and quantification of 12 benzodiazepines in serum using ultra-performance liquid chromatography (UPLC) coupled with tandem mass spectrometry (MS/MS). Analytes included alprazolam, temazepam, oxazepam, nordiazepam, clonazepam, lorazepam, diazepam, chlordiazepoxide, midazolam, flunitrazepam, 7-aminoclonazepam, and 7-aminoflunitrazepam. Sample pretreatment is simple consisting of protein precipitation using cold acetonitrile (ACN) mixed with the deuterated internal standards. Samples were capped and vortexed for 5 min to ensure maximum precipitation. Following a 5-min centrifugation period, 400 µL of the supernatant was transferred to a clean tube and evaporated down under nitrogen. Samples were reconstituted in 200 µL of a deionized water:ACN (80:20) mixture and transferred to appropriate vials for analysis. Chromatographic run time was 7.5 min, and the 12 analytes were quantified using multiple reaction monitoring (MRM) and 6-point calibration curves constructed for each analyte at concentrations covering a clinically significant range.

Key words: Benzodiazepines, mass spectrometry, protein precipitation, ultra-performance liquid chromatography.

1. Introduction

The benzodiazepines are a large family of psychoactive drugs with varying hypnotic, sedative, anxiolytic, anticonvulsant, muscle relaxant, and amnesic properties. Since their introduction in the early 1960s the benzodiazepines have become the most commonly prescribed psychoactive drugs in the world due to the large number of indications for the use of this drug class (1, 2). While the pharmacological properties of the benzodiazepines continue to

U. Garg, C.A. Hammett-Stabler (eds.), *Clinical Applications of Mass Spectrometry*, Methods in Molecular Biology 603,
DOI 10.1007/978-1-60761-459-3_10, © Humana Press, a part of Springer Science+Business Media, LLC 2010

find widespread therapeutic application, the same properties are commonly sought after by drug abusers worldwide (3). The pharmacological activity of the benzodiazepines results from their ability to modulate GABA$_A$ receptors in the central nervous system (CNS). Benzodiazepines have the ability to bind to GABA$_A$ receptors at the interface of the α- and γ-subunits. Once bound, the drug forces the receptor into a locked conformation conferring a higher affinity for the gamma-aminobutyric acid (GABA) neurotransmitter and leading to the observed sedatory or anxiolytic effects.

Traditionally, benzodiazepine use has been detected by using immunoassay-based screening assays followed by, when necessary, chromatography-based methods (i.e., GC/MS, HPLC, or LC/MS) for confirmation and quantitation. The simultaneous detection and quantification of multiple benzodiazepines in biological specimens is complicated by the wide range of therapeutic doses prescribed for the individual drugs, as well as by their diverse physicochemical properties including polarity which complicates their chromatographic separation (4). We describe a method using ultra-performance liquid chromatography coupled to tandem mass spectrometry that allows the simultaneous detection and quantification of 12 benzodiazepines in serum. The increased speed, resolution, sensitivity, and separation efficiency afforded by UPLC combined with the inherent specificity, sensitivity, and accuracy of tandem mass spectrometry allows for the accurate quantitation of all 12 analytes in a time and cost effective manner.

2. Materials

2.1. Samples

1. Blood should be collected using a non-gel barrier tube (*see* **Note 1**). Sample is stable for 6 weeks when stored at 4–8°C and for up to 1 year when stored at –20°C or colder.

2.2. Reagents

1. Mobile phase A: 0.1% (v/v) formic acid in deionized water. Stable for 3 months when stored at ambient temperature.

2. Mobile phase B: 0.1% (v/v) formic acid in acetonitrile. Stable for 3 months when stored at ambient temperature.

3. Adult bovine serum (Sigma Chemical Co) or drug-free serum.

4. Reconstitution solution: Mix deionized water and acetonitrile (HPLC grade or better) in 80:20 proportions (v/v). Stable for 3 months when stored at ambient temperature.

2.3. Standards and Calibrators

1. Stock Standards (Cerilliant, Round Rock, TX): Alprazolam (1.0 mg/mL), Temazepam (1.0 mg/mL), Oxazepam (1.0 mg/mL), Nordiazepam (1.0 mg/mL), Clonazepam (1.0 mg/mL), Lorazepam standard (1.0 mg/mL), Diazepam standard (1.0 mg/mL), Chlordiazepoxide standard (1.0 mg/mL), Midazolam standard (1.0 mg/mL), Flunitrazepam standard (1.0 mg/mL), 7-Aminoclonazepam standard (1.0 mg/mL), 7-Aminoflunitrazepam standard (1.0 mg/mL).

2. Working stock standard: Prepare according to **Table 10.1**. Stable for 1 year when stored at −20°C or colder.

3. Working standard 1: Add 20 µL of working stock standard to 980 µL of negative serum. Cap and rotate for approximately 5 min. This standard is prepared fresh on each day of use.

4. Working standard 2: Add 100 µL of working standard 1 to 900 µL of negative serum. Cap and rotate for approximately 5 min. This standard is prepared fresh on each day of use.

5. Calibrators: Prepare calibrators from working standard 1 and 2 according to **Table 10.2**. The calibrators are prepared fresh on each day of use.

Table 10.1
Preparation of working stock standard. Add following volumes of stock standards and bring the total volume to 4 mL using acetonitrile

Analyte	Target concentration (µg/mL)	Concentration of stock standard (mg/mL)	Volume (µL)
Alprazolam	5	1	20
Clonazepam	10	1	40
7-Aminoclonazepam	10	1	40
Flunitrazepam	20	1	80
7-Aminoflunitrazepam	20	1	80
Lorazepam	25	1	100
Midazolam	50	1	200
Diazepam	100	1	400
Nordiazepam	100	1	400
Oxazepam	100	1	400
Temazepam	100	1	400
Chlordiazepoxide	150	1	600

Table 10.2
Preparation of calibrators (ng/mL) from working standards 1 and 2

Calibrator level	Volume of working standards (µL)	Volume of negative serum (µL)	Final concentration of analyte											
			Alprazolam	Clona-zepam	7-Amino-clonazepam	Flunitra-zepam	7-Amino-flunitrazepam	Lorazepam	Midazolam	Diazepam	Nordi-azepam	Oxazepam	Temazepam	Chlordi-azepoxide
1	250 (Std 1)	0	100	200	200	400	400	500	1,000	2,000	2,000	2,000	2,000	3,000
2	125 (Std 1)	125	50	100	100	200	200	250	500	1,000	1,000	1,000	1,000	1,500
3	62.5 (Std 1)	187.5	25	50	50	100	100	125	250	500	500	500	500	750
4	250 (Std 2)	0	10	20	20	40	40	50	100	200	200	200	200	300
5	125 (Std 2)	125	5	10	10	20	20	25	50	100	100	100	100	150
6	62.5 (Std 2)	187.5	2.5	5	5	10	10	12.5	25	50	50	50	50	75
Neg	0	250	0	0	0	0	0	0	0	0	0	0	0	0

**2.4. Controls and
Internal Standards**

1. Stock controls: Prepare multidrug stock controls according to **Tables 10.3, 10.4, 10.5, 10.6,** and **10.7.** Stable for 1 year when stored at –20°C or colder.

2. Working controls: Prepare working controls according to **Table 10.8.** Single use aliquots are prepared and are stable for 1 year when stored at –20°C or colder.

3. Negative control: Drug-free serum.

4. Stock internal standards (Cerilliant, Round Rock, TX): Alprazolam d5 (1.0 mg/mL), Temazepam d5 (1.0 mg/mL), Oxazepam d5 standard (1.0 mg/mL), Nordiazepam d5 (1.0 mg/mL), Clonazepam d4 (100 µg/mL), Diazepam d5 (1.0 mg/mL), Midazolam d4 (100 µg/mL), Flunitrazepam d7 (100 µg/mL), 7-Aminoclonazepam d4 (100 µg/mL), and 7-Aminoflunitrazepam d7 (100 µg/mL).

5. Working internal standards: Prepare according to **Table 10.9.** Stable for 1 year when stored at –20°C or colder.

Table 10.3
Preparation of multidrug stock control for Alprazolam, 7-Aminoclonazepam, and Clonazepam. Add following volumes of stock standards and bring the total volume to 3 mL using acetonitrile

Analyte	Target concentration (µg/mL)	Concentration of standard (mg/mL)	Volume (µL)
Alprazolam	10	1	30
7-Aminoclonazepam	10	1	30
Clonazepam	10	1	30

Table 10.4
Preparation of multidrug stock control for Temazepam, Oxazepam, Nordiazepam, and Diazepam. Add following volumes of stock standards and bring the total volume to 1 mL using acetonitrile

Analyte	Target concentration (µg/mL)	Concentration of standard (mg/mL)	Volume (µL)
Temazepam	100	1	100
Oxazepam	100	1	100
Nordiazepam	100	1	100
Diazepam	100	1	100

Table 10.5
Preparation of stock control for Chlordiazepoxide. Add following volume of stock standard and bring the total volume to 1 mL using acetonitrile

Analyte	Target concentration (μg/mL)	Concentration of standard (mg/mL)	Volume (μL)
Chlordiazepoxide	100	1	100

Table 10.6
Preparation of stock control for Flunitrazepam and 7-Aminoflunitrazepam. Add following volumes of stock standards and bring the total volume to 3 mL using acetonitrile

Analyte	Target concentration (μg/mL)	Concentration of standard (mg/mL)	Volume (μL)
Flunitrazepam	10	1	30
7-Aminoflunitrazepam	10	1	30

Table 10.7
Preparation of multidrug stock control for Midazolam and Lorazepam. Add following volumes of stock standards and bring the total volume to 3 mL using acetonitrile

Analyte	Target concentration (μg/mL)	Concentration of standard (mg/mL)	Volume (μL)
Midazolam	20	1	60
Lorazepam	10	1	30

2.5. Supplies

1. Vials (Fisher Scientific or equivalent).
2. 11 mm blue cut septa snap cap (MicroLiter, Inc.).
3. 12 × 75 mm small glass culture tubes.
4. Column: Waters Acquity BEH Phenyl 2.1 × 100 mm; 1.7 μm particle size.
5. Desolvation gas: nitrogen (99.995% purity).
6. Collision gas: Ultra-pure argon (99.999% purity).

Table 10.8
Preparation of working quality control (QC) specimen using stock quality controls solutions. Add the following volumes of stock solutions and bring the total volume to 50 mL using drug-free bovine serum

Analytes	Target concentration (ng/mL)	Concentration of standard (μg/mL)	Volume (μL)
Alprazolam/Clonazepam/ 7-Aminoclonazepam	80	10	400
Flunitrazepam/ 7-Aminoflunitrazepam	120	10	600
Midazolam/Lorazepam	400/200	20/10	1000
Temazepam/Oxazepam/ Diazepam/Nordiazepam	800	100	400
Chlordiazepoxide	1200	100	600

Table 10.9
Preparation of working deuterated internal standards. Add following volumes of stock standards and bring the total volume to 500 mL using acetonitrile

Deuterated Standard	Target concentration (ng/mL)	Concentration of standard (μg/mL)	Volume (μL)
Alprazolam-d5	12	100	60
Clonazepam-d4	12	100	60
7-Aminoclonazepam-d4	12	100	60
Diazepam-d5	120	1	60
Nordiazepam-d5	120	1	60
Oxazepam-d5	120	1	60
Temazepam-d5	120	1	60
Chlordiazepoxide-d5	60	100	300
Midazolam-d4	30	100	150
Flunitrazepam-d4	24	100	120
7-Aminoflunitrazepam-d7	24	100	120

3. Methods

3.1. Sample Preparation

1. Add 250 µL of samples, controls, or calibrators to each appropriately labeled 12 × 75 mm tube.
2. Add 1 mL of the working internal standard to each tube.
3. Cap samples and vortex for 5 min.
4. Centrifuge samples for 5 min at 3000 × g.
5. Transfer 400 µL of the supernatant to a clean 12 × 75 mm test tube.
6. Evaporate to dryness under nitrogen at 40°C.
7. Reconstitute with 200 µL of the water:acetonitrile reconstituting solution.
8. Transfer to a snap cap autosampler vial for analysis.

3.2. Liquid Chromatography Conditions

Chromatographic separations were performed using an ultra-performance liquid chromatograph (UPLC) system (ACQUITY, Waters Corp., Milford, MA) that included a phenyl column (2.1 × 100 mm) packed with 1.7 µm bridged ethyl hybrid (BEH) particles (ACQUITY UPLC® Waters Corp., Milford, MA) maintained at 35°C. Analytes were eluted from the UPLC column using the following step-wise binary elution gradient: Initial mobile phase composition was 80% mobile phase A:20% mobile phase B. Initial conditions were held constant for 0.25 min after which the composition of mobile phase B was increased linearly to 35% over 2.25 min. Conditions were held constant for 2.50 min after which the composition of mobile phase B was ramped to 80% over 1 min. Conditions were returned to the initial composition of 80:20 over the next 0.01 min and held for 1.40 min to equilibrate the column before the next injection in the sequence. The total run time was 7.5 min (**Fig. 10.1**). Flow rates were maintained at 0.5 mL/min for the entire chromatographic separation. All flow was directed into the ESI source of the mass spectrometer. Samples were maintained at 7.5°C in the sample organizer and sample injection volumes were 5 µL for all analyses.

3.3. Mass Spectrometry

Mass spectrometric detection was performed using a triple quadrupole mass spectrometer (Waters Quattro Premier, Waters Corp., Milford, MA) equipped with an electrospray ionization (ESI) source operating in positive ion mode. MS/MS conditions were as follows: capillary voltage 0.60 kV, cone voltage 30 V, extractor voltage 6.0 V, RF lens voltage 0.1 V. The source temperature was 120°C while the desolvation temperature was set at 450°C. Cone gas was set at a flow of 50 L/h while the nitrogen desolvation gas flow was 900 L/h. The argon collision gas flow was set to 0.16 mL/min.

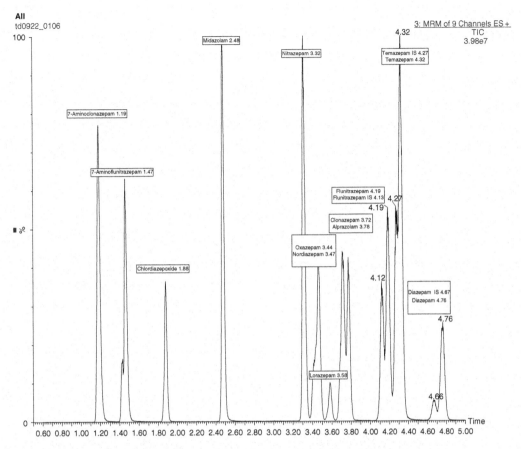

Fig. 10.1. Total ion chromatogram from the UPLC separation of 12 benzodiazepines in serum.

The mass transition from the protonated molecular ion $[M+H]^+$ to the most abundant product ion (**Table 10.9**) was designated the quantifying ion transition while the second most abundant mass transition was designated as the qualifying ion transition for each analyte (**Table 10.9**) (*see* **Notes 2 and 3**). The most abundant product ion for each deuterated internal standard was also monitored and used to calculate the response ratio between internal standards and analytes for all experiments. Following auto tuning of each analyte and internal standard, the optimized parameters were used to construct the MS/MS method, which was then used to acquire data in the MRM mode. **Table 10.10** reports the mass transitions, dwell times, cone voltages, and collision energies for each of the analytes and their deuterated internal standards.

3.4. Data Analysis   Analytical data was analyzed using Masslynx® version 4.1 software. Criteria for a positive result included accurate chromatographic retention time, presence of both the qualifying

Table 10.10
MS/MS parameters used for each analyte and deuterated internal standard

Compound	Mass transition	Purpose	Cone (V)	Collision (V)	Dwell (secs)
Alprazolam	309.30 > 274.0	Quantifying ion	45.0	24.0	0.010
Alprazolam	309.30 > 204.8	Qualifying ion	45.0	40.0	0.010
Alprazolam-d5	314.30 > 209.9	Quantifying ion	45.0	42.0	0.010
Temazepam	301.25 > 254.8	Quantifying ion	30.0	22.0	0.005
Temazepam	301.25 > 282.8	Qualifying ion	30.0	14.0	0.005
Temazepam-d5	306.30 > 259.9	Quantifying ion	30.0	22.0	0.005
Oxazepam	287.30 > 268.9	Quantifying ion	30.0	14.0	0.005
Oxazepam	287.30 > 240.9	Qualifying ion	30.0	22.0	0.005
Oxazepam-d5	292.30 > 273.9	Quantifying ion	30.0	16.0	0.005
Diazepam	285.30 > 153.7	Quantifying ion	40.0	26.0	0.005
Diazepam	285.30 > 192.8	Qualifying ion	40.0	32.0	0.005
Diazepam-d5	290.30 > 153.8	Quantifying ion	45.0	30.0	0.005
Lorazepam	321.30 > 302.9	Quantifying ion	30.0	16.0	0.005
Lorazepam	321.30 > 228.9	Qualifying ion	30.0	32.0	0.005
Lorazepam-d4	325.34 > 306.9	Quantifying ion	35.0	16.0	0.005
Clonazepam	316.40 > 269.9	Quantifying ion	40.0	26.0	0.005
Clonazepam	316.40 > 213.8	Qualifying ion	40.0	40.0	0.005
Clonazepam-d4	320.40 > 274.0	Quantifying ion	40.0	26.0	0.005

(continued)

Table 10.10 (continued)

Compound	Mass transition	Purpose	Cone (V)	Collision (V)	Dwell (secs)
Flunitrazepam	314.40 > 267.9	Quantifying ion	45.0	26.0	0.005
Flunitrazepam	314.40 > 239.1	Qualifying ion	45.0	34.0	0.005
Flunitrazepam-d7	321.00 > 274.8	Quantifying ion	45.0	26.0	0.005
Nordiazepam	271.20 > 139.7	Quantifying ion	45.0	28.0	0.005
Nordiazepam	271.20 > 164.7	Qualifying ion	45.0	30.0	0.005
Nordiazepam-d5	276.30 > 140.0	Quantifying ion	45.0	28.0	0.005
Midazolam	326.40 > 290.9	Quantifying ion	45.0	28.0	0.005
Midazolam	326.40 > 249.2	Qualifying ion	45.0	36.0	0.005
Midazolam-d4	330.40 > 295.0	Quantifying ion	45.0	28.0	0.005
Chlordiazepoxide	300.30 > 226.9	Quantifying ion	25.0	26.0	0.005
Chlordiazepoxide	300.30 > 282.9	Qualifying ion	25.0	14.0	0.005
Chlordiazepoxide-d5	305.20 > 231.8	Quantifying ion	25.0	26.0	0.005
7-Aminoclonazepam	286.36 > 120.7	Quantifying ion	40.0	32.0	0.005
7-Aminoclonazepam	286.36 > 221.9	Qualifying ion	40.0	24.0	0.005
7-Aminoclonazepam-d4	290.40 > 120.6	Quantifying ion	40.0	32.0	0.005
7-Aminoflunitrazepam	284.40 > 134.4	Quantifying ion	40.0	26.0	0.005
7-Aminoflunitrazepam	284.40 > 226.5	Qualifying ion	40.0	28.0	0.005
7-Aminoflunitrazepam-d7	291.20 > 137.5	Quantifying ion	40.0	26.0	0.005

product ion and the quantifying product ion, and product ion ratios within acceptable limits. (*see* **Note 4**) Retention times for each analyte were required to be within 5% of those determined with control samples. The ratio of the quantifying product ion peak area to the qualifying product ion peak area was required to be within +/− 20% of the ion ratio determined for calibrators.

Quantitation was performed using a working standard calibration curve and comparing the ratio of quantifying ion peak area to internal standard peak area. Patient samples were required to exhibit approximately the same internal standard peak areas as those in calibrators and controls in order to ensure an efficient extraction. The general acceptance criterion is that the internal standard response should be approximately 50–200% of the calibrator/control average. Calibration curves were required to comprise at least 50% of the original curve points and any specimens with quantitative values greater than the upper calibrator were required to be rerun at an appropriate dilution using negative serum. The analytical run is considered acceptable if the calculated concentrations of analyte/s in control samples are within 20% of the expected nominal value.

4. Notes

1. Many drugs have been shown to bind to gel-barrier phlebotomy tubes. Additionally, some of these types of tubes have been found to contain chemicals, which interfere with chromatography-based methods. The use of these tubes is not recommended; instead use plain redtops, gray top tubes (NaF containing), or royal blue (trace metal free) tubes for blood collection.

2. Appropriate quantifier and qualifier mass transitions were identified for each analyte by directly infusing a 10 μg/mL solution of each compound into the mass spectrometer ionization source at a flow rate of 20 μL/min. The flow path of the concentrated analyte solution was modified with a T-mixer to allow mixing of the solution with mobile phase and simultaneous infusion of the sample and initial mobile phase into the mass spectrometer.

3. Simultaneous infusion of the sample with the initial mobile phase was performed to ensure that any subsequently optimized tune page parameters were compatible with the initial mobile phase. By optimizing the tune page parameters while infusing a mixture of analyte and initial mobile phase, the

analytes are allowed to reach the ESI source under similar conditions and in a similar environment as those that would be encountered during the analysis of a positive sample. This ensures the most accurate optimization of ionization parameters specific to both the sample and the mobile phase. Concentrated (10 µg/mL) solutions of each benzodiazepine analyte and deuterated internal standard were individually infused concomitantly with a 80:20 (H_2O:ACN) composition of mobile phase at a flow rate of 0.5 mL/min. During infusion of each analyte, the collision gas was turned off to allow the protonated molecular ion of each compound to reach the detector and produce a recordable signal. Following identification of the molecular ion signal, an auto tune was completed for each analyte, which involved adjusting the capillary voltages, cone voltages, and collision energies to maximize the signal for both the precursor ions and the product ions generated in the collision cell. Auto tuning of the protonated molecular ion of each compound yielded information necessary to collect data in the MRM mode.

4. Chromatographic retention times are initially established for each analyte and deuterated internal standard during method validation through the analysis of reference standards. Retention times are updated following the analysis of calibration standards in each batch in order to account for minor drifts in daily retention times.

References

1. ElSohly, M.A., et al. 2007, *LC-(TOF) MS* analysis of benzodiazepines in urine from alleged victims of drug-facilitated sexual assault. *J Anal Toxicol*, 31(8): 505–514.

2. Baselt, R.C. 2008, *The Disposition of Toxic Drugs and Chemicals in Man.* 8th ed., Foster City: Biomedical Publications.

3. Peters, R.J.J., et al. 2007, Alprazolam (Xanax) use among southern youth: beliefs and social norms concerning dangerous rides on "handlebars". *J Drug Educ*, 37(4): 417–428.

4. Ngwa, G., et al. 2007, Simultaneous analysis of 14 Benzodiazepines in oral fluid by solid-phase extraction and LC-MS-MS. *J Anal Toxicol*, 31: 369–376.

Chapter 11

Determination of Benzoic Acid in Serum or Plasma by Gas Chromatography-Mass Spectrometry (GC/MS)

Jeff M. Knoblauch, David K. Scott, Laurie D. Smith, and Uttam Garg

Abstract

Nonketotic hyperglycinemia (NKH), a metabolic disorder due to defects in the glycine cleavage system, leads to the accumulation of toxic levels of glycine. Glycine levels in these patients may be lowered by sodium benzoate treatment. Benzoic acid binds to glycine to form hippurate, which is subsequently eliminated through the kidneys. At high concentrations, hippuric acid can crystallize in the kidneys and cause renal failure. Therefore, it is desirable to have benzoic acids concentrations within a therapeutic range.

In the gas chromatography method described, the drug from the acidified serum or plasma sample is extracted using ethyl acetate. The organic phase containing drug is separated and dried under a stream of nitrogen. After trimethylsilyl derivatization, benzoic acid analysis is done on a gas chromatograph mass spectrometer. Quantitation of the drug in a sample is achieved by comparing responses of the unknown sample to the responses of the calibrators using selected ion monitoring. Benzoic acid D_5 is used as an internal standard.

Key words: Benzoic acid, non-ketotic hyperglycinemia, cerebrospinal fluid, renal toxicity, mass spectrometry, gas chromatography.

1. Introduction

Nonketotic hyperglycinemia (NKH), also called glycine encephalopathy, is a metabolic disorder caused by a defect in the glycine cleavage system (GCS) (1, 2). The disease is inherited in an autosomal recessive pattern with a worldwide frequency of approximately 1 in 60,000. The most common form of NKH appears in infancy and is characterized by a progressive lethargy, hypotonia, myoclonic seizures, and apnea. Without treatment, death is likely. Surviving infants develop profound mental

U. Garg, C.A. Hammett-Stabler (eds.), *Clinical Applications of Mass Spectrometry*, Methods in Molecular Biology 603, DOI 10.1007/978-1-60761-459-3_11, © Humana Press, a part of Springer Science+Business Media, LLC 2010

retardation and intractable seizures. The diagnosis of the disease is based on the clinical findings and laboratory findings of increased glycine concentrations in plasma and cerebrospinal fluid (CSF) (3). Increased CSF to plasma glycine ratio has a higher diagnostic value than increased plasma or CSF glycine concentrations. Confirmation of the disease is done by measurement of GCS activity in liver tissue, fibroblasts, or lymphocytes (4). Molecular testing for detection of defects in GCS is also available.

The treatment of NKH includes oral administration of sodium benzoate, anticonvulsants to control seizures, and dextromethorphan (5–7). Sodium benzoate treatment improves clinical outcome by reducing both CSF and plasma glycine levels. In the circulation, the interaction between benzoic acid and glycine leads to the formation of hippurate which is relatively nontoxic and is excreted in the urine. However, in sufficient concentrations, hippuric acid can crystallize in the kidneys causing renal toxicity. Therefore, it is important that concentration of benzoic acid in blood be kept at ~250 µg/mL (7). Spectrophotometric, high performance liquid chromatography, and gas chromatographic methods are available for the determination of benzoic acid (8, 9). In this chapter, a gas chromatography-mass spectrometry method for the determination of benzoic acid in blood is described. The method involves acidic extraction of benzoic acid, preparation of trimethylsilyl derivatives, and analysis using selected ion monitoring.

2. Materials

2.1. Sample

Serum or plasma (heparin or EDTA tube) is acceptable samples for this procedure. Samples are stable for 1 week when refrigerated and 3 months when frozen at −20°C.

2.2. Reagents

1. Bis-(Trimethylsilyl)trifluoroacetamide (BSTFA) with 1% trimethylchlorosilane (TMCS) (UCT Inc., Bristol, PA).
2. Derivatization mixture: BSTFA with 1% TMCS: Pyridine (2:1).
3. Drug-free human serum (Utak Laboratories Inc, Valencia, CA) (see Note 1).
4. Sodium benzoate powder and deuterated benzoic acid D5 powder (Sigma Chemical Co., St. Louis, MO).
5. 1.0 N HCl: Add ~50 mL deionized water to a 100 mL volumetric flask. Add 8.3 mL concentrated HCl to the flask and bring to volume with deionized water (stable 1 year at room temperature).

2.3. Standards and Calibrators

1. Benzoic acid control stock standard (1 mg/mL): Add 118 mg Sodium Benzoate to a 100 mL volumetric flask. Add ∼50 mL deionized water to dissolve the powder (*see* **Note 2**). Add 830 μL concentrated (12 N) HCL then fill to volume with deionized water.

2. Prepare benzoic acid working calibrators using benzoic acid stock standard and blank serum in 10 mL volumetric flasks according to **Table 11.1**. The standards and calibrators are stable for 6 months stored at −20°C.

Table 11.1
Preparation of calibrators using 10 mL volumetric flasks. Fill to volume using drug-free serum

Stock solution (mL)	Final concentration (μg/mL)
0.1	10
0.5	50
1	100
5	500

2.4. Internal Standard and Quality Controls

1. Working internal standard (benzoic acid D_5 200 μg/mL): Add 20 mg benzoic acid D5 (Sigma) to a 100 mL volumetric flask. Add ∼50 mL deionized water to dissolve the powder (may need to sonicate or apply gentle heat). Add 830 μL concentrated (12 N) HCL, then fill to volume with deionized water. Stable 6 months stored at −20°C.

2. Quality control samples: In-house controls are used to validate sample analysis (*see* **Note 3**). The controls are made using benzoic acid stock standard and blank serum according to **Table 11.2**.

Table 11.2
Preparation of controls using 10 mL volumetric flasks. Fill to volume using drug-free serum

Stock solution (mL)	Final concentration (μg/mL)
0.25	25
2.5	250

2.5. Supplies

1. 13×100 mm test tubes and Teflon-lined caps (Fisher Scientific, Fair Lawn, NJ). These tubes are used for extraction and concentrating drug extracts.

2. Transfer pipettes (Samco Scientific, San Fernando, CA).

3. Autosampler vials (12×32 mm) with 0.3 mL limited volume inserts (P.J. Cobert Associates, Inc., St. Louis, MO).

4. GC column: Zebron ZB-1 with dimensions of $30\,m \times 0.25\,mm \times 0.25\,\mu m$ (Phenomenex, Torrance, California).

2.6. Equipment

1. TurboVap®1 V Evaporator (Zymark Corporation, Hopkinton, MA, USA)

2. A gas chromatograph/mass spectrometer (GC/MS), model 5890/5972, operated in electron impact mode (Agilent Technologies, CA)

3. Methods

3.1. Stepwise Procedure

1. Add 0.1 mL calibrator, control, or sample into a labeled 13×100 mm screw cap glass tube.

2. Add 1.0 mL of 1.0 N HCl.

3. Add 100 μL of working internal standard (200 μg/mL benzoic acid D5).

4. Add 4 mL ethyl acetate.

5. Cap and rock for 5 min.

6. Centrifuge at $\sim$1,600 $\times g$ for 5 min.

7. Transfer the upper organic layer to a 13×100 mm tube (*see* **Note 4**).

8. Evaporate the extract to dryness under stream of nitrogen at 45°C (*see* **Note 5**).

9. Add 150 μL of derivatization mixture.

10. Cap and heat the tubes in a dry block at 65°C for 20 min.

11. Remove the tubes from dry block and cool to the room temperature.

12. Transfer the contents to labeled auto sampler vials containing inserts.

3.2. Instrument Operating Conditions

The instrument's operating conditions are given in **Table 11.3**.

3.3. Data Analysis

1. A representative GC-MS chromatogram of benzoic acid is shown in **Fig. 11.1**. GC-MS selected ion chromatograms are shown in **Fig. 11.2**. An electron impact ionization mass

Table 11.3
GC/MS operating conditions

Head pressure	5 psi
Purge time	0.5 min
Injector temperature	250°C
Detector temperature	280°C
Initial temperature	70°C
Initial time	1.0 min
Ramp	20°C/min
Final temperature	280°C
Final time	3.0 min
MS mode	Electron impact at 70 eV, selected ion monitoring
MS tune	Autotune

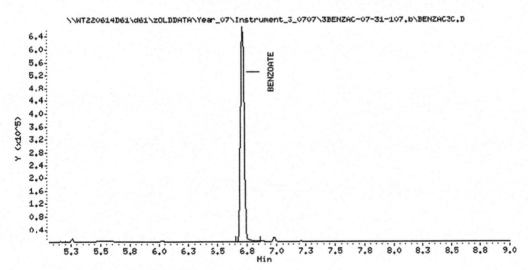

Fig. 11.1. GC-MS chromatogram of a 100 μg/mL calibrator.

spectrum of this compound is shown in **Fig. 11.3** (*see* **Note 6**). Ions used for identification and quantification are listed in **Table 11.4**.

2. Analyze data using Target Software (Thru-Put Systems, Orlando, FL) or similar software. The quantifying ions (**Table 11.4**) are used to construct standard curves of the peak area ratios (calibrator/internal standard pair) vs. concentration. These curves are then used to determine the concentrations of the controls and unknown samples.

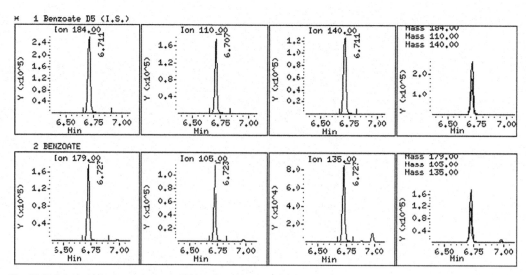

Fig. 11.2. Selected ion chromatograms of benzoic acid – monoTMS.

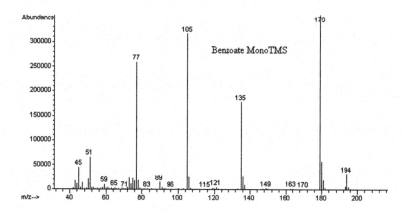

Fig. 11.3. Electron impact ionization mass spectrum of benzoic acid – monoTMS.

Table 11.4
The following ions are used in the quantitation of benzoic acid

	Quantitation ion	Qualifier ions
Benzoic acid	179	105, 135
Benzoic acid D5	184	110, 140

3. Standard curves should have a correlation coefficient (r^2) >0.99.

4. Typical intra- and inter-assay imprecision is <10%.

5. The linearity/limit of quantitation of the method is 10–500 μg/mL. Samples in which the drug concentrations exceed the upper limit of quantitation should be diluted with negative serum or plasma and retested.

6. Quality control: The run is considered acceptable if the calculated concentrations of drugs in the controls are within +/− 20% of target values. Quantifying ion in the sample is considered acceptable if the ratios of qualifier ions to quantifying ion are within +/− 20% of the ion ratios for the calibrators.

4. Notes

1. Certain lots of drug-free human serum may contain a small amount of benzoic acid. Test the material before putting into use.

2. Benzoic acid D5 and sodium benzoate powders are slow to dissolve. Be sure to completely dissolve the powder in deionized water first before adding the acid. Use sonication and/or gentle heat if needed.

3. Prepare calibrators and controls separately, preferably using different lots of drugs and analysts.

4. When transferring the ethyl acetate layer, make sure no aqueous layer is transferred. The aqueous portion will destroy the BSTFA and derivatization will not occur.

5. Do not over-dry. It will result in low recovery.

6. Electron impact ionization spectra are needed in the initial stages of method set up to establish retention times and later on, if there is a need for change in quantifying or qualifying ions. They are not needed for routine quantitation.

References

1. Hoover-Fong, J.E., Shah, S., Van Hove, J.L., Applegarth, D., Toone, J. and Hamosh, A. (2004) Natural history of nonketotic hyperglycinemia in 65 patients. *Neurology*, **63**, 1847–53.

2. Steiner, R.D., Sweetser, D.A., Rohrbaugh, J.R., Dowton, S.B., Toone, J.R. and Applegarth, D.A. (1996) Nonketotic hyperglycinemia: atypical clinical and biochemical manifestations. *J Pediatr*, **128**, 243–6.

3. Applegarth, D.A. and Toone, J.R. (2001) Nonketotic hyperglycinemia (glycine encephalopathy): laboratory diagnosis. *Mol Genet Metab*, **74**, 139–46.

4. Kure, S., Narisawa, K. and Tada, K. (1992) Enzymatic diagnosis of nonketotic hyperglycinemia with lymphoblasts. *J Pediatr*, **120**, 95–8.

5. Hamosh, A., McDonald, J.W., Valle, D., Francomano, C.A., Niedermeyer, E. and

Johnston, M.V. (1992) Dextromethorphan and high-dose benzoate therapy for nonketotic hyperglycinemia in an infant. *J Pediatr*, **121**, 131–5.

6. Hamosh, A., Maher, J.F., Bellus, G.A., Rasmussen, S.A. and Johnston, M.V. (1998) Long-term use of high-dose benzoate and dextromethorphan for the treatment of nonketotic hyperglycinemia. *J Pediatr*, **132**, 709–13.

7. Van Hove, J.L., Vande Kerckhove, K., Hennermann, J.B., Mahieu, V., Declercq, P., Mertens, S., De Becker, M., Kishnani, P.S. and Jaeken, J. (2005) Benzoate treatment and the glycine index in nonketotic hyperglycinaemia. *J Inherit Metab Dis*, **28**, 651–63.

8. Sioufi, A. and Pommier, F. (1980) Gas chromatographic determination of low concentrations of benzoic acid in human plasma and urine. *J Chromatogr*, **181**, 161–8.

9. Sarkissian, C.N., Scriver, C.R. and Mamer, O.A. (2007) Quantitation of phenylalanine and its trans-cinnamic, benzoic and hippuric acid metabolites in biological fluids in a single GC-MS analysis. *J Mass Spectrom*, **42**, 811–7.

Chapter 12

Quantification of Busulfan in Plasma Using Liquid Chromatography Electrospray Tandem Mass Spectrometry (HPLC-ESI-MS/MS)

Marion L. Snyder and James C. Ritchie

Abstract

Busulfan is a chemotherapy drug widely used as part of conditioning regimens for patients undergoing bone marrow transplantation (BMT). Challenges of busulfan treatment include a narrow therapeutic window and wide inter- and intra-patient variability. Inappropriately low drug levels lead to relapse and even graft rejection, while higher doses frequently have toxic and sometimes fatal consequences. Maintenance of plasma busulfan concentrations using repeated measurements and proper adjustment of dosage can reduce busulfan-related toxicity and improve treatment outcomes. We describe a rapid (2-minute total analysis time per sample) and simple method for accurate and precise busulfan concentration determination in plasma samples (100 μL) using high performance liquid chromatography combined with electrospray positive ionization tandem mass spectrometry (HPLC-ESI-MS/MS). Busulfan is isolated from plasma after internal standard (busulfan-D_8)–methanol extraction, dilution with mobile phase (ammonium acetate–formic acid–water), and centrifugation. The supernatant plasma is injected onto the HPLC-ESI-MS/MS and quantified using a six-point standard curve. The assay is linear from 0.025 μg/mL (∼0.1 μmol/L) to at least 6.2 μg/mL (∼25 μmol/L) with precisions of <5% over the entire range.

Key words: Busulfan, therapeutic drug monitoring, mass spectrometry, liquid chromatography, plasma, quantification.

1. Introduction

Busulfan, 1,4-butanediol dimethanesulfonate, is an alkylating chemotherapeutic agent widely used as a part of conditioning regimens for patients undergoing autologous and allogeneic bone marrow transplantation (BMT). The drug has been in clinical use since 1959, marketed in the U.S. under the brand name Myleran®

U. Garg, C.A. Hammett-Stabler (eds.), *Clinical Applications of Mass Spectrometry*, Methods in Molecular Biology 603, DOI 10.1007/978-1-60761-459-3_12, © Humana Press, a part of Springer Science+Business Media, LLC 2010

(GlaxoSmithKline) and more recently, as an IV formulation called Busulfex® (PDL BioPharma, Inc.). Optimal busulfan dosing is particularly challenging due to its narrow therapeutic window and pharmacokinetic variability. While under-dosing increases risk of disease recurrence and even graft rejection, higher doses are frequently associated with hepatic toxicity (1–4). Moreover, reports show up to 10-fold variability in area under the time–versus busulfan concentration curve (AUC) for patients on the same dosing regimen (1, 5). These challenges highlight the need for therapeutic drug monitoring (TDM) in optimizing busulfan treatment regimens. Studies show plasma busulfan concentrations correlate better with clinical effect and toxicity than the dose administered to the patient, and that maintenance of plasma busulfan concentrations using TDM may help decrease busulfan-related toxicity and improve treatment outcomes (1, 2).

Many of the available busulfan assays are complicated and time consuming, requiring extensive sample preparation/derivatization, and lack desired sensitivity or require large sample volumes (5, 6). The following chapter describes a simple, sensitive, accurate, and precise method to rapidly measure busulfan in plasma samples using high performance liquid chromatography-coupled electrospray positive ionization tandem mass spectrometry (HPLC-ESI-MS/MS). In this method, busulfan is isolated from 100 µL of plasma after internal standard (busulfan-D_8)-spiked methanol extraction, centrifugation, and dilution of the supernatant with mobile phase (ammonium acetate–formic acid–water) (6, 7). The diluted supernatant is injected into the HPLC-ESI-MS/MS and quantified using a six-point standard curve.

2. Materials

2.1. Samples

Heparinized plasma. Process and analyze within 4 h of collection to prevent analyte degradation or freeze at 4°C until analysis.

2.2. Solvents and Reagents

1. Formic acid, ACS grade.
2. Acetone, ACS grade.
3. Methanol, Optima® LC/MS grade.
4. Acetonitrile, Optima® LC/MS grade.
5. Ammonium acetate, ACS grade.
6. Ammonium acetate solution, 1 M, prepared with HPLC-grade water. Stable at room temperature, 18–24°C, up to 1 year.

7. Mobile Phase A and Purge Solvent (2 mM ammonium acetate/0.1% (v/v) formic acid in HPLC-grade water): Add 2 mL of 1 M ammonium acetate solution and 1 mL of formic acid to 1 L volumetric flask, bring to volume with water and mix. Stable at room temperature, 18–24°C, up to 1 year.

8. Mobile Phase B (2 mM ammonium acetate/0.1% (v/v) formic acid in methanol): Add 2 mL of 1 M ammonium acetate solution and 1 mL formic acid to 1 L volumetric flask; bring to volume with methanol and mix. Stable at room temperature, 18–24°C, up to 1 year.

9. Human drug-free pooled normal plasma.

2.3. Internal Standards and Standards

1. Primary standard: Busulfan (1,4-Butanediol dimethane sulfonate) (Sigma-Aldrich)

2. Primary internal standard (I.S.): Busulfan-D_8 (Tetramethylene) (Cambridge Isotope Laboratories)

3. Busulfan Standard Stock Solutions 1 and 2 (500 μg/mL primary standard in acetone) (*see* **Note 1**): Add 50 mg primary standard to 100 mL volumetric flask, bring to volume with acetone and mix. Stable at room temperature, 18–24°C, up to 1 year.

4. I.S. Working Solution/Extraction Solution (2.0 μg/mL busulfan-D_8 in methanol):
   a. Reconstitute primary I.S. by adding 10 mg to 100 mL volumetric flask; bring to volume with acetonitrile and mix. This intermediate solution is 100 μg/mL. Stable at room temperature, 18–24°C, up to 1 year.

   b. Add 2 mL of previous solution to 100 mL volumetric flask; fill to volume with methanol and mix. Stable at room temperature, 18–24°C, up to 1 year.

2.4. Calibrators and Controls

1. Calibrators: Prepare Calibrators 1–6 by making serial dilutions of the standard stock solution 1 according to **Table 12.1**. For each dilution step: Add appropriate amount of previous solution, as shown in **Table 12.1**, to 10.0 mL volumetric flask and fill to volume with heparinized plasma. Vortex mix after each dilution step. (*See* **Note 2**)

2. Controls: Prepare controls by making separate dilutions of the standard stock solution 2 according to **Table 12.2**, using procedure similar to preparation of Calibrators: For Controls, add appropriate amount of standard stock solution 2 to 10.0 mL volumetric flask and fill to volume with heparinized plasma. Vortex mix. (*See* **Note 2**)

Table 12.1
Preparation of calibrators

Calibrator	Volume of previous standard (mL)	Drug-free plasma (mL)	Final concentration (µg/mL)[a]
1	0.1 (Stock Solution 1)	9.9	5.0
2	6.0 (#1)	4.0	3.0
3	5.0 (#2)	5.0	1.5
4	5.0 (#3)	5.0	0.75
5	5.0 (#4)	5.0	0.38
6	2.0 (#5)	8.0	0.075

[a] 1 µg/mL busulfan = × 4.06 µmol/L.

Table 12.2
Preparation of quality controls

Control	Volume of standard stock solution 2 (µL)	Drug-free plasma (mL)	Final concentration (µg/mL)[a]
Low	5.0	9.995	0.25
Medium	25.0	9.975	1.25
High	75.0	9.925	3.75

[a] 1 µg/mL busulfan = × 4.06 µmol/L.

3. Check Standard: prepare by pooling several extracted calibrators and/or controls from previous run.

4. Assign instrumental operating parameters according to conditions presented in **Table 12.3A** and **B**. (Parameters are instrumentation-specific.)

2.5. Analytical Equipment and Supplies

1. Waters Alliance HT 2795 Separation Module with Micromass Quatro Micro API equipped with Masslink software

2. Analytical column: XTerra MS C-8, 35 µm, 21 × 50 mm

3. 1.5 mL microcentrifuge tubes

Table 12.3
HPLC-ESI-MS/MS operating conditions

A. HPLC[a]

Column temp.	55°C	
Flow rate	0.600 mL/min	
Gradient	Time (min)	Mobile Phase A (%)
	0.00	100
	1.25	45
	1.50	100
	3.00	100

B. MS/MS Tune Settings[b]

Capillary (kV)	0.40
Cone (V)	19.00
Source temp. (°C)	140
Desolvation temp (°C)	375
Cone gas flow (L/hr)	OFF
Desolvation gas flow (L/Hr)	800
LM 1 resolution	11.3
LM 2 resolution	8.7
HM 1 resolution	11.3
HM 2 resolution	8.7

[a]Optimized for Waters Alliance HT 2795 Separation Module equipped with XTerra MS C-8, 35 μm, 21 × 50 mm analytical column; Mobile phase A: 2 mM ammonium acetate-0.1% formic acid in water; Mobile phase B: 2 mM ammonium acetate-0.1% formic acid in methanol.
[b]Optimized for Micromass Quatro Micro API. Tune settings may vary slightly between instruments.

3. Methods

3.1. Stepwise Procedure

1. To labeled 1.5-mL microfuge tubes, pipette 100 μL plasma (calibrators, controls, or plasma samples) (*see* **Note 3**).

2. Add 200 μL I.S. Working Solution.

3. Cap and vortex mix tubes at maximum speed for 5 sec.

4. Centrifuge for 6 minutes at $12,000 \times g$.

5. Transfer supernatant into second set of labeled 1.5 mL micro tubes (*See* **Note 4**).

6. Add 1 mL Mobile Phase A to each tube, cap, and vortex mix for several seconds.

7. Remove tube caps.

8. Place sample tubes including Check Standard onto autosampler.

9. Inject 10 μL of sample onto HPLC-ESI-MS/MS (*see* **Note 5**). Sample HPLC-ESI-MS/MS ion chromatograms for busulfan and I.S. are shown in **Fig. 12.1**.

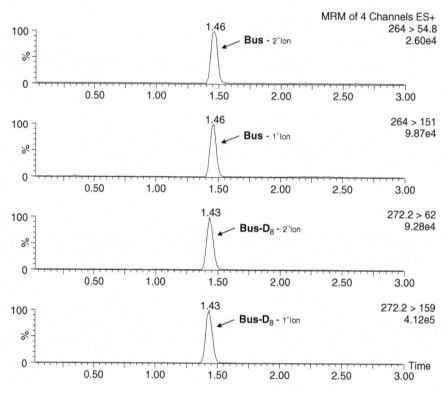

Fig. 12.1. HPLC-ESI-MS/MS ion chromatograms of busulfan and busulfan-D$_8$ (I.S.). Both primary (1°) and secondary (2°) ions are shown for both compounds.

3.2. Analysis

1. Instrumental operating parameters are given in **Table 12.3A** and **B**.

2. Data are analyzed using QuanLynx Software (Waters, Inc.).

3. Standard curves are generated based on linear regression of the analyte/I.S. peak-area ratio (y) versus analyte concentration (x) using the quantifying ions indicated in **Table 12.4**.

4. Acceptability of each run is confirmed if the calculated control concentrations fall within two standard deviations of the target values. Target values for controls are established in-house as the mean of twenty consecutive runs.

Table 12.4
Precursor and primary and secondary product ions for busulfan and busulfan-D$_8$

	Precursor ion $(M+NH_4)^{+1}$	Product ions(primary, secondary)	Dwell(sec)	Cone(V)	Collision(eV)	Delay(sec)
Busulfan	264[a]	151[a,b], 55[a]	0.2	19	13	0.030
Busulfan-D$_8$	272[a]	159[a,b], 62[a]	0.2	19	13	0.030

[a]Optimized m/z may change based on tuning parameters.
[b]Primary ions used for busulfan quantification.

5. Limits for the ratio of primary to secondary product ion areas are used to increase the specificity of the assay for busulfan and are set to 10%.

6. Liquid chromatography retention time window limits for busulfan and busulfan-D$_8$ are set at 1.45 and 1.42 ($\pm$ 0.05) min, respectively (*see* **Note 6**).

7. The assay is linear to at least 6.2 µg/mL ($\sim$25 µmol/L) and sensitive to at least 0.025 µg/mL ($\sim$0.1 µmol/L), with precisions of <5% over the entire range. *See* **Notes 7** and **8** for information regarding accuracy and ion suppression studies, respectively.

8. Typical coefficients of correlation of the standard curve are >0.99.

4. Notes

1. Prepare two separate busulfan standard stock solutions: Stock solution 1 is used to prepare calibrators and stock solution 2 is used to prepare the controls. These are ideally prepared as separate lots on separate days.

2. Individual sets of Calibrators 1–6 and high, medium, and low level Controls can be pre-aliquoted and frozen until use in each analytical run. For each set pipette 100 µL of calibrator/control solution into 1.5 mL microfuge tubes and freeze at –80°C until use. Thaw completely before use. Stable up to 1 year at –80°C.

3. A new standard curve (Calibrators 1–6) should be generated with each analytical run to optimize method performance.

4. If desired, a set volume of supernatant (e.g., 250 µL) can be transferred; however, we find that the pour-off method easier and quicker to perform and provides equivalent method precision and accuracy.

5. The Check Standard is tested prior to each analytical run to verify instrument function (peak shapes, retention times, etc.).

6. Retention times are instrumentation-specific and can vary somewhat with minor changes in instrumentation (different lengths of connective tubing, use of a column switching valve, etc.).

7. Correlation with an established GC-MS method revealed a slope of 0.95, an intercept of 0.1, and $r^2 = 0.99$.

8. Ion suppression effects were evaluated by sample infusion method. Interferences from several common anticoagulants, as well as hemolysis, lipemia, and icterus, were also evaluated. No significant interferences or ion suppression (>5%) were identified. For more information on potential interferences, *see* Kellog et al. (6).

References

1. Grochow, L. B., Jones, R. J., Brundrett, R. B., et al. (1989) Pharmacokinetics of busulfan: correlation with veno-occlusive disease in patients undergoing bone marrow transplantation *Cancer Chemother Pharmacol* **25**, 55–61.

2. Grochow, L. B. (1993) Busulfan disposition: the role of therapeutic monitoring in bone marrow transplantation induction regimens *Semin Oncol* **20**, 18–25.

3. Brodsky, R., Topolsky, D., Crilley, P., et al. (1990) Frequency of veno-occlusive disease of the liver in bone marrow transplantation with a modified busulfan/cyclophosphamide preparative regimen *Am J Clin Oncol* **13**, 221–5.

4. Hasegawa, S., Horibe, K., Kawabe, T., et al. (1998) Veno-occlusive disease of the liver after allogeneic bone marrow transplantation in children with hematolohgic malignancies: incidence, onset time and risk factors *Bone Marrow Transplant* **22**, 1191–7.

5. Oliveira dos Reis, E., Vianna-Jorge, R., Suarez-Kurtz, G., Luiz da Silva Lima, E., de Almeida Azevedo, D. (2005) Development of a rapad and specific assay for detection of busulfan in human plasma by high-performance liquid chromatography/electrospray ionization tandem mass spectrometry *Rapid Commun Mass Spectrom* **19**, 1664–74.

6. Kellog, M. D., Law, T., Sakamoto, M., Rifai, N. (2005) Tandem mass spectrometry method for the quantification of serum busulfan *Ther Drug Monit* **25**, 625–9.

7. Correspondence with Russell Potter, Department of Pathology, University of Alabama.

Chapter 13

Quantitation of Total 11-Nor-9-Carboxy-Delta 9-Tetrahydrocannabinol in Urine and Blood Using Gas Chromatography-Mass Spectrometry (GC-MS)

C. Clinton Frazee III, Michael Kiscoan, and Uttam Garg

Abstract

Marijuana, which is made from crushing the leaves, flowers, and sometimes the stems of the plant Cannabis sativa, contains more than 30 cannabinoids. The major psychoactive cannabinoid is delta-9-tetrahydro-cannabinol (THC). The major metabolite of THC, 11-nor-delta 9-carboxy-tetrahydrocannabionol (THC-COOH), is excreted in the urine primarily as a glucuronide conjugate and is commonly analyzed in biological specimens for detecting marijuana usage. The procedure described here involves the addition of deuterated internal standard THC-COOH-d9 into the sample followed by hydrolysis of conjugated THC-COOH by alkali. THC-COOH is extracted from urine or blood using liquid–liquid extraction followed by preparation of its trimethylsilyl derivatives. The analysis of derivatized THC-COOH is performed using gas-chromatography/mass spectrometry (GC/MS). Quantification of the drug in a sample is achieved by comparing the responses of the unknown sample to the responses of the calibrators using selected ion monitoring.

Key words: Marijuana, cannabinoids, THC, THC-COOH, mass spectrometry.

1. Introduction

Marijuana, which is smoked or administered orally, is one of the most commonly abused drugs. The main active ingredient of marijuana is tetrahydrocannabinol (THC). Its pharmacological effects appear within a few minutes of smoking and may last for several hours. Its behavioral effects include feelings of euphoria, relaxation, and altered time perception. Its major side effects may include lack of concentration, impaired memory, and paranoia (1–4). The use of marijuana or synthetic THC dronabinol to relieve nausea and vomiting associated with cancer chemotherapy

U. Garg, C.A. Hammett-Stabler (eds.), *Clinical Applications of Mass Spectrometry*, Methods in Molecular Biology 603,
DOI 10.1007/978-1-60761-459-3_13, © Humana Press, a part of Springer Science+Business Media, LLC 2010

has been described. Other uses of marijuana include control of intraocular pressure, reduction of muscle spasms, and relief from chronic pain (5). Though legal use of marijuana is highly debatable, its use is legal in some states (6).

The major metabolite of THC is 11-carboxy-THC and after glucuronide conjugation it is excreted in the urine where its presence is used as an indicator of marijuana use. Since the half-life of THC is 20–40 h, following smoking cessation, heavy marijuana use can be detected for several weeks (1, 7). The usual protocol in most laboratories for the detection of THC-COOH involves prescreening samples using an immunoassay followed by confirmation of positive samples using mass spectrometry. Both gas and liquid chromatography-mass spectrometry have been used for the confirmation of THC-COOH (8–12). Gas chromatography-mass spectrometry (GC-MS) remains the most commonly used method for the confirmation of THC-COOH. Since the majority of THC-COOH in urine exists in the conjugated form, the samples must be hydrolyzed prior to extraction of THC-COOH. Extraction methods involve either liquid–liquid extraction or solid phase extraction. The method described here involves alkaline sample hydrolysis, formation of trimethylsilyl derivatives, and quantitation by selected ion monitoring. This procedure, though quantitative, is intended as a confirmation for THC-COOH and does not attempt to offer interpretation of the levels.

2. Materials

2.1. Samples

Serum or plasma (heparin or EDTA) or randomly collected urine are acceptable samples for this procedure. Samples are stable for 1 week when refrigerated and 6 months when frozen at −20°C.

2.2. Reagents

1. Bis-(Trimethylsilyl)trifluoroacetamide (BSTFA) with 1% TMCS (trimethylchlorosilane) (United Chemical Technologies, Bristol, PA).

2. 11.8 N Potassium hydroxide: Add approximately 500 mL of deionized water to a 1 L volumetric flask. Slowly add 662 g of KOH pellets and bring the volume to 1 L with deionized water. Store in an amber bottle. Stable for 1 year at room temperature.

3. Hexanes: Ethyl acetate (8:2): Combine 800 mL hexanes with 200 mL of ethyl acetate. Store in an amber bottle. Stable for 1 year at room temperature.

4. 0.1 M acetic acid: Add approximately 400 mL of deionized water to a 500 mL volumetric flask. Slowly add 2.87 mL glacial acetic acid and bring the volume to 500 mL with deionized water. Stable for 6 months. at room temperature.

2.3. Standards and Calibrators

1. Primary standard: 100 µg/mL THC-COOH (Cerilliant Corporation, Round Top, TX).

2. Working standard, 10 µg/mL THC-COOH: 10 µg/mL THC-COOH. Add 1 mL primary standard to a 10 mL volumetric flask and bring the volume to the mark with methanol. Stable for 1 year in a refrigerator.

3. Prepare working calibrators according to **Table 13.1**.

Table 13.1
Preparation of calibrators. The total volume is made to 10 mL with drug-free urine or plasma

µL of working standard (10 µg/mL)	Concentration (ng/mL)
15	15
50	50
100	100
500	500

Calibrators are stable for 1 year at −20°C.

2.4. Quality Controls and Internal Standard

1. Quality control samples are prepared according to the manufacturer's instructions (Bio-Rad Laboratories, Hercules, CA).

1. Primary internal standard: 100 µg/mL THC-COOH-d9 (Cerilliant Corporation, Round Top, TX).

2. Working internal standard, 2 µg/mL: Add 1 mL primary internal standard to a 50 mL volumetric flask and bring the volume to the mark with methanol. Stable for 1 year in a refrigerator.

2.5. Supplies

1. 16 × 100 screw-cap glass tubes: These tubes are used for extraction.

2. 13 × 100 mm screw-cap glass tubes: These tubes are used for concentration of extracts.

3. Transfer pipettes (Samco Scientific, San Fernando CA).

4. Auto sampler vials (12 × 32 mm with crimp caps) with 0.3 mL limited volume inserts (P.J. Cobert Associates, St. Louis, MO).

5. GC column: Zebron ZB-1 with dimensions of 15 m × 0.25 mm × 0.25 µm. (Phenomenex, Torrance,California).

2.6. Equipment

1. A gas chromatograph/mass spectrometer system (GC/MS; 6890/5975 or 5890/5972) with autosampler and operated in electron impact mode (Agilent Technologies, Wilmington, DE).

2. TurboVap® IV Evaporator (Zymark Corporation, Hopkinton, MA, USA).

3. Methods

3.1. Stepwise Procedure

1. Prepare an unextracted standard by adding 100 µL working THC-COOH standard and 100 µL working THC-COOH-d9 internal standard to a 13×100 mm concentration tube. Set-aside until step 11.

2. Pipette 1 mL sample, calibrator or control, into appropriately labeled 16×100 mm culture tubes.

3. Add 4 mL deionized water to each tube.

4. Add 100 µL working internal standard to each tube.

5. Add 100 µL 11.8 N potassium hydroxide to each tube.

6. Vortex and allow to stand for 15 min.

7. Add 2 mL 0.1 M acetic acid to each tube (*see* **Note 1**).

8. Add 200 µL of glacial acetic acid to each tube.

9. Add 5 mL hexanes:ethyl acetate (8:2), cap and rock for 5 min.

10. Centrifuge the samples at ~$1,600 \times g$ for 5 min. and transfer organic layer to a concentration tube (*see* **Note 2**).

11. Concentrate the extract to dryness under nitrogen at 50°C (*see* **Note 3**).

12. Reconstitute the residue in 100 µL BSTFA with 1% TMCS.

13. For derivatization, cap the tubes and incubate them in a dry bath for 10 min. at 65°C.

14. Cool the derivatized sample and transfer to appropriately labeled autosampler vials and inject 1 µL onto GC/MS for analysis.

3.2. Instrument Operating Conditions

The instrument's operating conditions are given in **Table 13.2**.

3.3. Data Analysis

1. A representative GC-MS chromatogram of the TMS derivatives of THC-COOH and THC-COOH-d9 is shown in **Fig. 13.1**. GC-MS selected ion chromatograms are shown in **Fig. 13.2**. Electron impact ionization mass spectrum

Table 13.2
GC/MS operating conditions

Column pressure	5 psi
Injector temp.	250°C
Purge time on	1.0 min
Detector temp.	280°C
Initial oven temp.	120°C
Initial time	0.5 min
Temperature ramp	30°C/min
Final oven temp.	280°C
Final time	6.0 min
MS source temp.	230°C
MS mode	Electron impact at 70 eV, selected ion monitoring
MS tune	Auto-tune

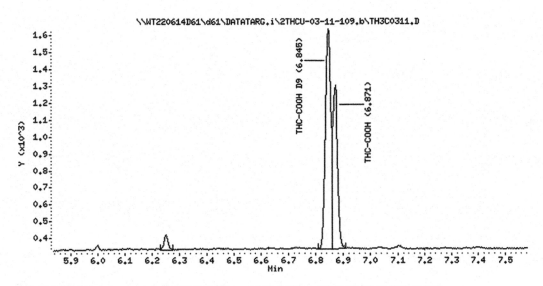

Fig. 13.1. GC-MS chromatogram of TMS-derivatives of THC-COOH and THC-COOH-d9 (100 ng/mL).

of Di-TMS THC-COOH is shown in the **Fig. 13.3** (*see* **Note 4**). The ions used for identification and quantification is listed in **Table 13.3**.

2. Analyze data using Target Software (Thru-Put Systems, Orlando, FL) or similar software. The quantifying ions (**Table 13.3**) are used to construct standard curves of the

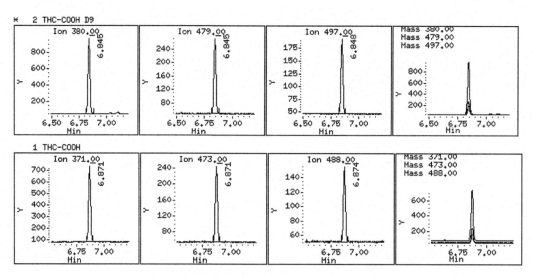

Fig. 13.2. Selected ion chromatograms of TMS-derivatives THC-COOH and THC-COOH-d9.

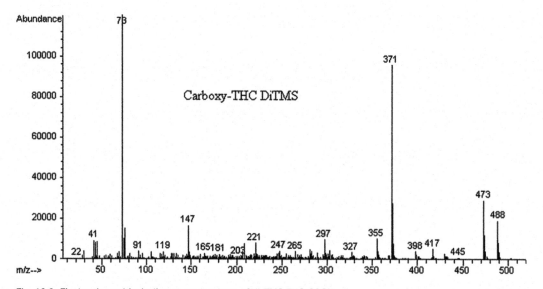

Fig. 13.3. Electron impact ionization mass spectrum of di-TMS THC-COOH.

Table 13.3
Quantifying and qualifying ions

Compound	Quantitation ion	Qualifier ions
THC-COOH	371	473, 488
THC-COOH-d9	380	479, 497

peak area ratios (calibrator/internal standard pair) vs. concentration. These curves are then used to determine the concentrations of the controls and unknown samples.

3. The linearity/limit of quantitation of the method is 15–500 ng/mL. Samples in which the drug concentrations exceed the upper limit of quantitation should be diluted with negative urine, serum, or plasma and retested.

4. A typical calibration curve has correlation coefficient (R^2) of >0.99.

5. Typical intra- and inter-assay imprecision is <10%.

6. Quality control: The analytical run is considered acceptable if the calculated concentrations of drugs in the controls are within +/− 20% of target values. The quantifying ion in the sample is considered acceptable if the ratios of qualifier ions to quantifying ion are within +/− 20% of the ion ratios for the calibrators.

4. Notes

1. This step is particularly important for serum or plasma samples. The purpose of this step is to acidify the sample slowly. Rapid acidification of the serum or plasma samples results in rapid protein precipitation and low recovery.

2. Be careful not to transfer any aqueous phase into the concentration tubes. This will result in low recovery, as the drug will not derivatize.

3. Do not over-dry the extract. This will result in poor recovery and failed run.

4. Electron impact ionization spectra are needed in the initial stages of the method set up to establish retention times, and later on if there is a need for change in quantifying or qualifying ions. However, these spectra are not needed for routine quantification.

Acknowledgment

We acknowledge the help of David Scott in preparing the figures.

References

1. Baselt, R.C. (2008) Cannabinoids. In *Disposition of Toxic Drugs and Chemicals in Man*, Foster City, pp. 1513–8.

2. Degenhardt, L. and Hall, W.D. (2008) The adverse effects of cannabinoids: implications for use of medical marijuana. *CMAJ*, **178**, 1685–6.

3. O'Leary, D.S., Block, R.I., Koeppel, J.A., Schultz, S.K., Magnotta, V.A., Ponto, L.B., Watkins, G.L. and Hichwa, R.D. (2007) Effects of smoking marijuana on focal attention and brain blood flow. *Hum Psychopharmacol*, **22**, 135–48.

4. Nixon, P.J. (2006) Health effects of marijuana: a review. *Pac Health Dialog*, **13**, 123–9.

5. Haney, M., Gunderson, E.W., Rabkin, J., Hart, C.L., Vosburg, S.K., Comer, S.D. and Foltin, R.W. (2007) Dronabinol and marijuana in HIV-positive marijuana smokers. Caloric intake, mood, and sleep. *J Acquir Immune Defic Syndr*, **45**, 545–54.

6. Seamon, M.J. (2006) The legal status of medical marijuana. *Ann Pharmacother*, **40**, 2211–5.

7. Huestis, M.A., Henningfield, J.E. and Cone, E.J. (1992) Blood cannabinoids. I. Absorption of THC and formation of 11-OH-THC and THCCOOH during and after smoking marijuana. *J Anal Toxicol*, **16**, 276–82.

8. Weinmann, W., Goerner, M., Vogt, S., Goerke, R. and Pollak, S. (2001) Fast confirmation of 11-nor-9-carboxy-Delta(9)-tetrahydrocannabinol (THC-COOH) in urine by LC/MS/MS using negative atmospheric-pressure chemical ionisation (APCI). *Forensic Sci Int*, **121**, 103–7.

9. Weinmann, W., Vogt, S., Goerke, R., Muller, C. and Bromberger, A. (2000) Simultaneous determination of THC-COOH and THC-COOH-glucuronide in urine samples by LC/MS/MS. *Forensic Sci Int*, **113**, 381–7.

10. Scurlock, R.D., Ohlson, G.B. and Worthen, D.K. (2006) The detection of Delta9-tetrahydrocannabinol (THC) and 11-nor-9-carboxy-Delta9-tetrahydrocannabinol (THCA) in whole blood using two-dimensional gas chromatography and EI-mass spectrometry. *J Anal Toxicol*, **30**, 262–6.

11. Moeller, M.R., Doerr, G. and Warth, S. (1992) Simultaneous quantitation of delta-9-tetrahydrocannabinol (THC) and 11-nor-9-carboxy-delta-9-tetrahydrocannabinol (THC-COOH) in serum by GC/MS using deuterated internal standards and its application to a smoking study and forensic cases. *J Forensic Sci*, **37**, 969–83.

12. Gustafson, R.A., Moolchan, E.T., Barnes, A., Levine, B. and Huestis, M.A. (2003) Validated method for the simultaneous determination of Delta 9-tetrahydrocannabinol (THC), 11-hydroxy-THC and 11-nor-9-carboxy-THC in human plasma using solid phase extraction and gas chromatography-mass spectrometry with positive chemical ionization. *J Chromatogr B Analyt Technol Biomed Life Sci*, **798**, 145–54.

Chapter 14

Quantitation of Cocaine, Benzoylecgonine, Ecgonine Methyl Ester, and Cocaethylene in Urine and Blood Using Gas Chromatography-Mass Spectrometry (GC-MS)

Steven W. Fleming, Amitava Dasgupta, and Uttam Garg

Abstract

Cocaine, a stimulant, is a commonly abused drug. Cocaine and its metabolites are measured in various biological specimens for clinical and forensic purposes. Urine or plasma or serum is spiked with deuterated internal standards cocaine-d3, benzoylecgonine-d3, ecgonine methyl ester-d3, and cocaethylene-d3 and buffered with phosphate buffer. The drugs in the sample are extracted by cation-exchange solid phase extraction. The drugs from the solid phase cartridge are eluted and the eluent is dried under the stream of nitrogen. The residue is incubated with pentafluoropropionic acid anhydride and pentafluoropropanol to form pentafluoropropionyl derivatives of ecgonine methyl ester and benzoylecgonine. Cocaine and cocaethylene are refractory to derivatization. The extract is dried, reconstituted in ethyl acetate, and injected into gas chromatography mass-spectrometry analyzer. Quantitation of the drugs in the samples is made, using selected ion monitoring, from a 3-point calibration curve.

Key words: Cocaine, benzoylecgonine, ecgonine methyl ester, cocaethylene, mass spectrometry, gas chromatography, Solid Phase Extraction (SPE).

1. Introduction

Cocaine is a naturally occurring alkaloid found in *Erythroxylon coca*, which primarily grows in the northern South American Andes. It is a potent central nervous system stimulant that is widely abused (1, 2). Its mechanism of action is through blocking uptake of neurotransmitters norepinephrine, dopamine, and serotonin. Cocaine is generally administered by drug users intranasally, intravenously, or by smoking (free base form known as "crack"). Its bioavailability by oral intake is very low due to first-pass effect. Once in the body, cocaine is rapidly metabolized in the body to several metabolites.

U. Garg, C.A. Hammett-Stabler (eds.), *Clinical Applications of Mass Spectrometry*, Methods in Molecular Biology 603, DOI 10.1007/978-1-60761-459-3_14, © Humana Press, a part of Springer Science+Business Media, LLC 2010

The two major metabolites are benzoylecgonine and ecgonine methyl ester. If cocaine and ethanol are co-consumed, a toxic metabolite cocaethylene (ethylcocaine) is produced in the body by transesterification of cocaine by ethanol (3, 4). The half-lives for cocaine, benzoylecgonine, ecgonine methyl ester, and cocaethylene are 0.5–1.5, 4–7, 3–4, and 2.5–6 h, respectively (5).

Cocaine and its metabolites are measured in the biological specimens for toxicological and forensic purposes. Samples are generally screened by immunoassays followed by confirmation of positive samples by gas chromatography mass-spectrometry. The drugs in the samples are extracted using liquid or solid phase extraction (6–8). Deuterated internal standards are available and are frequently employed in the assay of these drugs. To increase volatility and stability, the extracted drugs are frequently derivatized (9–12). The derivatized drugs are separated by gas chromatography and quantified by mass-spectrometry.

Almost all immunoassays for screening of cocaine in urine specimens recognize benzoylecgonine, the major metabolite of cocaine. Usually immunoassays for determining the presence of benzoylecgonine in urine have good specificity. However, most immunoassays are specific for benzoylecgonine and have poor cross-reactivity with cocaine. If in a forensic scenario, enough cocaine is not metabolized, urinary screen for cocaine metabolite in a deceased may be negative despite the fact that the death is due to cocaine overdose. In a recent report, the authors observed that the urine drug screen using EMIT assay for detecting benzoylecgonine was negative in a person who died from cocaine overdose. The heart–blood concentrations of cocaine and benzoylecgonine were 18,330 and 8,640 ng/mL, respectively. However, the urine concentrations of cocaine and benzoylecgonine were 75 and 55 ng/mL, respectively as determined by GC/MS. These values explain why urine assay for benzoylecgonine was negative because the cut-off of the assay was 300 ng/mL. The authors concluded that individuals utilizing results of drug screening by immunoassay must be aware of limitations of this methodology (13).

Because benzoylecgonine is not conjugated in urine, no hydrolysis step is needed. Benzoylecgonine can be extracted from urine using liquid extraction or solid phase extraction using extraction column. Benzoylecgonine, being a relatively polar molecule requires derivatization prior to analysis. In the literature, silylation, perfluoroalkylation, and alkylation are the most frequently described derivatives although benzoylecgonine can also be derivatized with diazomethane. In this process, benzoylecgonine is converted back into cocaine which is relatively less polar and more volatile (14). Although silylation is a popular method for confirmation of benzoylecgonine using GC/MS, fluconazole interferes in this method because trimethylsilyl derivative of fluconazole is co-eluted with trimethylsilyl benzoylecgonine. Interestingly, fluconazole does not interfere with immunoassay screening for benzoylecgonine (15). This problem can be circumvented by using pentafluoropropionyl derivative of benzoylecgonine because fluconazole derivative elutes after elution of derivatized benzoylecgonine (16). Pentafluoropropionyl derivative is a relatively easy derivative to prepare for GC/MS confirmation of benzoylecgonine. Cocaethylene if present can also be analyzed by this method although cocaethylene does not have any appropriate function group that undergoes derivatization.

2. Materials

2.1. Sample

Serum or plasma (heparin or EDTA) or randomly collected urine is acceptable samples for this procedure. Samples are stable for 3 days when refrigerated and 3 months when frozen at $-20°C$.

2.2. Reagents and Buffers

1. 100 mM Sodium phosphate buffer pH 6.0. Buffer is stable at room temperature for 6 months.

2. 100 mM HCl. Stable at room temperature for 6 months.

3. Elution solvent: dichloromethane:isopropanol (80:20). Store in an amber bottle at room temperature and is stable for 1 year.

4. Working elution solvent: Make fresh with each analytical run by adding 0.200 mL ammonium hydroxide per 10 mL elution solvent, which makes a working elution solvent of dichloromethane:isopropanol:ammonium hydroxide (78:20:2). The pH of the working elution solvent should be 11.0 and if necessary should be adjusted with ammonium hydroxide.

5. Derivatizing agents: Pentafluoropropionic acid anhydride (PFAA) and Pentafluoropropanol (PFPOH) (United Chemical Technologies, Inc., Bristol, PA).

6. Human drug-free serum and urine (UTAK Laboratories, Inc., Valencia, CA).

2.3. Standards and Calibrators

1. 1 mg/mL primary combo standards: cocaine, benzoylecgonine, ecgonine methyl ester, and cocaethylene (Cerilliant Corporation, Round Rock, TX).

2. 100 μg/mL secondary combo standards: Prepared by transferring 1.0 mL of primary standard to a 10 mL volumetric flask and diluting with acetonitrile. Secondary combo standard is stable for 1 year at $-20°C$.

3. 10 μg/mL tertiary combo standards: Prepared by transferring 1.0 mL of secondary combo standard to a 10 mL volumetric flask and diluting with acetonitrile. Tertiary combo standard is stable for 1 year at $-20°C$.

4. Working calibrators are made according to **Table 14.1** using 10 mL volumetric flasks. The calibrators are stable for 6 months when stored at −20°C.

Table 14.1
Preparation of calibrators

Calibrator	Drug-free urine (mL)	Tertiary combo standard (mL)	Secondary combo standard (mL)	Concentration (ng/mL)
0	10.00	0	0	0
1	9.85	0.15	0	150
2	9.90	0	0.10	1,000
3	9.50	0	0.50	5,000

2.4. Internal Standards and Quality Controls

1. 100 µg/mL primary internal standards: Cocaine-d3, benzoyl-lecgonine-d3, ecgonine methyl ester-d3, and cocaethylene-d3 (Cerilliant Corporation, Round Rock, TX).

2. 10 µg/mL working internal standards: Prepared by transferring 1.0 mL of primary internal standard to a 10 mL volumetric flask and diluting with acetonitrile. Secondary combo standard is stable for 1 year at −20°C.

3. Quality control samples: Both in-house and commercial controls are used to valid samples analysis. The in-house controls are made according to **Table 14.2** using 10 mL volumetric flasks (*see* **Note 1**). The controls are stable for 6 months when stored at −20°C. Commercial controls are Bio-Rad Liquichek™ Urine Toxicology Controls C3 and C4 (Bio-Rad Inc., Irvine, CA). The target values for these two quality control samples are established in-house.

Table 14.2
Preparation of in-house controls

Control	Drug-free urine (mL)	Tertiary combo standard (mL)	Secondary combo standard (mL)	Concentration (ng/mL)
Low	9.50	0.50	0	500
High	9.80	0	0.20	2,000

2.5. Supplies

1. 10 mL BondElut® LRC Certify solid phase extraction (SPE) cartridges with a 130 mg sorbent bed (Varian Corporation, Harbor City, CA).

2. Auto sampler vials (12 × 32 mm; clear snap top) with 0.300 mL limited volume inserts (P.J. Cobert Associates, Inc., St. Louis, MO).

3. 16 × 100 mm glass tubes (Fisher Scientific, Fair Lawn, NJ)

4. Concentration vials (Fisher Scientific, Fair Lawn, NJ).

5. GC column: Zebron ZB-1 with dimensions of 15 m × 0.25 mm × 0.25 μm (Phenomenex, Torrance, CA).

2.6. Equipment

1. Visiprep™ SPE vacuum manifold, 24-port model (Supelco, St. Louis, MO).

2. A gas chromatograph/mass spectrometer system (GC/MS; 6890/5975 or 5890/5972 (Agilent Technologies, Wilmington, DE) utilizing electron impact mode.

3. TurboVap® IV Evaporator (Zymark Corporation, Hopkinton, MA, USA)

3. Methods

3.1. Stepwise Procedure

1. Prepare an unextracted standard by adding 100 μL of the 10 μg /mL tertiary combo standard and 100 μL of the working internal standard to a concentration tube. Dry down under nitrogen and set aside until derivatization step 17.

2. Add 1 mL urine, 1 mL serum, or 1 mL plasma to appropriately labeled 16 × 100 mm glass tubes.

3. Add 100 μL of internal standard to each tube.

4. Add 3 mL of 100 mM phosphate buffer and vortex. (*see* **Note 2**)

5. Centrifuge samples at ∼1,600 × g for at least 5 min.

6. Label and insert a SPE cartridge into the Visiprep™ SPE vacuum manifold box for every calibrator, control, and patient sample.

7. Start a vacuum and draw 2 mL of methanol through each column at ∼ 1–2 mL/min.

8. Load and draw 2 mL of 100 mM phosphate buffer through each column being careful not to let the cartridge go dry.

9. Load sample and draw through at ∼ 1–2 mL/min. (*see* Note 3)

10. Add 6 mL of deionized water and draw through at ∼ 1–2 mL/min.

11. Add 3 mL of 100 m*M* HCl and draw through at $\sim$ 1–2 mL/min.

12. Dry the cartridges for 5 min under vacuum.

13. Add 6 mL methanol and draw through under vacuum.

14. Wipe tips off on the Visiprep™ SPE vacuum manifold box and place appropriately labeled concentration tubes in appropriate position for eluate collection.

15. Add 2 mL elution solvent and collect eluate at $\sim$ 1–2 mL/min.

16. Concentrate eluate to dryness under N_2 in a water bath at 40°C (*see* **Note 4**).

17. Add 100 µL pentafluoropropionic acid anhydride (PFAA) and 50 µL pentafluoropropanol (PFPOH) to each tube.

18. Cap and incubate at 65°C for 15 min.

19. Concentrate each tube to dryness under N_2 in a water bath at 40°C.

20. Reconstitute with 100 µL ethyl acetate and inject 1 µL on GC/MS for analysis.

3.2. Instrument Operating Conditions

The instrument's operating conditions are given in **Table 14.3**.

Table 14.3
GC operating conditions

Initial oven temp.	90°C
Initial time	1.0 min
Ramp 1	32°C/min
Temp. 2	170°C
Time	2.0 min
Ramp 2	20°C/min
Final temp.	260°C
Injector temp.	250°C
Detector temp.	280°C
Purge time on	0.5 min
Column pressure	5 psi
MS source temp.	230°C
MS mode	Electron Impact at 70 eV, selected ion monitoring
MS tune	Autotune

3.3. Data Analysis

1. Representative GC-MS chromatogram of pentafluoropropionyl derivatives of ecgonine methyl ester and benzoylecgonine, and underivatized cocaine and cocaethylene is shown in **Fig. 14.1**. GC-MS selected ion chromatograms are shown in **Fig. 14.2**. Electron impact ionization mass spectra of these compounds are shown in the **Fig. 14.3** (*see* **Note 5**). Ions used for identification and quantification are listed in **Table 14.4**.

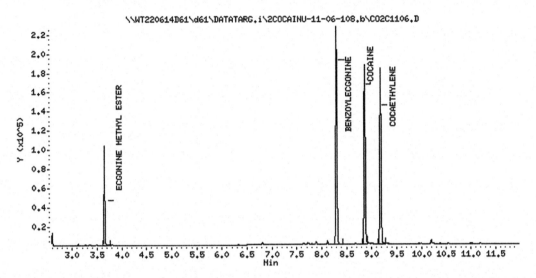

Fig. 14.1. GC-MS chromatogram of pentafluoropropionic acid derivatives of ecgonine methyl ester and benzoylecgonine, and underivatized cocaine and cocaethylene (1,000 ng/mL each). Deuterated internal standards co-elute with each compound.

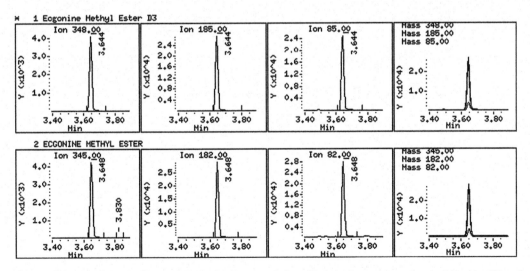

Fig. 14.2. Selected ion chromatograms of pentafluoropropionic acid derivatives of ecgonine methyl ester (**A**) and benzoylecgonine (**B**), and underivatized cocaine (**C**) and cocaethylene (**D**). Also shown are selected ion chromatograms of deuterated compounds.

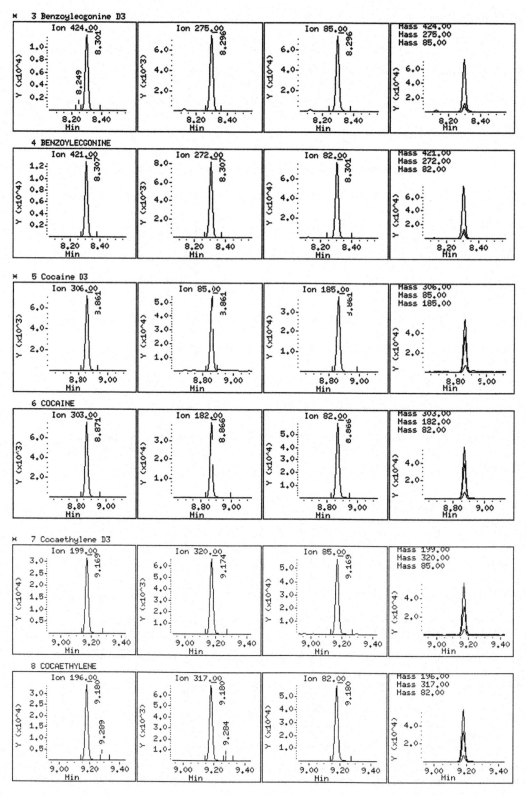

Fig. 14.2. (continued)

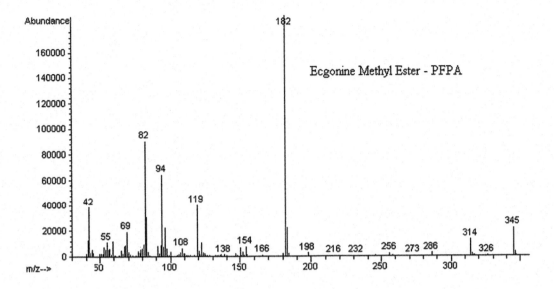

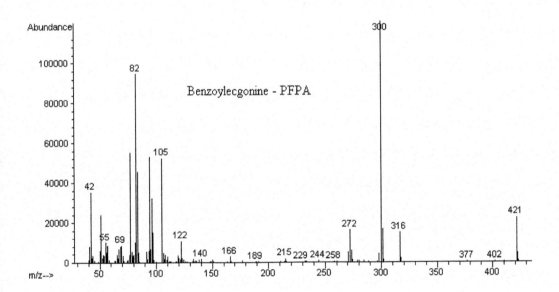

Fig. 14.3. Electron impact ionization mass spectra of pentafluoropropionic acid derivatives of ecgonine methyl ester (**A**) and benzoylecgonine (**B**), and underivatized cocaine (**C**) and cocaethylene (**D**).

2. Analyze data using Target Software (Thru-Put Systems, Orlando, FL) or similar software. The quantifying ions (**Table 14.4**) are used to construct standard curves of the peak area ratios (calibrator/internal standard pair) vs. concentration. These curves are then used to determine the concentrations of the controls and unknown samples (*see* **Note 6**).

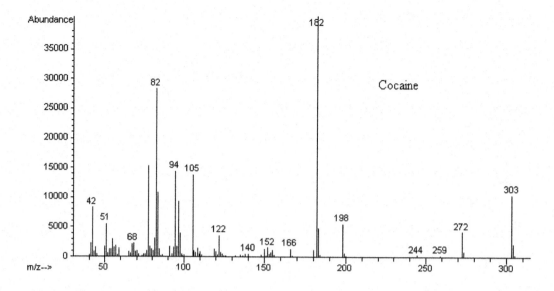

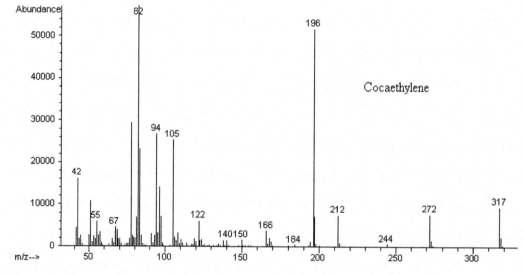

Fig. 14.3. (continued)

3. The linearity/limit of quantitation of the method is 150–5,000 ng/mL. Samples in which the drug concentrations exceed the upper limit of quantitation should be diluted with negative serum or plasma and retested.

4. Standard curves should have a correlation coefficient (r^2) >0.99.

5. Typical intra- and inter-assay imprecision is <10%.

6. Quality control: The run is considered acceptable if calculated concentrations of drugs in the controls are within +/−20% of target values. Quantifying ion in the sample is considered acceptable if the ratios of qualifier ions to quantifying ion are within +/− 20% of the ion ratios for the calibrators.

Table 14.4
Quantitation and qualifying ions for cocaine and its metabolites

	Quantitation ion	Qualifier ions
Cocaine	303	182, 82
Cocaine d3	306	85, 185
Benzoylecgonine	421	272, 82
Benzoylecgonine d3	424	275, 85
Ecgonine methyl ester	345	182, 82
Ecgonine methyl ester d3	348	185, 85
Cocaethylene	196	317, 82
Cocaethylene d3	199	320, 85

4. Notes

1. When possible, calibrators and controls should be made from different batches of drugs at another time by different analysts.

2. pH of samples should be 6.0 ± 0.5 (Adjust by adding more phosphate buffer if necessary.).

3. Be careful not to transfer any pellets at the bottom of the tube.

4. Do not over-dry the extract. This will result in poor recovery and failed run.

5. Electron impact ionization spectra are needed in the initial stages of method set up to establish retention times and later on, if there is a need for change in quantifying or qualifying ions. They are not needed for routine quantitation.

6. The use of urine calibrators and controls for serum testing has been validated for this procedure.

References

1. Isenschmid, D.S. (2002) Cocaine: Effects on human performance and behavior. *Forensic Science Review*, **14**, 61–100.

2. Blaho, K., Logan, B., Winbery, S., Park, L. and Schwilke, E. (2000) Blood cocaine and metabolite concentrations, clinical findings, and outcome of patients presenting to an ED. *Am J Emerg Med*, **18**, 593–8.

3. Casale, J.F. (2007) Cocaethylene as a component in illicit cocaine. *J Anal Toxicol*, **31**, 170–1.

4. Harris, D.S., Everhart, E.T., Mendelson, J. and Jones, R.T. (2003) The pharmacology of cocaethylene in humans following cocaine and ethanol administration. *Drug Alcohol Depend*, **72**, 169–82.

5. Porter, W.H. (2006) Clinical Toxicology. In Burtis, C.A., Ashwood, E.R. and Bruns, D.E. (eds.), *Tietz Textbook of Clinical Chemistry and Molecular Diagnostics.* Elsevier Saunders, St. Louis, pp. 1287–1390.

6. Bermejo, A.M., Lopez, P., Alvarez, I., Tabernero, M.J. and Fernandez, P. (2006) Solid-phase microextraction for the determination of cocaine and cocaethylene in human hair by gas chromatography-mass spectrometry. *Forensic Sci Int*, **156**, 2–8.

7. Yonamine, M. and Saviano, A.M. (2006) Determination of cocaine and cocaethylene in urine by solid-phase microextraction and gas chromatography-mass spectrometry. *Biomed Chromatogr*, **20**, 1071–5.

8. Cardona, P.S., Chaturvedi, A.K., Soper, J.W. and Canfield, D.V. (2006) Simultaneous analyses of cocaine, cocaethylene, and their possible metabolic and pyrolytic products. *Forensic Sci Int*, **157**, 46–56.

9. Aderjan, R.E., Schmitt, G., Wu, M. and Meyer, C. (1993) Determination of cocaine and benzoylecgonine by derivatization with iodomethane-D3 or PFPA/HFIP in human blood and urine using GC/MS (EI or PCI mode). *J Anal Toxicol*, **17**, 51–5.

10. Pichini, S., Pacifici, R., Altieri, I., Pellegrini, M. and Zuccaro, P. (1999) Determination of opiates and cocaine in hair as trimethylsilyl derivatives using gas chromatography-tandem mass spectrometry. *J Anal Toxicol*, **23**, 343–8.

11. Thompson, W.C. and Dasgupta, A. (1995) Confirmation and quantitation of cocaine, benzoylecgonine, ecgonine methyl ester, and cocaethylene by gas chromatography/mass spectrometry. Use of microwave irradiation for rapid preparation of trimethylsilyl and T-butyldimethylsilyl derivatives. *Am J Clin Pathol*, **104**, 187–92.

12. Cone, E.J., Hillsgrove, M. and Darwin, W.D. (1994) Simultaneous measurement of cocaine, cocaethylene, their metabolites, and "crack" pyrolysis products by gas chromatography-mass spectrometry. *Clin Chem*, **40**, 1299–305.

13. Baker, J.E. and Jenkins, A.J. (2008) Screening for cocaine metabolite fails to detect an intoxication. *Am J Forensic Med Pathol*, **29**, 141–4.

14. Yonamine, M. and Silva, O.A. (2002) Confirmation of cocaine exposure by gas chromatography-mass spectrometry of urine extracts after methylation of benzoylecgonine. *J Chromatogr B Analyt Technol Biomed Life Sci*, **773**, 83–7.

15. Wu, A.H., Ostheimer, D., Cremese, M., Forte, E. and Hill, D. (1994) Characterization of drug interferences caused by coelution of substances in gas chromatography/mass spectrometry confirmation of targeted drugs in full-scan and selected-ion monitoring modes. *Clin Chem*, **40**, 216–20.

16. De Giovanni, N. and Strano Rossi, S. (1994) Simultaneous detection of cocaine and heroin metabolites in urine by solid-phase extraction and gas chromatography-mass spectrometry. *J Chromatogr B Biomed Appl*, **658**, 69–73.

Chapter 15

Identification and Quantitation of Cocaine, Benzoylecgonine, and Cocaethylene in Blood, Serum, and Plasma Using Ultra-Performance Liquid Chromatography Coupled to Tandem Mass Spectrometry (UPLC-MS/MS)

Scott Kriger, Josh Gunn, and Andrea R. Terrell

Abstract

Cocaine is a widely abused stimulant. Numerous methods exist for the identification of the drug, or more commonly, one of its metabolites in urine. Urine testing is useful for most cases, but it is necessary to use other matrices in forensic situations and when subjects are anuric. We describe a novel method for the analysis of cocaine, benzoylecgonine, and cocaethylene in blood, serum, and plasma utilizing ultra-performance liquid chromatography coupled to tandem mass spectrometry (UPLC-MS/MS). Sample preparation has been minimized to a simple deproteinization step in which each specimen is mixed with an acetonitrile–internal standard mixture. The method has excellent precision across the linear range of 25–2,000 ng/mL for each analyte. With a run-time of 4 min, this method provides a significant improvement over traditional GC/MS methods.

Key words: Cocaine, benzoylecgonine, cocaethylene, tandem mass spectrometry, ultra-performance liquid chromatography, protein precipitation.

1. Introduction

Cocaine is a naturally occurring central nervous system stimulant that interferes with the actions of dopamine, norepinephrine, and serotonin in functioning nerves. Cocaine is classified as a sympathomimetic agent due to its ability to activate the sympathetic nervous system both centrally and peripherally (1). Once administered, cocaine is rapidly metabolized to several products. One of the major metabolites, benzoylecgonine, is widely believed to be inactive. In contrast, cocaethylene, an active, toxic metabolite, is

U. Garg, C.A. Hammett-Stabler (eds.), *Clinical Applications of Mass Spectrometry*, Methods in Molecular Biology 603, DOI 10.1007/978-1-60761-459-3_15, © Humana Press, a part of Springer Science+Business Media, LLC 2010

formed with the co-administration of cocaine and ethanol. This transesterification product is considered to be equally or more toxic than cocaine itself.

Cocaine is available on the street in either the base form (referred to as crack or free base) or as the hydrochloride salt. Although both forms of the drug are available in high purity for a similar street value, the free-base 'crack' cocaine is predominately used for smoking while cocaine hydrochloride is mainly used for intravenous injection and nasal insufflation.

The prevalence of cocaine use has increased substantially over the past two decades. In 1989, ~50 million Americans were estimated to have used cocaine at least once and ~8 million people were regular users of the drug (2). More recent findings from the 2006 National survey on drug use and health (NSDUH) estimate 2,700 people per day initiate cocaine use based on the 977,000 persons aged 12 years or older who admitted to using cocaine for the first time within the past twelve months (3). The NSDUH estimates that in addition to the 2.4 million frequent (i.e., at least twice weekly) users of cocaine, there are approximately 4.6 million occasional users (i.e., once a month or less) of the drug, excluding individuals already in prison (3).

Urine is the most common matrix for the detection of cocaine, but there are situations in which other matrices are useful. For example, blood is useful in medical examiner investigations and when subjects are anuric. We describe a novel method for the analysis of cocaine, benzoylecgonine, and cocaethylene in blood, serum, and plasma utilizing ultra-performance liquid chromatography coupled to tandem mass spectrometry (UPLC-MS/MS).

2. Materials

2.1. Specimens

1. Whole Blood or plasma: Collect using a NaF/Potassium Oxalate or NaF/EDTA (gray top) tube.

2. Serum: Collect without preservatives or additives (plain red or royal blue). (*see* **Note 1**)

3. Samples are stable for 6 weeks when refrigerated at 4–8°C and up to 1 year when frozen at −20°C or colder.

2.2. Reagents and Solutions

1. Mobile phase solvent A: 0.1% (v/v) Formic acid in deionized water

2. Mobile phase solvent B: 0.1% (v/v) Formic acid in acetonitrile

3. Adult bovine serum (Sigma Chemical Co) or drug-free serum

2.3. Standards and Calibrators

1. Stock standards: benzoylecgonine, cocaine, and cocaethylene (1 mg/mL each), (Cerilliant, Round Rock, TX).

2. Working stock standards, 100 ug/mL: Prepare by diluting 1 mg/mL stock standard with acetonitrile. Stable for 1 year when stored at −20°C or colder.

3. Working standard 1: Prepare the 2,000 ng/mL working standard by adding 1.96 mL of adult bovine serum to a 12 × 75 mm small culture tube. Add 40 μL of the 100 μg/mL working stock standard. Cap, vortex well, and centrifuge for 5–10 min to remove any debris. This standard is prepared fresh on each day of use.

4. Working standard 2: Prepare the 200 ng/mL working standard 2 by adding 1.8 mL of adult bovine serum to a 12 × 75 mm small culture tube. Add 200 μL of the 2,000 μg/mL working standard 1. Cap, vortex well, and centrifuge for 5–10 min to remove any debris. This standard is prepared fresh on each day of use.

5. Calibrators: Prepare the calibrators according to **Table 15.1**. Calibrators are prepared fresh on each day of use.

Table 15.1
Preparation of calibrators

Final concentration (ng/mL)	Volume working standards (μL)	Volume of negative serum (μL)
2000	200 (Std1)	0
1000	100 (Std1)	100
500	50 (Std1)	150
200	200 (Std2)	0
100	100 (Std2)	100
50	50 (Std2)	150
25	25 (Std2)	175
NEGATIVE	0	200

2.4. Controls and Internal Standards

1. High quality control, 800 ng/mL: Prepare by diluting 200 μL of a 100 μg/mL working stock standard to 25 mL with acetonitrile. Single use aliquots are prepared and are stable for 1 year when stored at −20°C or colder.

2. Low quality control, 200 ng/mL: Prepare by diluting 50 µL of the 100 µg/mL working stock standard to 25 mL with acetonitrile. Single use aliquots are prepared and are stable for 1 year when stored at −20°C or colder.

3. Negative control: Adult bovine serum.

4. Stock internal standards, 1 mg/mL each: d3 benzoylecgonine and d3 cocaine (Cerilliant, Round Rock, TX). Cocaethylene is quantified using the Cocaine d3 internal standard.

5. Working internal standard, 100 ng/mL: Prepare by diluting 1 mg/mL stock internal standard with acetonitrile. Stable for 1 year when stored at −20°C or colder (*see* **Note 2**).

2.5. Supplies and Equipment

1. 12 × 75 mm glass test tubes.

2. Plastic GC autosampler vials.

3. 11 mm Blue Cut Septa Snap Cap.

4. Waters ACQUITY™ ultra performance liquid chromatograph (UPLC) (Waters Corp., Milford, MA, USA).

5. Waters Quattro Premier triple quadruple mass spectrometer (Waters Corp., Milford, MA, USA).

6. Column: Aquity UPLC HSS T3 2.1 × 100 mm (1.7 µm particle size) (Waters Corp., Milford, MA, USA).

3. Methods

3.1. Stepwise Standard and Calibration Curve Preparation Procedure

1. Add 200 µL of sample, control or calibrator in each appropriately labeled 12 × 75 mm glass tube.

2. Add 2.0 mL of *cold* working internal standard solution to each tube (*see* **Note 2**).

3. Cap each sample with non-vented caps.

4. Vortex each sample for 5–10 sec. Inspect to assure sample is thoroughly mixed.

5. Place on the block vortex for 5 min. Thorough mixing is critical to this analytical method.

6. Centrifuge for 5 min at 3,000 × g.

7. Transfer 200 µL of supernatant to a new culture tube.

8. Add 600 µL of deionized water to each sample and vortex well.

9. Transfer 200 µL to an appropriately labeled injection vial for analysis.

**3.2. Instrument
Operating Conditions**

1. **Ultra-Performance Liquid Chromatography (UPLC):** Chromatographic separations were performed using an ACQUITY™ ultra-performance liquid chromatograph (UPLC) (Waters Corp., Milford, MA, USA) that included an ACQUITY UPLC® HSS T3 (2.1 × 100 mm) column packed with 1.7 μm particles maintained at 35°C. Analytes were eluted from the UPLC column using a binary elution gradient: Initial mobile phase composition was 75% mobile phase A:25% mobile phase B held constant for 0. 50 min after which the composition of mobile phase B was increased linearly to 50% over 2.5 min. Conditions were returned to their initial composition of 75:25 over the next 0.01 min and held for 1.00 min to equilibrate the column before the next injection in the sequence. The total run-time was 4.0 min. Flow rates were maintained at 0.5 mL/min for the entire chromatographic separation and increased to 0.6 mL/min used for post-run column equilibration. *See* **Fig. 15.1** for illustration. All flow was directed into the ESI source of the mass spectrometer. Samples were maintained at 7.5°C in the sample organizer and sample injection volumes were 5 μL for all analyses.

2. **Tandem Mass Spectrometry:** Mass spectrometric detection was performed using a Waters Quattro Premier triple quadrupole mass spectrometer (Waters Corp., Milford, MA, USA) equipped with an electrospray ionization (ESI) source operating in positive ion mode. MS/MS conditions were as

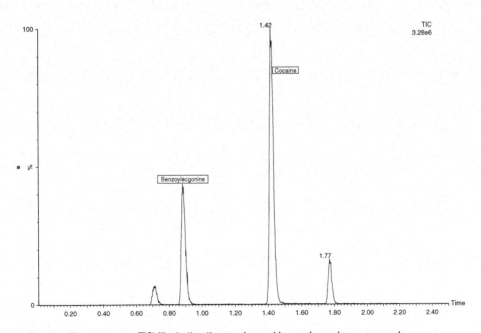

Fig. 15.1. Total Ion Chromatogram (TIC) illustrating the cocaine and benzoylecgonine components.

follows: capillary voltage 0.80 kV, cone voltage 27 V, extractor voltage 3.0 V, RF lens voltage 0.3 V. The source temperature was 120°C while the desolvation temperature was set at 350°C. Cone gas was set at a flow rate of 50 L/h while the desolvation gas flow was 900 L/h. The collision gas flow was set to 0.10 mL/min. Nitrogen (99.995% purity) was used as the desolvation gas, and ultra-pure argon (99.999% purity) was used as the collision gas.

The mass transition from the protonated molecular ion $[M+H]^+$ to the most abundant product ion (**Table 15.2**) was designated the quantifying ion transition, while the second most abundant mass transition was designated as the qualifying ion transition for each analyte (**Table 15.2**). (*see* **Notes 3** and **4**). The most abundant product ion for each deuterated internal standard was also monitored and used to calculate the response ratio between internal standards and analytes for all experiments. Following auto tuning of each analyte and internal standard, the optimized parameters were used to construct the MS/MS method, which was then used to acquire data in the MRM mode. **Table 15.2** reports the mass transitions, dwell times, cone voltages, and collision energies for each of the analytes and their deuterated internal standards.

Table 15.2
MS/MS parameters used for each analyte and deuterated internal standard

Compound	Mass transition	Purpose	Cone (V)	Collision (V)	Dwell (secs)
Cocaine	304.14→182.10	Quantifying	40	20	0.010
Cocaine	304.14→150.16	Qualifying	40	20	0.010
Cocaine d3	307.15→184.96	Quantifying	30	20	0.010
Benzoylecgonine	290.08→168.24	Quantifying	40	20	0.010
Benzoylecgonine	290.08→104.78	Qualifying	40	40	0.010
Benzoylecgonine d3	293.11→170.98	Quantifying	40	20	0.010
Cocaethylene	318.18→196.08	Quantifying	30	20	0.010
Cocaethylene	318.18→149.90	Qualifying	30	20	0.010

3.3. Data Analysis

The data was processed using the MassLynx v4.1(Waters, Inc.) software. The results for the batch are considered accurate, reliable, and acceptable if the quantitation of the quality control samples agrees within ± 20% of the expected values. The calibration curve is considered acceptable if $r^2 \geq 0.98$ and each individual data point agrees to within ± 20% of the expected value. Analyte quantitation was accomplished using the standard calibration curve and the ratio of the peak area of the analyte-quantifying ion to the peak area of the internal standard quantifying ion.

4. Notes

1. Many drugs have been shown to bind to gel-barrier phlebotomy tubes. Additionally, some of these types of tubes have been found to contain chemicals, which interfere with chromatography-based methods. The use of these tubes is not recommended; instead use either plain redtop, gray top (containing NaF/EDTA), or royal blue top (trace metal free) tubes for blood collection.

2. It is important that the working internal standard solution be cold to adequately precipitate blood and serum proteins. Remove from the freezer just prior to use.

3. Appropriate quantifier and qualifier mass transitions were identified for each analyte by directly infusing a 10 μg/mL solution of each compound into the mass spectrometer ionization source at a flow rate of 20 μL/min. The flow paths of the concentrated analyte solutions was modified with a T-mixer to allow mixing of the solution with mobile phase and simultaneous infusion of the sample and initial mobile phase into the mass spectrometer.

4. Simultaneous infusion of the sample with the initial mobile phase was performed to ensure that any subsequently optimized tune page parameters were compatible with the initial mobile phase. By optimizing the tune page parameters while infusing a mixture of analyte and initial mobile phase, the analytes are allowed to reach the ESI source under similar conditions and in a similar environment as those that would be encountered during the analysis of a positive sample. This ensures the most accurate optimization of ionization parameters specific to both the sample and the mobile phase. Concentrated (10 μg/mL) solutions of each analyte and deuterated internal standard were individually infused concomitantly with a 80% mobile phase A:20% mobile phase B composition at a flow rate of 0.5 mL/min. During infusion of each analyte, the collision gas was turned off to allow the protonated molecular ion of each compound to reach the detector and produce a recordable signal. Following identification of the molecular ion signal, an auto tune was completed for each analyte, which involved adjusting the capillary

voltages, cone voltages, and collision energies to maximize the signal for both the precursor ions and the product ions generated in the collision cell. Auto tuning of the protonated molecular ion of each compound yielded information necessary to collect data in the MRM mode.

References

1. Ross, D.L. and T.C. Chan, 2006, *Sudden Deaths in Custody*. Totowa, NJ: Humana Press.
2. Dixon, S.D. 1989, Effects of transplacental exposure to cocaine and methamphetamine on the neonate. *West J Med*, **150**(4): 436–442.
3. *Results from the 2006 National Survey on Drug Use and Health: National Findings.*, Substance Abuse and Mental Health Services Administration.

Chapter 16

Detection and Quantification of Cocaine and Benzoylecgonine in Meconium Using Solid Phase Extraction and UPLC/MS/MS

Josh Gunn, Scott Kriger, and Andrea R. Terrell

Abstract

The simultaneous determination and quantification of cocaine and its major metabolite, benzoylecgonine, in meconium using UPLC-MS/MS is described. Ultra-performance liquid chromatography (UPLC) is an emerging analytical technique which draws upon the principles of chromatography to run separations at higher flow rates for increased speed, while simultaneously achieving superior resolution and sensitivity. Extraction of cocaine and benzoylecgonine from the homogenized meconium matrix was achieved with a preliminary protein precipitation or protein 'crash' employing cold acetonitrile, followed by a mixed mode solid phase extraction (SPE). Following elution from the SPE cartridge, eluents were dried down under nitrogen, reconstituted in 200 µL of DI water:acetonitrile (ACN) (75:25), and injected onto the UPLC/MS/MS for analysis. The increased speed and separation efficiency afforded by UPLC, allowed for the separation and subsequent quantification of both analytes in less than 2 min. Analytes were quantified using multiple reaction monitoring (MRM) and six-point calibration curves constructed in negative blood. Limits of detection for both analytes were 3 ng/g and the lower limit of quantitation (LLOQ) was 30 ng/g.

Key words: Cocaine, benzoylecgonine, mass spectrometry, Solid Phase Extraction (SPE).

1. Introduction

Cocaine belongs to the tropane alkaloid family and is obtained from the leaves of the plant *Erythroxylon coca*. Leaves of the *E. coca* and other related species have been used by the Peruvian Indians for centuries to increase endurance and improve well-being. By the mid-twentieth century the recreational use of cocaine had become a significant concern across socioeconomic lines. Cocaine is available on the street in either the base form (crack, free base) or as the hydrochloride salt. Although both forms of the drug are available

U. Garg, C.A. Hammett-Stabler (eds.), *Clinical Applications of Mass Spectrometry*, Methods in Molecular Biology 603,
DOI 10.1007/978-1-60761-459-3_16, © Humana Press, a part of Springer Science+Business Media, LLC 2010

in high purity for a similar street value, the free-base 'crack' cocaine is predominately used for smoking while cocaine hydrochloride is mainly employed for intravenous injection and nasal insufflations.

Cocaine is a naturally occurring central nervous system (CNS) stimulant that interferes with the actions of dopamine, norepinephrine, and serotonin in functioning nerves. Cocaine is classified as a sympathomimetic agent due to its ability to activate the sympathetic nervous system both centrally and peripherally (1). Central stimulation of the sympathetic nervous system results from cocaine's ability to selectively bind dopamine reuptake transporters (DATs) in the brain (2). Clearance of dopamine at the synapse and subsequent termination of dopaminergic neurotransmission is achieved through reuptake into the presynaptic neuron which is mediated by DATs[3]. By binding to DATs, cocaine impairs the reuptake of dopamine into the presynaptic neuron resulting in elevated dopamine levels in the central nervous system. Direct stimulation of the central nervous system by cocaine and other sympathomimetic agents also results in increased norepinephrine release from peripheral synapses. Cocaine not only facilitates increased release of norepinephrine peripherally but also acts to inhibit its reuptake at the synapses, causing it to remain in the synaptic cleft for a prolonged period of time. As a result of cocaine's ability to inhibit the reuptake of dopamine centrally and norepinephrine peripherally, the natural effect of these neurotransmitters is amplified. Excessive levels of CNS dopamine and peripheral norepinephrine account for the feelings of euphoria and increased alertness associated with cocaine use (1).

Although it is difficult to accurately determine the number of women who use cocaine during pregnancy, there is evidence to suggest that cocaine represents a significant proportion of the illicit substances, which 4% of pregnant women admitted using during pregnancy in the 2006 National Survey of Drug Use and Health (NSDUH) (3–7). Rates of infants exposed to cocaine prenatally have been estimated to be between 2.6 and 11% of all live births (7). Prenatal cocaine use has been associated with premature labor, placental abruption, low birth parameters (weight, head circumference, length), microcephaly, congenital malformations, increased risk of sudden infant death syndrome (SIDS), spontaneous abortion, acute hypoxic-ischemic encephalopathy, cerebral hemorrhage or infarction, abnormal neonatal behavior, and limb deformities (8–11).

2. Materials

2.1. Specimen

1. Collect meconium in a plastic urine collection cup or conical vial by scraping the material from the diaper. The preferred sample size is 10–20 g. The specimen should be stored at 2–8°C.

2. Samples are stable for 1 week at ambient temperature, 1 month when refrigerated at 2–8°C, and up to 3 years when frozen at –20°C or colder.

2.2. Reagents and Solutions

1. Negative blood: Gently mix 500 mL of expired, packed-red cells with 550 mL of 0.9% saline. Add sodium fluoride to a concentration of 5 mg/mL. Prior to use, analyze to validate that the material is free of the drugs of interest. Stable for 1 year when stored at 2–8°C.

2. Dichloromethane (HPLC grade).

3. 2-Propanol (HPLC grade).

4. 50:50 Methanol:deionized water.

5. 0.1 M Phosphate buffer pH 6.0.

6. 1.0 M Hydrochloric acid.

7. 75:25 Deionized water:acetonitrile

8. Elution solvent (78:20:2 Dichloromethane:2-propanol:ammonium hydroxide)

9. Mobile phase A:0.1% (v/v) formic acid in DI water

10. Mobile phase B: 0.1% (v/v) formic acid in acetonitrile

2.3. Standards and Calibrators

1. Stock standards, 1 mg/mL: benzoylecgonine and cocaine (Cerilliant, Round Rock, TX).

2. Stock standard dilution, 10 µg/mL: Prepare by diluting 1 mg/mL stock standard with methanol (*see* **Table 16.1**). Stable for 1 year when stored at −20°C or colder.

3. Working standard, 500 ng/mL: In a 12 × 75 mm small culture tube add 4.75 mL of deionized water and add 250 µL of cocaine/benzoylecgonine stock standard dilution (10 µg/mL). Cap and vortex for 10–15 sec. The working standard is prepared fresh on each day of use.

4. Calibrators: Prepare the calibrators according to **Table 16.2**. Calibrators are prepared fresh on each day of use.

Table 16.1
Volumes and analyte concentrations in cocaine/benzoylecgonine standard dilution solution.

Analyte	Target Concentration (µg/mL)	Concentration of STD (mg/mL)	Volume (µL)
Cocaine	10	1	30
Benzoylecgonine	10	1	30

Table 16.2
Preparation and concentrations of working standard curve calibrators

Standard concentration (ng/mL)	Volume of working standard (µL)	Volume of neg blood (µL)
250	500	500
100	200	800
50	100	900
25	50	950
10	20	980
5	10	990
NEG	0	1,000

2.4. Controls and Internal Standards

1. Positive quality control, 10 ng/mL: Prepare by diluting 25 µL of the cocaine/benzoylecgonine standard dilution solution (10 µg/mL) in 25 mL negative blood. Single use aliquots are prepared and are stable for 1 year when stored at −20°C or colder.

2. Negative quality control: Negative blood.

3. Stock internal standards, 1 mg/mL each: d3 benzoylecgonine and d3 cocaine (Cerilliant, Round Rock, TX).

4. Working internal standard, 500 ng/mL: Prepare by diluting 1 mg/mL stock internal standard with acetonitrile. Stable for 1 year when stored at −20°C or colder (*see* **Table 16.3**).

2.5. Supplies and Equipment

1. 13 × 100 mm large glass culture tube.
2. 12 × 75 mm small glass test tubes.
3. 16 × 125 mm large glass screw-top tube.

Table 16.3
Preparation of working internal standard solution, final volume of 100 mL

Internal standard	Target concentration (ng/mL)	Concentration of STD (mg/mL)	Volume (µL)
Cocaine-d3	500	1	50
Benzoylecgonine-d3	500	1	50

4. CLEAN-SCREEN solid phase extraction cartridges (United Chemical Technologies, Bristol, PA).

5. Plastic GC vials.

6. 11 mm blue cut septa snap cap (Microliter Analytical Supplies, Suwanee, GA).

7. Column: Acquity UPLC HSS T3 2.1 × 100 mm (1.8 µm particle size) (Waters Corporation, Milford, MA).

3. Methods

3.1. Stepwise Sample and Calibration Curve Preparation Procedure

1. Add 250 µL of sample, control or calibrator to appropriately labeled 13 × 100 mm glass test tubes.

2. Add 100 µL of internal standard working solution to each tube.

3. Vortex for 10–20 sec.

4. Add 2 mL of cold acetonitrile (*see* **Note 1**).

5. Place on block vortex for 5 min.

6. Centrifuge sample for 10 min at 3,000 × *g*.

7. Transfer organic layer to labeled large screw-top test tubes.

8. Add 3 mL of 0.1 M phosphate buffer, pH 6.0.

9. Add 1 mL of concentrated ammonium hydroxide.

10. Vortex for 10–15 sec, then centrifuge for 5 min at 3,000 × *g*.

3.2. Solid Phase Extraction

1. Add 3 mL of methanol into each purple UCT CLEAN-SCREEN ZSDAU020 column using a re-pipette dispenser.

2. Add 3 mL of deionized water into each column using a re-pipette dispenser.

3. Use the re-pipette dispenser to add 1 mL of 0.1 M phosphate buffer, pH 6.0.

4. Allow solvents to drip through slowly–change waste tubes/containers, accordingly.

5. Pour samples onto the columns – allow to drip through for 15 min before applying any vacuum pressure.

6. Turn valves off as liquid drips through.

7. Add 1 mL of deionized water into each column using a re-pipette dispenser.

8. Add 1 mL of 1.0 M HCL into each column.

9. Add 3 mL of methanol into each column using a re-pipette dispenser.

10. Allow to drip through, then apply vacuum pressure for 5–10 min.

11. Turn pressure off and replace waste containers with 12 × 75 mm small glass test tubes.

12. Add 3 mL of 78:20:2 dichloromethane:2-propanol:ammonium hydroxide (*see* **Note 2**).

13. Dry down samples under nitrogen.

14. Reconstitute with 200 μL of 75:25 acetonitrile:deionized water

15. Transfer into appropriately labeled plastic autosampler vials for analysis.

3.3. Instrument Operating Conditions

1. **Ultra-performance Liquid Chromatography (UPLC)**: Separations were performed on a Waters ACQUITY™ ultra-performance liquid chromatograph (UPLC) (Waters Corp., Milford, MA, USA). Separations were achieved on an ACQUITY UPLC® HSS T3 column (2.1 × 50 mm) packed with 1.8 μm bridged ethyl hybrid (BEH) particles and maintained at 35°C. The mobile phase consisted of deionized water containing 0.1% formic acid (solvent A), and acetonitrile containing 0.1% formic acid (solvent B). Analytes were eluted from the UPLC column using the following stepwise binary elution gradient: Initial mobile phase composition was 75:25 (H_2O:ACN). Initial conditions were held constant for 0.5 min after which the composition of solvent B was linearly increased to 50% over 1.5 min, finally conditions were returned to their initial composition of 75:25 (H_2O:ACN) over the next 0.01 min and held for 0.49 min to equilibrate the column before the next injection in the sequence. The total run-time was 2.50 min (**Fig. 16.1**). Samples were maintained at 7.5°C in the sample organizer and sample injection volumes were 5 μL for all analyses. Flow rates were maintained at 0.5 mL/min for the first 0.50 min after which they were increased to 0.6 mL/min for the remainder of the chromatographic separation. All flow was directed into the ESI source of the mass spectrometer.

2. **Tandem Mass Spectrometry**: Mass spectrometric detection was performed using a Waters TQD triple quadrupole mass spectrometer (Waters Corp., Milford, MA, USA) equipped with an electrospray ionization (ESI) source operating in positive ion mode. MS/MS conditions were as follows: capillary voltage 0.80 kV, cone voltage 20 V, extractor voltage 3.0 V, RF lens voltage 0 V. The source temperature was 120°C while the desolvation temperature was set at 350°C. Cone gas was set at a flow of 100 L/Hr while the desolvation

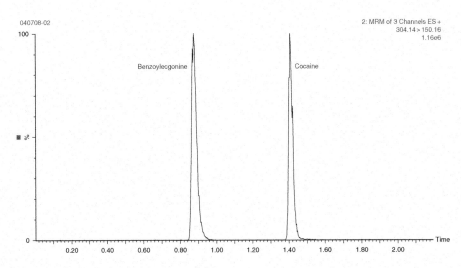

Fig. 16.1. TIC from the UPLC separation of cocaine and benzoylecgonine in meconium.

gas flow was 900 L/h. The collision gas flow was set to 0.10 mL/min. Nitrogen (99.995% purity) was used as the desolvation gas, and ultra-pure argon (99.999% purity) was used as the collision gas. Appropriate quantifier and qualifier mass transitions were identified for each analyte by directly infusing a 10 μg/mL solution of each compound into the mass spectrometer ionization source at a flow rate of 20 μL/min. The flow paths of the concentrated analyte solutions was modified with a T-mixer to allow mixing of the solution with mobile phase and simultaneous infusion of the sample and initial mobile phase into the mass spectrometer. Simultaneous infusion of the sample with the initial mobile phase was performed to ensure that any subsequently optimized tune page parameters were compatible with the initial mobile phase. By optimizing the tune page parameters while infusing a mixture of analyte and initial mobile phase, the analytes are allowed to reach the ESI source under similar conditions and in a similar environment as those that would be encountered during the analysis of a positive sample. This ensures the most accurate optimization of ionization parameters, specific to both the sample and the mobile phase. Concentrated (10 μg/mL) solutions of cocaine, benzoylecgonine, cocaine-d_3, and benzoylecgonine-d_3 were individually infused concomitantly with a 75:25 (H_2O:ACN) composition of mobile phase at a flow rate of 0.5 mL/min. During infusion of each analyte, the collision gas was turned off to allow the protonated molecular ion of each compound to reach the detector and produce a recordable signal. Following identification of the molecular ion signal, an auto tune was completed for each analyte which involved adjusting the capillary voltages, cone

Table 16.4
MS/MS parameters used for each analyte and deuterated internal standard

Compound	Mass transition	Purpose	Cone (V)	Collision (V)	Dwell (sec)
Cocaine	304.14 > 182.10	Quantifying ion	40.0	20.0	0.010
Cocaine	304.14 > 150.16	Qualifying ion	40.0	20.0	0.010
Cocaine-d3	307.15 > 184.96	Quantifying ion	30.0	20.0	0.010
Benzoylecgonine	290.08 > 168.24	Quantifying ion	40.0	20.0	0.010
Benzoylecgonine	290.08 > 104.78	Qualifying ion	40.0	40.0	0.010
Benzoylecgonine-d3	293.11 > 170.98	Quantifying ion	40.0	20.0	0.010

voltages, and collision energies to maximize the signal for both the precursor ions and the product ions generated in the collision cell. Auto tuning of the protonated molecular ion of each compound yielded information necessary to collect data in the MRM mode. The mass transition from the protonated molecular ion $[M+H]^+$ to the most abundant product ion (**Table 16.4**) was designated the quantifying ion transition, while the second most abundant mass transition was designated as the qualifying ion transition for each analyte (**Table 16.4**). The most abundant product ion for each deuterated internal standard was also monitored and used to calculate the response ratio between internal standards and analytes for all experiments. Following auto tuning of each analyte and internal standard, the optimized parameters were used to construct the MS/MS method, which was then used to acquire data in the MRM mode. **Table 16.4** reports the mass transitions, dwell times, cone voltages, and collision energies for each of the analytes and their deuterated internal standards (**Table 16.4**).

3. **Data Analysis**: Analytical data was analyzed using Masslynx version 4.1 software. Criteria for a positive result included accurate chromatographic retention time, presence of both the qualifying product ion and the quantifying product ion, and product ion ratios within acceptable limits. Chromatographic retention times are initially established for each analyte and deuterated internal standard during method validation through the analysis of reference standards. Retention times are updated following the analysis of calibration standards in each batch in order to account for minor drifts in daily retention times. Retention times for each analyte of interest were required to be within 5% of those determined with control samples. The ratio of the quantifying product ion

peak area to the qualifying product ion peak area was required to be within +/− 20% of the ion ratio determined for calibrators. Quantitation was performed using a working standard calibration curve and comparing the ratio of quantifying ion peak area to internal standard peak area. Authentic samples were required to exhibit approximately the same internal standard peak areas as those in calibrators and controls in order ensure an efficient extraction. For matrices other than urine the general acceptance criterion is that the internal standard response should be approximately 10–200% of the calibrator/control average. Calibration curves were required to comprise at least 50% of the original curve points and any specimens with quantitative values above/greater than the upper calibrator were required to be rerun at an appropriate dilution using negative serum. The analytical run is considered acceptable if the calculated concentrations of analyte/s in control samples are within 20% of the expected nominal value.

4. Notes

1. It is important that the acetonitrile be cold to adequately precipitate proteins. Remove from the freezer just prior to use.

2. The 78:20:2 dichloromethane:2-propanol:ammonium hydroxide, used in step 12 of the solid phase extraction must be made fresh on each day of use.

References

1. Ross, D.L. and T.C. Chan 2006. *Sudden Deaths in Custody*. Totowa, NJ: Humana Press.

2. Karch, S.B. 2002. *Karch's Pathology of Drug Abuse*. Boca Raton, FL: CRC Press.

3. Di Maio, T.G. and V.J.M. Di Maio 2006. *Excited Delirium Syndrome Cause of Death and Prevention*. Boca Raton: Taylor & Francis Group.

4. *Results from the 2006 National Survey on Drug Use and Health: National Findings*. Substance Abuse and Mental Health Services Administration.

5. Ostrea, E.M., et al. 1992. Drug screening of newborns by meconium analysis: a large-scale, prospective, epidemiologic study. *Pediatrics* 89(1): 107–113.

6. Ebrahim, S.H. and J. Gfroerer 2003. Pregnancy-related substance use in the United States during 1996–1998. *Obstet Gynecol Clin North Am* 101(2): 374–379.

7. Birchfield, M., J. Scully, and A. Handler 1995. Perinatal screening for illicit drugs: policies in hospitals in a large metropolitan area. *J Perinatol* 15(3): 208–214.

8. Rosenberg, E., 2003. The potential of organic (electrospray- and atmospheric pressure chemical ionization) mass spectrometric techniques coupled to liquid-phase separation for speciation analysis. *J Chromatography A*, 1000: 841–889.

9. Browne, S., et al. 1994. Detection of cocaine, norcocaine, and cocaethylene in

the meconium of premature neonates. *J Forensic Sci* **39**(6):. 1515–1519.

10. l. 1992. Detection of cocaine and its metabolites in human amniotic fluid. *J Analytical Toxicol* **16**(5): 328–331.

11. Sandberg, J.A. and G.D. Olsen 1992. Cocaine and metabolite concentrations in the fetal guinea pig after chronic maternal cocaine administration. *J Pharmacol Exp Ther* **260**(2): 587–591.

Chapter 17

Diagnosis of Creatine Metabolism Disorders by Determining Creatine and Guanidinoacetate in Plasma and Urine

Qin Sun and William E. O'Brien

Abstract

Creatine metabolism disorders include a creatine transporter deficiency, as well as, deficiencies of two enzymes involved in creatine synthesis, arginine–glycine amidinotransferase (AGAT) and guanidinoacetate methyltransferase (GAMT). Laboratory diagnosis of these disorders relies on the determination of creatine and guanidinoacetate in both plasma and urine. Here we describe a rapid HPLC/MS/MS method for these measurements using a normal phase HILIC column after analyte derivatization.

Key words: Creatine, guanidinoacetate, arginine–glycine amidinotransferase, AGAT, guanidinoacetate methyltransferase, GAMT, creatine transporter, mass spectrometry.

1. Introduction

Creatine and phosphocreatine are critical for temporal and spatial regulation of the intracellular ATP energy pool (1–3). Humans acquire creatine from dietary sources and endogenously. The latter provides up to 50% of daily creatine requirements through a short two-step synthesis pathway involving the enzymes arginine–glycine amidinotransferase (AGAT) and guanidinoacetate methyltransferase. In the first step, arginine–glycine amidinotransferase catalyzes the synthesis of ornithine and guanidinoacetate (GAA) from glycine and arginine. The subsequent conversion of GAA to creatine by guanidinoacetate methyltransferase requires S-adenosylmethionine (SAM) as a methyl donor and is the largest SAM consuming reaction. The majority of creatine synthesis occurs in the kidney and is later taken up by creatine membrane transporters (CRT) (4).

U. Garg, C.A. Hammett-Stabler (eds.), *Clinical Applications of Mass Spectrometry*, Methods in Molecular Biology 603, DOI 10.1007/978-1-60761-459-3_17, © Humana Press, a part of Springer Science+Business Media, LLC 2010

Disorders have been described involving both creatine synthesis and creatine membrane transporters (5–10). All are rare with only two cases of AGAT deficiency; over a hundred cases of GAMT deficiency and few case of CRTR deficiency described to date, but it is likely that as more will be identified with increased awareness. Patients present in infancy and childhood with development delay, hypotonia, and in some cases seizures. For patients with GAMT deficiency, creatine is low to within normal limits in infancy but declines by adolescence if the proband is on strict vegetarian diet. High plasma GAA concentrations are considered pathognomonic for GAMT deficiency. Magnetic resonance spectroscopy of the brain of these patients reveals creatine depletion and guanidinoacetate phosphate accumulation (11). With AGAT deficiency, biochemical testing reveals extremely low plasma creatine and creatinine, as well as low GAA concentrations. Prenatal diagnosis is possible by measuring guanidinoacetate in amniotic fluids (12).

The creatine transporter, important in providing adequate energy to the brain, skeletal, and cardiac muscle, is also critical for renal reabsorption of creatine and responsible for cerebral distribution and uptake of GAA (13, 14). The transporter is encoded by an X chromosome gene (SLC6A8). Mutations of this gene are also found in patients with X-linked mental retardation (6, 7, 15–17). Biochemical changes for the initial case included an elevated urinary creatine to creatinine ratio. Interestingly, changes in guanidinoacetate and/or creatine excretion are also reported in disorders in which one-carbon methyl group metabolism is disturbed, for example, in S-adenosylhomocysteine hydrolase deficiency and cobalamin deficiency (18, 19).

Several guanidinoacetate and creatine determination methods have been reported using techniques such as HPLC, GC/MS, and HPLC/tandem MS (20–23). While high guanidinoacetate and creatine concentrations in urine were quantified without difficulty using these methods, the methods were not sufficiently sensitive to accurately measure the low plasma guanidinoacetate essential for diagnosing AGAT deficiency. We describe an improved tandem MS protocol utilizing derivatized analytes and a hydrophilic interaction column (HILIC). HILIC is a variation of normal-phase chromatography designed for separation of very polar analytes. The method is sufficiently sensitive to permit the measurement of low concentrations of guanidinoacetate, and allows simultaneous analysis of both urine and plasma samples.

2. Materials

2.1. Specimen Requirements

1. Plasma samples (sodium heparin): Immediately centrifuge sample and transfer plasma (0.5–1.0 mL) to labeled tube. Store frozen at −16°C or colder. Samples should be analyzed

within two weeks of collection. Samples should be transported between collection site and laboratory on dry ice (*see* **Note 1**).

2. Urine samples: Freeze 2–5 ml of urine. Specimens may be stored frozen for up to 2 weeks. Samples should be transported between collection site and laboratory on dry ice (*see* **Note 1**).

2.2. Reagents

1. 1% Ammonium Formic Acid: Add 10 ml of formic acid to 800 ml of HPLC grade H_2O, adjust pH to 4.0 with concentrated ammonium hydroxide and bring total volume to 1 liter. Store at room temperature. Stable for up to 2 years.

2. HPLC Running Buffer: 95% acetonitrile containing 0.05% ammonium formate. Add 50 ml of the 1% ammonium formate solution to 950 ml of acetonitrile. Store at room temperature. Stable for up to 2 years.

3. Derivatizing reagent, 3 N HCl in n-Butanol (Regis Technologies, Inc., Morton Grove, IL).

2.3. Standards and Calibrators

1. Creatine primary stock standard, 6 mM: Add 44.75 mg of creatine into 50 ml of acetonitrile:H_2O (50:50). Store frozen ($-16°C$ or colder) for up to 2 years.

2. Creatine secondary stock solution, 600 μM: Prepare by diluting 1 ml of the stock solution with 9 ml acetonitrile:H_2O (50:50) for effective concentration of 600 μM. Store frozen ($-16°C$ or colder) for up to 1 year.

3. Creatine quality control solution, 7.6 mM: Add 50 mg of creatine into 50 ml of acetonitrile:H_2O (50:50).

4. Guanidinoacetate primary stock standard, 5 mM: Add 29.3 mg of guanidinoacetate into 50 ml of acetonitrile:H_2O (50:50). Store the stock solution frozen ($-16°C$ or colder) for up to 2 years.

5. Guanidinoacetate secondary stock standard, 50 μM: Dilute the stock solution by mixing 100 μl with 9.9 ml of acetonitrile:H_2O (50:50). Store in freezer ($-16°C$ or colder) for up to 1 year.

6. Guanidinoacetate quality control solution, 850 μM: Add 25 mg of GAA into 250 ml of acetonitrile:H_2O (50:50).

7. Prepare working creatine and guanidinoacetate standards as follows (*see* **Note 1**):
   a. GAMT A standard solution, 300 μM creatine and 25 μM GAA – Mix equal volumes of 600 μM creatine and 50 μM guanidinoacetate secondary stock standards.

   b. GAMT B standard solution, 30 μM creatine and 2.5 μM GAA – Dilute GAMT A standard solution 1:10 with 95% acetonitrile.

8. Prepare 6 calibrators according to **Table 17.1**. Mix all calibrators thoroughly and dry using SpeedVac. Drying time is approximately 30 min.

Table 17.1
Standard solutions for GAA and CRT

	GAMT A µl	GAMT B µl	Internal standard µl	CRT final concentration (µM)	GAA final concentration (µM)
G1	30	–	10	300	25
G2	15	–	10	150	12.5
G3	10	–	10	100	8.33
G4	–	30	10	30	2.5
G5	–	15	10	15	1.25
G6	–	10	10	5	0.4

2.4. Internal Standard and Quality Controls

1. Internal Standard (D_3-Creatine): Prepare by adding 10 mg D3-creatine to 20 ml of acetonitrile:H_2O (50:50) to give a solution of approximately 3.7 mM. The stock solution is stable when stored frozen at $-16°C$ or colder for up to 5 years. The working internal standard is prepared by diluting the stock solution 1/500 into acetonitrile: H_2O (50:50) for the working solution (about 6.35 µM). Stable at $-16°C$ or colder for up to 1 year.

2. Urine controls: Pooled urine from volunteers or drug-free urine may be used. Dilute pool 1:10 using water to prepare a "normal" urine control. Use undiluted urine as a "high" control. Aliquot control materials and store frozen at $-16°C$ or colder.

3. Plasma controls: Fresh frozen plasma or similar material is used to prepare controls. A portion of the pool should be aliquoted to prepare a "normal" control. Prepare a "high" control by adding 100 µl of the 7.6 mM creatine and 100 µl of the 850 µM guanidinoacetate quality control standard solutions to 4 ml of fresh frozen plasma.

2.5. Equipments

1. Micromass Quattro Micro Spectrometer (Waters Corp., Milford, MA).

2. Waters 2695 HPLC (Waters Corp., Milford, MA).

3. Atlantis HILIC (Hydrophilic Interaction Chromatography) HPLC column, 2.1 × 100 mm, 5 µm (Waters Corp. Milford, MA).

4. Masslynx analytical software (Waters Corp., Milford, MA).

5. SPD1010 SpeedVac (Thermo Fisher Scientific, Waltham, MA).

3. Methods

3.1. Preparation of Plasma Samples

1. Thaw the sample and mix thoroughly. Sample may be maintained at room temperature while processing.

2. Transfer 30 μl of the plasma sample and add 10 μl of the internal standard working solution to an appropriately labeled 1.5 ml microfuge tube.

3. Add 300 μl of acetonitrile into above mixture and vortex to precipitate the protein.

4. Centrifuge for 10 min at $10,500 \times g$ in a desktop microfuge.

5. Transfer the supernatant to a new, labeled tube and dry liquid completely in a SpeedVac with heating temperature of 45°C for 45 min.

3.2. Preparation of Urine Samples

1. Determine the creatinine concentration of urine samples. Creatinine values are used to adjust the appropriate urine dilution in subsequent steps. The assay is not valid for samples with creatinine less than 0.1 mg/ml (*see* **Note 2**).

2. Dilute samples 1:10 with deionized water if creatinine is 0.1–0.4 mg/ml and 1:75 if creatinine is above 0.4 mg/ml (*see* **Note 3**).

3. Dilute normal and high urine controls 1:10 times with deionized water.

4. Transfer 10 μl of the diluted urine sample and control into a labeled 1.5 mL microfuge tube. Add 10 μl of the internal standard working solution and vortex to mix.

5. Dry liquid completely using a SpeedVac with heating temperature of 45°C for 45 min. Trace amount of liquid does not affect final results.

3.3. Sample Preparation for HPLC/MS/MS Analysis

1. Add 70 μl of the 3 N HCl in n-Butanol to all dried tubes, including standard solutions, controls, and the samples and mix thoroughly.

2. Place the tubes in the 65°C heating block for 20 min. Vortex at least once during the time period.

3. Remove the tubes and dry the contents in the SpeedVac with heating temperature of 45°C. Drying time is approximately 30 min. Trace amount of liquid does not affect final results.

4. Add 1 ml of 95% acetonitrile to each tube and mix thoroughly.

5. Centrifuge the tubes at $10,500 \times g$ for 6 min to remove any particulate matter.

6. Transfer 150 μl supernatant to 150 μl sample vials for chromatographic analysis.

3.4. Setup for HPLC and Tandem Mass Spectrometer (see Note 4)

1. The chromatographic separation of hydrophilic CRT and GAA is performed at room temperature using an Atlantis HILIC (Hydrophilic Interaction Chromatography) HPLC column, 2.1 × 100 mm, 5 μm. The mobile phase is maintained at a flow rate of 0.2 mL/min.

2. The settings for the mass spectrometer are as follows: Positive ion mode of electrospray (ESI⁺), capillary voltage 3.5 kV, source cone voltage 25 V, extract voltage 4 V, source temperature 130°C, desolvation temperature 350°C, and analyzer collision 15. Nitrogen gas flows at 90 and 600 l/h for nebulizer and desolvation, respectively.

3. Multiple Reaction monitoring (MRM) transitions are utilized: Guanidinoacetate 174>101, creatine 188>90, and D3-Creatine 191>93. A typical chromatogram is seen in **Fig. 17.1**.

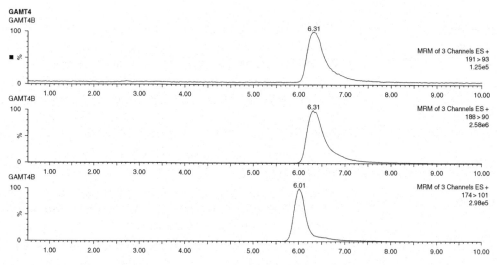

Fig. 17.1. Chromatogram of LC-MS/MS analyses of a sample. The individual traces of MRM transitions are as follows: *top panel*, D3-Creatine (*m/z* 191>93); *middle*, Creatine (*m/z* 188>90) at about 6.3 min; and *bottom*, GAA (*m/z* 174>101) at 6.0 min.

3.5. Data Analysis

1. Concentrations of unknowns and quality control samples are calculated from standard curves using Masslynx software. The concentration values in **Table 17.2** takes into account the sample volume being used in the assay. Urine dilution factors need to be entered to allow Masslynx to calculate the value. No further calculations are necessary as long as the sample volume does not change.

2. The coefficient of correlation (r^2) for the standard curve should be above 0.97.3.

3. For results to be acceptable, the control values should be within ± 2 SD of the established control means. Examples of quality control ranges are shown in **Tables 17.3** and **17.4**.

Table 17.2
Dilution factor calculation for urine samples

Creatinine mg/ml	Dilution in H$_2$O	Urine sample volume analyzed (μL)	Standard sample volume analyzed (μL)	Extra dilutions due to volume Dif.	Final dilution factor
0.1–0.4	10	10	30	3	30 (10 × 3)
0.4 and above	75	10	30	3	225 (75 × 3)

Table 17.3
Example of urine quality control ranges

Compound	High tolerance limits (μM)	Normal tolerance limits (μM)
Creatine	660–1260	50–165
Guanidinoacetate	160–330	10–40

Table 17.4
Example of plasma quality control ranges

Compound	High tolerance limits (μM)	Normal tolerance limits (μM)
Creatine	75–240	10–55
Guanidinoacetate	8–45	0.5–2.5

4. If the values of clinical samples are above the upper limit of standard curve the sample must be diluted and re-analyzed.

3.6. Result Interpretation

1. Reference ranges for both plasma and urine are shown in **Table 17.5**. These ranges should only serve as a guide and should be established by each laboratory. The concentration of guanidinoacetate was reported to decline with age in urine, but increases with age in plasma (20). Others have similarly reported age-related changes in these reference values (24). However this trend is not reflected in our reference range (**Table 17.5**). This could be due to either differences in age groups or sample population as our reference ranges are established in-house.

Table 17.5
Reference ranges for CRT and GAA

	Plasma (μM)		Urine (mmoles/mole creatinine)	
	0–2 years	Above 2 years	1–5 years	Above 5 years
CRT	28–102	20–110	20–900	12–500
GAA	0.3–1.5	0.3–2.8	8–130	8–150

2. **Abnormalities in creatine deficiencies**: Biochemical findings for these patients are highly dependent on differences in methodology. The following general guidelines are used to interpret patient results.

It is important to remember that diet and supplements have a major impact on creatine concentrations. Adults may obtain up to 50% of creatine from diet, mostly from meat and fish. Thus patients on western diet rarely have low creatine concentrations. In addition, creatine is the most common nutritional supplement used by athletes (25). A patient's diet and any history of supplement usage are important information for interpreting guanidinoacetate and creatine results. Since patients who have AGAT or GAMT deficiency are often found to have creatine concentrations within or only slightly above the reference range, plasma guanidinoacetate concentrations (**Table 17.6**) are more significant and should be sufficient for diagnoses. It is interesting that plasma guanidinoacetate is suggested to increase in combined methylmalonic aciduria and homocystinuria due to cobalamin deficiencies (18). Creatinuria is also found following trauma, likely due to muscle damage (26). Such patients should be retested at a later date.

Table 17.6
Interpretation of results

	Plasma (μM)		Urine (mmoles/mole creatinine)	
	GAA	CRT	GAA	CRT
AGAT deficiency	Low	Low	Low	Normal/low
GAMT deficiency	High	Low	Normal/high	Normal/low
CRT transporter defect	Normal	Normal/low	Normal	High

4. Notes

1. GAA and creatine are unstable. Plasma and urine samples should be stored frozen at $-16°C$ or colder. Use dry ice when shipping from collection site to laboratory. Samples should be analyzed within 2 weeks of collection. Plasma samples that are hemolyzed are unacceptable due to contamination from erythrocyte creatine. Stock solutions are stable up to one year when kept frozen. Working solutions are stable for 6 weeks when stored at 4°C.

2. The urinary excretion of creatine and guanidinoacetate varies considerably even in normal populations. To minimize the numerous issues associated with a random urine collection, it is critical that these results are normalized to the urinary creatinine concentration before making any interpretations. Extremely low creatinine levels are problematic and tend to falsely elevate the creatinine normalized creatine and guanidinoacetate levels. It is recommended to cancel urine samples with very low creatinine (<0.1 mg/ml) for this reason. When possible, a first morning void or 24-hour collection is preferred. These limitations underscore the importance of plasma creatine and GAA in diagnoses of AGAT and GAMT deficiency.

3. Because the urine volume (10 μl) analyzed is less than that of standard solution and plasma samples (30 μl), another 3x dilution factor must be taken into account. As a result, the final dilution factors are 30 for samples with 0.1–0.4 mg/ml creatine and 225 for those with creatinine >0.4 mg/ml, respectively. These dilution factors are entered in running sequence list in Masslynx software for later calculation. *See* **Table 17.2** for dilution factor calculation for urine samples.

4. The sensitivity of this tandem mass spectrometry method is improved compared to previous methods (23). Shifting retention times often indicate deterioration of the column. Changing the guard column and/or washing HILIC column with running buffer are usually sufficient to solve the problem. For persistent column-related problems, it may be useful to perform the following: wash the column for 30 min with 5% acetonitrile (95% water), continue for an additional 30 min with 95% acetonitrile, and conclude with 30 min of 5% acetonitrile (95% water) again. The time for each step could vary from 30 min to few hours depending on different situations. It is also recommended that the column be washed with 5–10 column volumes of running buffer before each routine run to assure equilibration and improve consistency of retention times.

References

1. Monge, C., Beraud, N., Kuznetsov, A.V., Rostovtseva, T., Sackett, D., Schlattner, U., Vendelin, M. and Saks, V.A. (2008) Regulation of respiration in brain mitochondria and synaptosomes: restrictions of ADP diffusion in situ, roles of tubulin, and mitochondrial creatine kinase. *Mol Cell Biochem*, **318**, 147–165.

2. Saks, V., Kaambre, T., Guzun, R., Anmann, T., Sikk, P., Schlattner, U., Wallimann, T., Aliev, M. and Vendelin, M. (2007) The creatine kinase phosphotransfer network: thermodynamic and kinetic considerations, the impact of the mitochondrial outer membrane and modelling approaches. *Subcell Biochem*, **46**, 27–65.

3. Saks, V., Kuznetsov, A., Andrienko, T., Usson, Y., Appaix, F., Guerrero, K., Kaambre, T., Sikk, P., Lemba, M. and Vendelin, M. (2003) Heterogeneity of ADP diffusion and regulation of respiration in cardiac cells. *Biophys J*, **84**, 3436–3456.

4. Braissant, O., Bachmann, C. and Henry, H. (2007) Expression and function of AGAT, GAMT and CT1 in the mammalian brain. *Subcell Biochem*, **46**, 67–81.

5. Ganesan, V., Johnson, A., Connelly, A., Eckhardt, S. and Surtees, R.A. (1997) Guanidinoacetate methyltransferase deficiency: new clinical features. *Pediatr Neurol*, **17**, 155–157.

6. Item, C.B., Stockler-Ipsiroglu, S., Stromberger, C., Muhl, A., Alessandri, M.G., Bianchi, M.C., Tosetti, M., Fornai, F. and Cioni, G. (2001) Arginine:glycine amidinotransferase deficiency: the third inborn error of creatine metabolism in humans. *Am J Hum Genet*, **69**, 1127–1133.

7. Salomons, G.S., van Dooren, S.J., Verhoeven, N.M., Cecil, K.M., Ball, W.S., Degrauw, T.J. and Jakobs, C. (2001) X-linked creatine-transporter gene (SLC6A8) defect: a new creatine-deficiency syndrome. *Am J Hum Genet*, **68**, 1497–1500.

8. Stockler, S., Hanefeld, F. and Frahm, J. (1996) Creatine replacement therapy in guanidinoacetate methyltransferase deficiency, a novel inborn error of metabolism. *Lancet*, **348**, 789–790.

9. Stockler, S., Marescau, B., De Deyn, P.P., Trijbels, J.M. and Hanefeld, F. (1997) Guanidino compounds in guanidinoacetate methyltransferase deficiency, a new inborn error of creatine synthesis. *Metabolism*, **46**, 1189–1193.

10. van der Knaap, M.S., Verhoeven, N.M., Maaswinkel-Mooij, P., Pouwels, P.J., Onkenhout, W., Peeters, E.A., Stockler-Ipsiroglu, S. and Jakobs, C. (2000) Mental retardation and behavioral problems as presenting signs of a creatine synthesis defect. *Ann Neurol*, **47**, 540–543.

11. Schulze, A., Hess, T., Wevers, R., Mayatepek, E., Bachert, P., Marescau, B., Knopp, M.V., De Deyn, P.P., Bremer, H.J. and Rating, D. (1997) Creatine deficiency syndrome caused by guanidinoacetate methyltransferase deficiency: diagnostic tools for a new inborn error of metabolism. *J Pediatr*, **131**, 626–631.

12. Cheillan, D., Salomons, G.S., Acquaviva, C., Boisson, C., Roth, P., Cordier, M.P., Francois, L., Jakobs, C. and Vianey-Saban, C. (2006) Prenatal diagnosis of guanidinoacetate methyltransferase deficiency: increased guanidinoacetate concentrations in amniotic fluid. *Clin Chem*, **52**, 775–777.

13. Braissant, O. and Henry, H. (2008) AGAT, GAMT and SLC6A8 distribution in the central nervous system, in relation to creatine deficiency syndromes: A review. *J Inherit Metab Dis*.

14. Guimbal, C. and Kilimann, M.W. (1993) A Na(+)-dependent creatine transporter in rabbit brain, muscle, heart, and kidney. cDNA cloning and functional expression. *J Biol Chem*, **268**, 8418–8421.

15. Clark, A.J., Rosenberg, E.H., Almeida, L.S., Wood, T.C., Jakobs, C., Stevenson, R.E., Schwartz, C.E. and Salomons, G.S. (2006) X-linked creatine transporter (SLC6A8) mutations in about 1% of males with mental retardation of unknown etiology. *Hum Genet*, **119**, 604–610.

16. Rosenberg, E.H., Almeida, L.S., Kleefstra, T., deGrauw, R.S., Yntema, H.G., Bahi, N., Moraine, C., Ropers, H.H., Fryns, J.P., deGrauw, T.J. et al. (2004) High prevalence of SLC6A8 deficiency in X-linked mental retardation. *Am J Hum Genet*, **75**, 97–105.

17. Salomons, G.S., van Dooren, S.J., Verhoeven, N.M., Marsden, D., Schwartz, C., Cecil, K.M., DeGrauw, T.J. and Jakobs, C. (2003) X-linked creatine transporter defect: an overview. *J Inherit Metab Dis*, **26**, 309–318.

18. Bodamer, O.A., Sahoo, T., Beaudet, A.L., O'Brien, W.E., Bottiglieri, T., Stockler-Ipsiroglu, S., Wagner, C. and Scaglia, F. (2005) Creatine metabolism in combined

methylmalonic aciduria and homocystinuria. *Ann Neurol*, **57**, 557–560.

19. Buist, N.R., Glenn, B., Vugrek, O., Wagner, C., Stabler, S., Allen, R.H., Pogribny, I., Schulze, A., Zeisel, S.H., Baric, I. et al. (2006) S-adenosylhomocysteine hydrolase deficiency in a 26-year-old man. *J Inherit Metab Dis*, **29**, 538–545.

20. Arias, A., Ormazabal, A., Moreno, J., Gonzalez, B., Vilaseca, M.A., Garcia-Villoria, J., Pampols, T., Briones, P., Artuch, R. and Ribes, A. (2006) Methods for the diagnosis of creatine deficiency syndromes: a comparative study. *J Neurosci Methods*, **156**, 305–309.

21. Bodamer, O.A., Bloesch, S.M., Gregg, A.R., Stockler-Ipsiroglu, S. and O'Brien, W.E. (2001) Analysis of guanidinoacetate and creatine by isotope dilution electrospray tandem mass spectrometry. *Clin Chim Acta*, **308**, 173–178.

22. Struys, E.A., Jansen, E.E., ten Brink, H.J., Verhoeven, N.M., van der Knaap, M.S. and Jakobs, C. (1998) An accurate stable isotope dilution gas chromatographic-mass spectrometric approach to the diagnosis of guanidinoacetate methyltransferase deficiency. *J Pharm Biomed Anal*, **18**, 659–665.

23. Young, S., Struys, E. and Wood, T. (2007) Quantification of creatine and guanidinoacetate using GC-MS and LC-MS/MS for the detection of cerebral creatine deficiency syndromes. *Curr Protoc Hum Genet*, **Chapter 17**, Unit 17 13.

24. Valongo, C., Cardoso, M.L., Domingues, P., Almeida, L., Verhoeven, N., Salomons, G., Jakobs, C. and Vilarinho, L. (2004) Age related reference values for urine creatine and guanidinoacetic acid concentration in children and adolescents by gas chromatography-mass spectrometry. *Clin Chim Acta*, **348**, 155–161.

25. Derave, W., Marescau, B., Vanden Eede, E., Eijnde, B.O., De Deyn, P.P. and Hespel, P. (2004) Plasma guanidino compounds are altered by oral creatine supplementation in healthy humans. *J Appl Physiol*, **97**, 852–857.

26. Threlfall, C.J., Maxwell, A.R. and Stoner, H.B. (1984) Post-traumatic creatinuria. *J Trauma*, **24**, 516–523.

Chapter 18

Broad Spectrum Drug Screening Using Electron-Ionization Gas Chromatography-Mass Spectrometry (EI-GCMS)

Judy Stone

Abstract

A liquid–liquid extraction (LLE) of drugs and internal standard (promazine) is performed by mixing urine at basic pH with 1-chlorobutane. There are no hydrolysis or derivatization steps. After centrifugation the organic (upper) layer is transferred to another tube and evaporated. The dried extract is reconstituted with ethyl acetate and 1 μL is injected onto the GCMS. Drugs are volatilized in the GC inlet and separated on a capillary column. In the EI source drugs become positively charged and fragment. Mass analysis of ionized fragments occurs with a single quadrupole. The resulting full scan mass spectra are automatically searched against three libraries.

Key words: Broad spectrum drug screening, urine liquid–liquid extraction, Full Scan Electron-Ionization Gas Chromatography-Mass Spectrometry (EI-GCMS), NIST mass spectral library, library search algorithms.

1. Introduction

Full scan EI-GCMS has been the reference method for clinical broad spectrum drug screening for many years (1, 2). However, the number of clinical laboratories performing the test has decreased because it is labor intensive, there are cost constraints, shortages of labor and expertise, and only a small number of patients benefit from such testing (3). All such methods are "home-brew" or "lab-developed tests" with additional effort required for validation compared to FDA approved assays (4). There is no single protocol for EI-GCMS drug screening that is universally practiced (5).

Until recently, an alternative used by some clinical laboratories was an automated drug screening system incorporating liquid chromatography with UV detection, the REMEDI HS® from

U. Garg, C.A. Hammett-Stabler (eds.), *Clinical Applications of Mass Spectrometry*, Methods in Molecular Biology 603, DOI 10.1007/978-1-60761-459-3_18, © Humana Press, a part of Springer Science+Business Media, LLC 2010

Bio-Rad Laboratories (Hercules, CA). This system combined ease of use (e.g. online extraction), FDA approval (for a subset of drugs), kit reagents and columns, an extensive library of drug and metabolite UV spectra, and excellent technical support from Bio-Rad. Disadvantages were a relative lack of sensitivity, specificity, and resolution compared to capillary EI-GCMS. Support for the instrument has been discontinued, and laboratories that wish to continue broad spectrum drug testing must consider EI-GCMS and/or LC-MS or LC-MSMS as alternatives.

The advantage of EI-GCMS is the highly reproducible nature of the spectra. Good matches can be obtained between spectra acquired on different models and even different generations of an instrument, from different laboratories, and relatively independent of specimen preparation and chromatographic procedures (1). A number of commercial libraries of EI mass spectra are available such as the NIST 08 EI-GCMS database with >200,000 spectra of 191,436 unique compounds (6). Although manual examination of any spectral match is basic good laboratory practice, such review is simplified by the robust performance of automated library search and deconvolution algorithms that exist for EI-GCMS spectra (7, 8). With the exception of selected drugs/drug classes that have sparse spectra dominated by a single low mass ion (e.g. amitriptyline, amphetamine, etc.) we have found automated library search of EI spectra combined with capillary GC and the use of RRt reference ranges to yield very few false positives. Each laboratory must establish reporting criteria for the list of drugs in their screening method based on chromatographic and spectral characteristics and interference studies (9, 10).

In order to detect polar and heat-labile compounds such as drug metabolites, sample preparation for GCMS must include hydrolysis to remove glucuronides/sulfates and derivatization to make compounds GC compatible, i.e., more volatile and heat stable (1, 2). A well-documented EI-GCMS drug screening procedure with a menu of over 2000 detectable drugs has been described by Maurer et al. (5). Their "systematic toxicological analysis" (STA) method includes acid hydrolysis, liquid–liquid extraction, and microwave-assisted acetylation (5). The use of customized software for spectral deconvolution, library searching, and peak integration becomes important with such methods as data interpretation is quite complex when the menu of GCMS detectable chemicals, including endogenous and environmental compounds, is greatly expanded via the hydrolysis and derivatization steps (11).

The EI-GCMS method described here does *not* include hydrolysis or derivatization and so is better characterized as broad spectrum rather than STA or a general unknown or comprehensive drug screening method (GUS or CDS). The extraction is simple – but is vulnerable to operator variability if not monitored. Many of the commonly encountered drugs found in emergency toxicology (100–150 compounds) can be detected at therapeutic levels within approximately 90 min. The only software necessary for reporting and interpretation is the Chemstation software package supplied with GCMS instruments purchased from Agilent Technologies (Wilmington, DE) or comparable products from other vendors. Broad spectrum drug screening with LC-MSMS is a good complement to this EI-GCMS procedure, as certain polar and heat labile compounds – such as risperidone and metabolite – that are not detected by this GCMS method are easily identified with LC-MSMS.

At San Francisco General Hospital (SFGH) the broad spectrum drug screen by GCMS in tandem with the broad spectrum drug screen by LC-MSMS are used to identify drugs in urine that are not detected by automated immunoassays. It can be informative in the setting of substance abuse for over-the-counter or prescription compounds like carisoprodol, ketamine, and dextromethorphan, and for synthetic opioids such as tramadol (Ultram®), meperidine (pethidine, Demerol®), and propoxyphene (Darvon®). It can be employed in the setting of overdose to identify a wide range of clinically significant drugs or poisons (e.g. antidepressants, antipsychotics, anticonvulsants, beta and calcium channel blockers, etc.) when a patient does not respond to supportive care, remains obtunded, has intractable seizures or when the clinical impression is not consistent with history. There are significant toxins that are not detectable or are poorly/variably detected with both this method and LC-MSMS screening – such as acetaminophen, salicylate, digoxin, opiates, amphetamine and benzodiazepines. This is not a concern at SFGH as testing for these compounds is done by immunoassay and enzymatic methods on automated, high throughput chemistry analyzers that can provide STAT turnaround times on a 24 h/7-day basis. Targeted, semi-quantitative mass spectrometry methods are also available, although usually are not needed for emergency toxicology cases to confirm and identify individual opiates, amphetamines, and benzodiazepines.

2. Materials

2.1. Reagents and Buffers

1. Human drug-free urine (UTAK Laboratories, INC, Valencia, CA).

2. Compressed nitrogen (high purity 99.997%).

3. Drug standards, 1.0 and 0.1 mg/mL in methanol or as powder, (Cerilliant Corp., Round Rock, TX; Grace Davison/Alltech, Deerfield, IL; Sigma-Aldrich, St. Louis, MO; and Lipomed Inc., Cambridge, MA).

2.2. Library Reference Standards (Calibrators/ Calibration)

1. This is a qualitative method; no calibrators are used and no concentrations are reported.

2. Instead of a lower limit of quantitation (LLOQ), the criteria necessary for reporting a drug as "detected" or "positive" are

   a) Peak area must be ≥4% of the internal standard peak area

   b) Quality of the library match factor must be ≥90 and visual comparison of the unknown versus library spectra must pass laboratory-defined criteria

   c) Relative retention time (RRt) must be within +/− 0.02 of the library standard RRt and

   d) the drug must not be identified in the blank (drug-free urine) control.

3. Reference relative retention times (RRt) and typical retention times (Rt) for drugs reportable with this method are established by adding approximately 50 μg of drug standards (e.g. 50 μL of 1 mg/mL standard in methanol) and 50 μL of internal standard working solution to a 13 × 100 mm glass tube and evaporating the solvent with a stream of nitrogen in a water or dry bath at 30–35°C. The drug and internal standard are reconstituted by adding 50 μL of ethyl acetate to the tube, vortex-mixing vigorously for 20 sec and transferring the contents to an autosampler vial with deactivated glass micro-insert using a glass Pasteur pipet (*see* **Note 1**). The vial is capped and analyzed using the GCMS parameters described.

4. From the same analyses of pure drug standards as in (3), full scan spectra for drugs are acquired, reviewed, and when acceptable used to create an in-house spectral library. Entering the name of the drug in the spectral library record followed by the reference RRt, e.g., "Cocaine RRt = 0.91" is helpful when reviewing library search reports.

2.3. Internal Standard and Quality Control

1. Internal standard: promazine HCl (Sigma Chemical Co., St. Louis, MO).

2. Internal Standard stock solution: promazine, 20 mg/mL, in methanol. Prepare by weighing 225.64 mg of promazine HCl. Dissolve in methanol in a 10 mL volumetric flask. Mix well. Store in an amber vial with a PTFE-lined cap. Stable at −8 to −20°C for 2 years.

3. Internal Standard working solution: promazine (0.05 mg/mL) in methanol. Prepare by pipetting 50 μL of the internal standard stock solution and 20 mL of acetonitrile into an amber glass vial with a PTFE-lined cap. Mix well. Stable at 15–30°C for 3 months.

4. Quality control samples: Drug-free urine (blank or negative control) and a commercial custom positive urine control containing 11 drugs (UTAK 11 Drug Custom Urine CDS QC; UTAK Laboratories, INC Valencia, CA) are used to validate performance daily. The drugs in the positive control, their concentrations, typical Rt, RRt, and peak areas are shown in **Table 18.1**.

Table 18.1
Drugs present, their nominal concentrations, and typical Rt, RRt, and peak areas for the positive control analyzed with the method as described

Drug	Concentration (ng/mL)	Typical Rt (min) and (RRt)	Typical peak area __e+6
Internal standard (Promazine)	500	14.35 (1.00)	5.9
1. Amitriptyline	500	12.88 (0.90)	5.9
2. Clozapine	500	19.59 (1.38)	4.6
3. Cocaine	500	12.99 (0.91)	4.8
4. Codeine	500	14.94 (1.05)	2.9
5. Diazepam	500	15.54 (1.09)	5.1
6. Meperidine (Pethidine)	500	8.10 (0.56)	3.8
7. Methadone	500	12.32 (0.86)	6.7
8. dl-Methadone primary metab.	100	11.13 (0.78)	1.2
9. MDMA	1000	6.22 (0.43)	1.1
10. Oxycodone	250	16.43 (1.16)	1.8
11. Phencyclidine	500	9.62 (0.67)	4.1

2.4. Extraction Tubes

1. Rinse 16 × 125 mm glass extraction tubes (with PTFE-lined screw caps) with methanol and air dry inverted. Tubes and caps are not reused (*see* **Note 1**).

2. Weigh 1+/− 0.2 g of sodium bicarbonate and pour into each tube, then cap. Stable for 6 months.

2.5. Supplies

1. 16 × 125 mm borosilicate glass tubes with PTFE-lined screw caps

2. Plain 13 × 100 mm borosilicate glass test tubes

3. Autosampler vials, 100 µL deactivated glass inserts, and caps with PTFE-lined septa (2 mL Amber wide opening glass vial with screw cap) (Agilent Technologies, Santa Clara, CA).

4. Amber glass vials with PTFE-lined caps for drug and internal standard solutions were purchased from Grace Davison/Alltech (Deerfield, IL).

5. Volumetric glassware

6. Repipet® dispensers

7. Positive displacement pipets (micro-dispensers) with glass capillaries and PTFE tips

8. GC capillary columns (HP-5MS, 0.250 mm × 30 m, 0.25 µm film) (Agilent Technologies, Santa Clara, CA).

9. Merlin microseal high-pressure septa (Merlin Instrument Co., Half Moon Bay, CA).

10. GC inlet liners (gooseneck, splitless, w/o glass wool 4 × 6.5 × 78.5 mm for Agilent GCs, IP deactivated).

11. GC inlet gold plated seals, 0.8 mm (Restek Chromatography Products, Bellefonte, PA).

2.6. Equipment

1. GCMS (e.g. Agilent Technologies model 6890 GC interfaced to a model 5973 MS).

2. One or more EI GCMS mass spectral libraries (purchase from the GCMS vendor for software compatibility): (a) MPW (formerly PMW) Mass Spectral Library of Drugs, Poisons, Pesticides, Pollutants and their Metabolites, 2007 edition with 7,840 EI spectra (including a two-volume text set, Mass Spectral and GC Data of Drugs, Poisons, Pesticides, Pollutants and their Metabolites, 3rd Edition by Hans H. Maurer, Karl Pfleger, and Arnin A. Weber) and/or (b) NIST 08 Mass Spectral Library with 191,436 EI spectra (http://www.nist.gov/srd/nist1a.htm) and/or (c) Wiley Registry™ of Mass Spectral Data, 8th edition with 399,383 EI spectra (John Wiley & Sons, Inc., Hoboken, NJ).

3. In addition to the library search function included in the GCMS vendor's software (e.g. Agilent Probability Based Matching [PBM]) – also available from GCMS vendors are the NIST Mass Spectral Search Program (Version 2.0f) and a separate utility – Automated Mass Spectrometry Deconvolution and Identification System (AMDIS) (see **Note 2**).

4. Heated evaporator module for drying down the liquid–liquid extraction (LLE) organic phase (e.g. N-VAP™ 112 and water bath from Organomation Assoc., Berlin, MA).

5. Centrifuge, rocking mixer for LLE tubes, vortex-mixer, adjustable pipets (100–1,000 µL and 1–5 mL).

3. Methods

3.1. Stepwise Procedure

1. Centrifuge samples at $\cong 2000 \times g$ for at least 10 min to remove particulate.

2. Label extraction tubes containing sodium bicarbonate (*see* **Section 2.4**) for the negative control, positive control, and each sample.

3. Add 5 mL of urine to the extraction tube.

4. Add 50 μL of internal standard working solution with a positive displacement pipet. Correct technique to avoid cross-contamination is critical (*see* **Note 3**).

5. Use Repipet dispenser to add 5 mL of 1-chlorobutane.

6. Cap the tubes firmly; and alternately tap or shake (gently) the bottom of the tubes to break up the sodium bicarbonate powder and then invert to mix the salt with the liquids. Repeat (3–5 times) until the salt and liquid have formed a uniform slurry and immediately put the tubes on a rocking mixer for 5 min (*see* **Note 4**).

7. After 5 min of mixing, centrifuge the tubes for 10 min at $\cong 2,000 \times g$ to separate the aqueous and organic layers.

8. Transfer the upper organic (1-chlorobutane) layer from each centrifuged tube to a labeled 13 × 100 mm tube using a glass transfer pipet. Some of the organic layer should remain in the original tube as transferring any of the debris at the organic/aqueous interface must be avoided.

9. Evaporate the 1-chlorobutane under a stream of nitrogen while heating the tubes to 30–35°C (*see* **Note 5**).

10. Add 50 μL of ethyl acetate to each dried 13 × 100 mm tube using a positive displacement pipet. Correct technique to avoid cross-contamination is critical (*see* **Note 2**). Vortex-mix vigorously for at least 20 sec.

11. Transfer the reconstituted extracts to a 100 μL deactivated, glass microinsert in an autosampler vial using a glass transfer pipet (*see* **Note 1**). Cap, using a PTFE-lined septa, and place on the autosampler.

3.2. Instrument Operating Conditions

1. Capillary GC column used: 0.25 mm i.d. × 30 m, 250 μm film, HP5-MS (5% phenyl-methyl silicone cross-linked to reduce bleed).

2. MS source used: EI at 70 eV.

3. Carrier gas is helium (*see* **Note 6**).

4. GC and MS parameters are given in **Tables 18.2** and **18.3**.

Table 18.2
GC operating conditions (Agilent 6890 model GC)

Component	Parameter	Details
Oven	Initial temp: 70 °C Initial time: 3.50 min Ramp 1: Ramp 2: Run time: 27.71 min	 Rate: 35.0/min Final temp: 200 Final time: 2.00 Rate: 8.0/min Final temp: 300 Final time: 6.00
Inlet	Mode: Splitless Initial temp: 250°C Pressure: 8.39 psi Gas saver: On Gas type: Helium	Purge flow: 50.0 mL/min Purge time: 1.00 min Total flow: 53.8 mL/min Saver flow: 20.0 mL/min Saver time: 2.00 min
Column	Capillary column Max temp 350°C Nominal length:30 m 30 m Mode: Constant flow Initial flow: 1 mL/min Outlet: MSD	Agilent 19091S-433, HP5-MS, 0.25 mm*30m*0.25um Nominal diameter: 250.00 um Nominal film thickness: 0.25 um Nominal initial pressure: 8.39 psi Average velocity: 36 cm/sec Outlet pressure: vacuum
Thermal	Aux 2 Initial temp: 280°C	Use: MSD transfer line heater
Injector	Sample washes: 1 Sample pumps: 4 Injection volume:1 uL Viscosity delay: 0 sec Plunger speed: fast	PreInj solvent A washes:4 (Solvent A=methanol) PreInj solvent B washes:4 (Solvent B=ethyl acetate) PostInj solvent A washes: 2 PostInj solvent B washes: 2 Pre- and post-injection dwell: 0.00 min

Table 18.3
MS operating conditions (Agilent model 5973)

Parameter	Setting	Details
Tune file	Atune.u	
Acquisition mode	Scan	
Solvent delay	3.00 min	
Low mass	40.0	
High mass	750.0	
Sample #	2	A/D samples: 4
MS Source temp	230°C	
MS quad temp	150°C	

5. As needed or on a schedule, the GC inlet liner and the gold-plated seal at the base of the inlet are changed. The number of specimens injected is directly related to the frequency with which the liner and gold seal must be replaced. Peak tailing, Rt shifts, and decreasing peak areas of drugs in the positive control are indicators that the liner and/or the gold seal may need to be changed.

6. As needed, the inlet end of the column is cut to remove the first few centimeters that become contaminated with strongly retained matrix components. The number of specimens injected is directly related to the frequency with which the column must be cut. Use of an inlet liner containing glass wool will protect the column and reduce the frequency for column cutting. However, the recovery of certain drugs is reduced when using a liner containing glass wool. If a clean liner does not resolve problems with peak shape, peak area, and Rt, the next logical step is to cut the column and replace the gold seal. The peak areas of oxycodone and codeine are particularly sensitive to a dirty inlet and column and to any active sites that are introduced by a poorly cut column end (*see* **Note** 7).

3.3. Data Analysis

1. Relative retention time (RRt) is calculated as drug Rt/internal standard Rt. A corrected RRt calculation takes into account the void volume of the column. The uncorrected RRt used in this method is less accurate but robustly distinguishes between closely eluting compounds and is easier to calculate.

2. As described in **Section 2.2**, building an in-house spectral library by injecting drug standards is the first phase of method development for broad spectrum drug screening by GCMS (*see* **Note 2**). Vendor software typically permits automated searching of more than one spectral library. A common approach would be to search the in-house library first and display a user-defined number of hits (e.g., 3), if the match factor is above a user-defined threshold (e.g., 75). If there are <3 hits from the in-house library above the threshold, the algorithm will then search the next user-defined library (e.g., MPW, NIST, or Wiley) and display hits with match factor >75, and then the next library, etc.

3. The data are typically analyzed with a summary and a detailed report. The summary report at a minimum displays the total ion chromatogram (TIC) and lists the peaks detected in Rt order with peak area and/or peak height. Additional features that make review easier are the inclusion of RRt, the best library match for the peak (if any) and its library match factor, and the percent of the peak area compared to the internal standard peak area.

4. The detailed report displays in Rt order the unknown spectrum for each peak and the spectrum from the library with the highest match factor and associated peak parameters (Rt, RRt, peak area, etc.).

5. The run is considered acceptable if
   (a) for the negative control: no drugs are identified other than the internal standard (*see* **Note 8**), and

   b) for the positive control: (i) if peak areas are above thresholds (established in-house), (ii) if RRts for target compounds in the control are within $+/-$ 0.02 of the reference compounds in the library, (iii) if all target compounds are identified with a match factor ≥ 90, (iv) if visual comparison of unknown and library spectra pass reviewing criteria (*see* **Note 2**), and (v) only target compounds are identified with a match factor ≥ 90 (*see* **Note 8**). A typical total ion chromatogram (TIC) for the positive control is shown in **Fig. 18.1**.

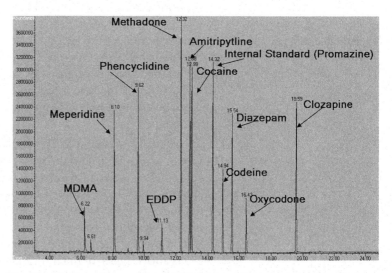

Fig. 18.1. The total ion chromatogram (TIC) is shown for an extract of the 11-drug positive control analyzed with this method.

6. Automated library search of EI spectra is usually robust and false positives are not common. Nonetheless, review of the match between each unknown spectra and library spectra by trained analysts using in-house developed criteria is necessary (*see* **Note 2**).

7. False negatives in patient specimens are more common, often caused by co-elution of drugs with each other, matrix components, and environmental contaminants.

Background subtraction of the interfering spectrum followed by a manual library search of the subtracted spectrum will often result in an acceptable spectrum with a match factor above 90. Using the extracted ion chromatogram (EIC or XIC) function of the software can be helpful to determine the best scans to select for subtraction (*see* **Fig. 18.2**).

8. False negatives may occur for certain drugs that have sparse (few ions) spectra, such as amitriptyline and methadone. The correct library match for the drug may be displayed but with a match factor below 90.

9. If target compounds with sparse spectra in the positive control are frequently matched with a factor <90 – review the library spectrum (may need to be updated), consider cleaning the ion source, check peak areas for adequate abundance. If all of these causes have been ruled out – it may be necessary to establish a threshold for the match factor of <90 for that drug. In that case, it may be appropriate to permit reporting of that drug in patient samples only if also detected by a second technique (e.g. LC-MSMS) or if a metabolite as well as the parent drug is detected, before reporting a positive (*see* **Note 2**).

10. A second common cause for false negatives is missed or incorrect integration. A peak must be integrated in order to trigger a library search. Drugs present at high concentrations, as may occur with an overdose, can have a bizarre peak shape (column overload) and may not be integrated. An example is shown in **Fig. 18.3**. A peak occurring as a shoulder on a larger peak may not be integrated separately and therefore not searched. It is common for the software to search the scan at the peak apex, after automatic background subtraction of a scan from the start or tail of the peak. Informed manual selection of spectra and search of the appropriate scan can usually produce the correct match with a >90 factor.

11. In some cases very high concentrations or very low concentrations of drug may result in distorted spectra that do not match well with automated or manual library search and are visually aberrant. Such samples can be diluted (with drug-free urine) or concentrated (evaporation of two extracts in a single tube) in an attempt to improve spectral characteristics.

12. Carryover is an ever-present hazard with broad spectrum drug screening as some drugs such as cocaine can be detected at a concentration of 50ng/mL but may be present in patient

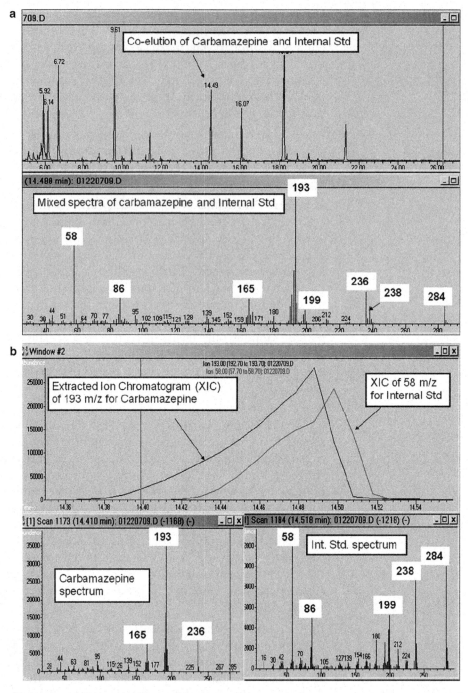

Fig. 18.2. **(A)** The top pane displays the TIC for a patient specimen extract in which the peaks of the internal standard (promazine) and of carbamazepine (an anticonvulsant) overlap and appear as one peak at 14.49 min. The lower pane shows the spectrum from the scan at the 14.49 min peak apex that contains fragment ions from both drugs. Neither compound can be identified. **(B)** The extracted ion chromatogram (XIC or EIC) function of the software was used to select and display the signal for a single, abundant, distinctive fragment ion for each drug so that the two overlapping peaks can be distinguished as seen in the upper pane. Scans for each compound can then be manually selected at the start (14.41 min for carbamazepine) and end (14.52 min for promazine internal standard) of the 14.49 min peak in order to display in the lower pane good spectra for each drug that have library search match factors >90.

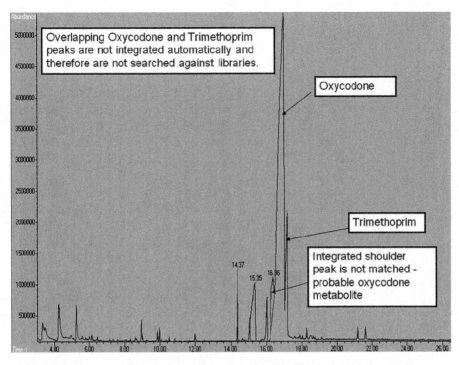

Fig. 18.3. The TIC for a patient specimen extract is shown with large, partially overlapping peaks for oxycodone and trimethoprim. The abnormal peak shapes were not recognized by the automated integration algorithm. Integration of a peak is the trigger for the library search, so these two peaks were not searched or identified or listed in the summary report. Careful review of the TIC and manual selection and library searching of scans from these peaks produced spectra with match factors >90 for both drugs. A false negative will occur if a missed integration is not noticed.

urines at concentrations >1 mg/mL (1,000,000 ng/mL). Helpful tactics to avoid reporting false positives caused by carryover include,

a. Rigorous pre- and post-injection wash of the autosampler needle (*see* **Table 18.2**).

b. Evaluation at validation and subsequently at intervals defined in the procedure of carryover for drugs frequently found at high concentrations in urine with this protocol – for example, cocaine, oxycodone, methadone, and EDDP (methadone metabolite).

c. Review of all samples in a batch by one analyst in retention time order.

d. Use of a review form that correlates data from immunoassays, broad spectrum drug screening, and targeted confirmation assays so that discordant results (e.g., broad spectrum screen positive for cocaine, but cocaine metabolite immunoassay is negative) are easily observed.

4. Notes

1. Minimize exposure of organic solvents to any form of plastic (tubes, vials, pipets, transfer pipets, parafilm, pipet tips, etc.). Phthalate contamination from plastics is common and can be significant. Use polytetrafluoroethylene (PTFE)-lined caps for glass containers and PTFE-lined septa for glass ALS vials. Use positive displacement pipets with glass capillaries to accurately measure stock solutions of drug standards in methanol. Glassware, water, solvents, sodium bicarbonate, drug standard solutions, GCMS columns and consumables (inlet liners, ferrules, gold seals, etc.) that are exposed to air, soap residue, dust, fingerprints, latex gloves, and other environmental contaminants can also cause major interferences.

2. Informed review and selection of spectra from the analysis of reference drug standards is critical for building a robust in-house library (7–11). Equally important is establishing appropriate criteria for reviewing and accepting spectra from library search reports for patient samples. There is not a single consensus standard about the minimum number of ions at a defined percentage of the base peak that must be present in a full scan EI spectra (e.g. 3 or 4) for definitive identification. References *7-11* provide some guidance. Laboratories should have a policy for spectral review, for training of clinical laboratory scientists in analysis of library search reports, and for retrospective review of every QC- and patient-report by a supervisor or lead CLS with advanced training and experience. Traditional forensic practice for broad spectrum screening requires that a drug be identified by two different chemical techniques in order to report a positive. For clinical emergency toxicology, some toxicologists would recommend a two-tier policy; reporting as "presumptive positive" when a drug has been identified with one technique and as "confirmed positive" when identified with two techniques or if both a parent drug and metabolite are present. Laboratories should create and periodically update a list of drugs and metabolites that can be reported using their own procedure, based on consultation with clinicians and the chemistry, chromatography, mass spectra, pharmacology, library search characteristics, and clinical toxicology of the compounds.

3. Positive displacement pipets are necessary to accurately measure small volumes of organic solvent. They can be highly precise if used correctly but there are more operator-dependent variables than with air displacement pipets. Careful adherence to the manufacturer's instructions is important, including appropriate wiping of the capillary before, and of

the capillary and Teflon tip after dispensing and touching the end of the capillary to the receiving vessel to remove the last of the sample.

4. Lengthy or too vigorous mixing can cause an emulsion, preventing good separation of organic and aqueous layers. Too little mixing causes poor and irreproducible extraction recovery. Use of a timer and a rocking mixer for 5 min consistently will minimize between analyst variance.

5. The drying step is the phase of testing that is most prone to operator error. Optimization for the evaporator module in use, and then rigorous compliance with the maximum temperature and length of time that the tube is heated is critical for acceptable recovery of heat labile compounds. Peak areas for MDMA, meperidine, and phencyclidine (PCP) decrease rapidly with exposure to heat in a dried tube, for example, we observed a 30–40% decrease in peak areas for these three compounds when extracts remained in the water bath for 20 min versus 15 min or at temperatures >35°C.

6. Hydrogen may be used as a carrier gas instead of helium, at lower cost and with a shorter run-time that yields similar resolution. Consult the GCMS manufacturer for appropriate safeguards before using hydrogen. Rts will change and reference compound Rts, RRts, and spectra would need to be re-verified if switching from helium to hydrogen.

7. The ideal concentration for drugs in the positive control is only 20–30% above their limits of detection (variable by drug) to better observe any extraction errors and loss of GCMS sensitivity.

8. This method detects caffeine to the degree that it may prove difficult to obtain a caffeine-free blank urine. Identification of cholesterol is also ubiquitous. Phthalates with characteristic retention times, antiseptics (e.g., chloroxylenol), soap (e.g. triclosan), and numerous small-to-moderate sized unidentified peaks – presumably environmental contaminants of the extraction – routinely occur in drug-free urine even with good compliance to trace analysis precautions.

References

1. Maurer HH. (2006) Hyphenated mass spectrometric techniques-indispensable tools in clinical and forensic toxicology and in doping control. *J Mass Spectrom* **4**, 1399–413. Review.

2. Polettini A. (1999) Systematic toxicological analysis of drugs and poisons in biosamples by hyphenated chromatographic and spectroscopic techniques. *J Chromatogr B Biomed Sci App* **733**, 47–63. Review.

3. Greller HA, Barrueto F Jr. (2004) Comprehensive Drug Screening. *Emerg Med J* **21**, 646.

4. Current CLIA Regulations, CDC http://wwwn.cdc.gov/clia/regs/toc.aspx.

5. Peters FT, Drvarov O, Lottner S, Spellmeier A, Rieger K, Haefeli W, Maurer HH. (2009) A systematic comparison of four different workup procedures for systematic toxicological analysis of urine samples using gas chromatography – mass spectrometry. *Anal Bioanal Chem* **393**, 735–45.

6. NIST Scientific and Technical Databases, NIST/EPA/NIH Mass Spectral Library (NIST 08) http://www.nist.gov/srd/nist1.htm.

7. Stein SE, Heller DN. (2006) On the risk of false positive identification using multiple ion monitoring in qualitative mass spectrometry: large scale intercomparisons with a comprehensive mass spectral library. *J Am Soc Mass Spectrom* **17**, 823–85.

8. Aebi B, Berhnard W. (2002) Advances in the use of mass spectral libraries for forensic toxicology. *J Anal Toxicol* **26**, 149–56.

9. Bethem R, Boison J, Gale J, Heller D, Lehotay S, Loo J, Musser S, Price P and Stein S. (2003) Establishing the fitness for purpose of mass spectrometric methods. *J Am Soc Mass Spectrom* **14**, 528–41.

10. Chase D et al. (2007) Mass Spectrometry in the Clinical Laboratory: General Principles and Guidance; Approved Guideline C50-A. *Clinical and Laboratory Standards Institute*, Wayne, PA.

11. Stimpfl T, Demuth W, Varmuz K, Vycudilik W. (2003) Systematic toxicological analysis:computer assisted identification of poisons in biological materials. *J Chromatrogr B* **789**, 3–7.

Chapter 19

Broad Spectrum Drug Screening Using Liquid Chromatography-Hybrid Triple Quadrupole Linear Ion Trap Mass Spectrometry

Judy Stone

Abstract

Centrifuged urine, internal standard (promazine), and ammonium formate buffer are mixed in an auto-sampler vial to achieve a 10-fold dilution of the specimen. Without additional pretreatment, 10 µL of the sample is injected onto a C18 reverse phase column for gradient analysis with ammonium formate/acetonitrile mobile phases. Drugs in the column eluent become charged in the ion source using positive electrospray atmospheric pressure ionization. Pseudomolecular drug ions are analyzed by a hybrid triple quadrupole linear ion trap mass spectrometer operated with a 264-drug selected ion monitoring (SRM) acquisition method that includes an information-dependant acquisition (IDA) algorithm.

Key words: Broad Spectrum Drug Screening, Urine Dilution, Positive Mode Electrospray, Hybrid Triple Quadrupole Linear Ion Trap, Liquid Chromatography-Tandem Mass Spectrometry (LC-MSMS).

1. Introduction

The toxicologic process of testing body fluids for the widest possible range of drugs and poisons has been described as Comprehensive Drug Screening (CDS), General Unknown Screening (GUS), and Systematic Toxicological Analysis (STA) (1–2). Regardless of the name applied, no method has the capability to detect all possible chemicals of interest. The degree to which different chromatographic-mass spectrometry techniques approach the goal depends on sample preparation, chromatographic ionization, and mass analyzer characteristics (3, 4, 5). Traditionally, the use of two orthogonal methods – gas chromatography-mass spectrometry

U. Garg, C.A. Hammett-Stabler (eds.), *Clinical Applications of Mass Spectrometry*, Methods in Molecular Biology 603, DOI 10.1007/978-1-60761-459-3_19, © Humana Press, a part of Springer Science+Business Media, LLC 2010

(GCMS) and liquid chromatography with an ultraviolet detector (LC-UV) has identified the broadest possible menu of drugs (4, 5). GCMS has higher resolution, greater specificity, and lower detection limits than does LC-UV but requires hydrolysis and derivatization steps in order to detect polar and/or heat-labile compounds, such as glucuronide/sulfate metabolites (4, 5).

The advent of liquid chromatography/atmospheric pressure ionization (API) coupled to tandem mass spectrometry appeared to offer the potential for improving comprehensive screening (2, 3). Unfortunately, the broad range of compounds that can be ionized with electrospray ionization sources (ESI) includes many endogenous compounds and environmental contaminants. Even with selective sample preparation, a highly complex mélange of ions occurs when biological samples such as urine or serum undergo ESI (2, 3). As a result, it has proved challenging to tease out all ions that represent compounds of interest from the overwhelming background noise. The enhanced selectivity of tandem mass analysis that uses a "multi-targeted" acquisition strategy, as described in this method, is one approach for dealing with this problem (1).

In this method, drugs in the column eluent become charged in the ion source using positive electrospray atmospheric pressure ionization. Pseudomolecular drug ions are analyzed using a hybrid triple quadrupole linear ion trap mass spectrometer operated with a 264-drug selected ion monitoring (SRM) acquisition method that includes an information-dependant acquisition (IDA) algorithm. The IDA algorithm will select the three most abundant precursor ions from the list of 264 SRM transitions for each signal with a peak height above a defined threshold. In the subsequent dependant scan(s), the selected precursor ion(s) are passed, one m/z per scan, through the first quadrupole (Q1), fragmented with three different collision energies in the collision cell (Q2), and the resulting product ions are collected in the third quadrupole (Q3) operating as a linear ion trap (LIT). The ions are scanned out of the LIT to produce an enhanced (EPI) or full mass range product ion spectrum (1). Each EPI spectrum is the product of precursor ions of a defined m/z and is linked to a peak defined by the associated SRM transition from the survey scan. The spectra are automatically searched against a 1262-drug library for identification.

A multi-targeted method looks for only those compounds previously defined and ignores all other ions resulting from the sample matrix or as environmental background noise (1). An advantage of the approach is that there is a greater degree of certainty that the targeted compounds, when present, will be detected (6), but it should be recognized that any drug not included in the defined target list will remain undetected.

2. Materials

2.1. Reagents and Buffers (see Note 1)

1. 1 mM Ammonium formate. Stable for 1 year at 4°C.

2. Specimen Preparation Buffer: 10 mM ammonium formate, pH 3.00 +/− 0.03. To avoid contaminants, do *not* expose the buffer to the pH meter probe to check/adjust pH. Test the pH of an aliquot, add formic acid or 1 mM ammonium formate to the volumetric flask, if needed, to adjust pH, remix, and test a new aliquot. Filter using a 0.22 μm pore size membrane under vacuum. Stable at 15–30°C for 3 months.

3. Mobile Phase A: 0.5 mM ammonium formate buffer, pH 3.00 +/− 0.03. Use the same precautions to check/adjust pH as for the specimen preparation buffer. Filter using a 0.22 μm pore size membrane under vacuum. Stable at 15–30°C for 3 months.

4. Mobile Phase B: Acetonitrile: Specimen Preparation Buffer (90:10). Measure the acetonitrile and buffer in separate graduated cylinders and then mix. Filter using a 0.22 μm pore size membrane under vacuum. Stable at 15–30°C for 3 months.

5. Pump A Wash Solution: HPLC water:acetonitrile (50:50). Filter using a 0.22 μm pore size membrane under vacuum. Stable at 15–30°C for 1 year.

6. Pump B Wash Solution: 100% acetonitrile. Filter using a 0.22 μm pore size membrane under vacuum. Stable at 15–30°C for 1 year.

7. Human drug-free urine (UTAK Laboratories, Valencia, CA).

2.2. Library Reference Standards (see Note 2)

1. Drug standards, 1.0 and 0.1 mg/mL in methanol or powders, were purchased from Cerilliant (Round Rock, TX), Grace Davison/Alltech (Deerfield, IL), Sigma-Aldrich (St. Louis, MO), and Lipomed Inc. (Cambridge, MA). Standards are prepared by spiking desired compounds into drug-free urine at concentrations of 5,000 ng/mL and testing to confirm or acquire spectra for the library.

2.3. Internal Standard and Quality Controls (see Note 1)

1. Promazine internal standard stock solution: Prepare by weighing 11.28 mg of promazine HCl (Sigma-Aldrich, St. Louis, MO). Dissolve in methanol in a 10 mL volumetric flask. Mix well. Store in an amber vial with a PTFE-lined cap. Stable at −8 to −20°C for 2 years.

2. Promazine internal standard working solution, 99 mg/L in acetonitrile: Prepare by pipetting 50 μL of the internal standard stock solution and 5 mL of acetonitrile into an amber vial with a PTFE-lined cap. Mix well. Stable at 15–30 °C for 3 months.

3. Quality control samples: Drug-free urine (blank or negative control) and a commercial custom-positive urine control containing 11 drugs (UTAK 11 Drug Custom Urine CDS QC; UTAK Laboratories, Valencia, CA) are used to validate performance daily. The drugs in the positive control, their concentrations, typical Rt, RRt, and peak areas are shown in **Table 19.1**. The acceptable thresholds and variances for the peak areas of the drugs in the control are established in-house (**Fig. 19.4**).

Table 19.1
Drugs present, their nominal concentrations, and typical Rt, RRt, and peak areas for the positive control analyzed with the method as described

Drug	Concentration (ng/mL)	Typical Rt (min) and RRt	Typical peak area__e+5
1. Amitriptyline	500	7.5 and 1.04	1.2
2. Clozapine	500	6.6 and 0.91	2.6
3. Cocaine	500	6.0 and 0.83	4.9
4. Codeine	500	4.4 and 0.61	0.3
5. Diazepam	500	9.4 and 1.30	0.5
6. Meperidine (Pethidine)	500	6.0 and 0.84	3.2
7. Methadone	500	7.5 and 1.04	8.3
8. dl-Methadone Primary Metab. (EDDP)	100	7.1 and 0.99	0.6
9. MDMA	1,000	4.9 and 0.68	4.1
10. Oxycodone	250	4.7 and 0.65	0.8
11. Phencyclidine	500	6.5 and 0.90	2.5

2.4. Supplies

1. Autosampler vials and caps with polytetrafluoroethylene (PTFE)-lined septa (2 mL Amber wide-opening glass vial with screw cap) (Agilent Technologies, Santa Clara, CA).

2. Amber vials with PTFE-lined caps for drug and internal standard solutions (Grace Davison/Alltech, Deerfield, IL).

3. HPLC columns (XTerra® MS C18, 3.5 μm, 2.1 × 100 mm) and guard columns (XTerra® MS C18, 5 μm, 2.1 × 10 mm) (Waters Co., Milford, MA).

4. Pre-column filters (0.5 μm) (MAC-MOD Analytical, Inc., Chadds Ford, PA).

2.5. Equipment

1. An HPLC system compatible with Life Technologies/ Applied Biosystems Analyst software (e.g., Agilent Model 1200) with (a) microdegasser, (b) binary pump with solvent switching valve, (c) temperature-controlled autosampler, (d) thermal column compartment with column switching valve (Agilent Technologies, Santa Clara, CA).

2. A 3200 QTrap® hybrid triple quadrupole linear ion trap mass spectrometer (Life Technologies/Applied Biosystems, Foster City, CA).

3. Cliquid 1.0 Drug Screen and Quant software (Life Technologies/Applied Biosystems, Foster City, CA).

3. Methods

3.1. Stepwise Procedure

1. Centrifuge samples at ~3,000 × g for 10 min to remove particulate.

2. Label autosampler (ALS) vials for the negative control (drug-free urine), the positive control (UTAK 11 Drug Custom Urine CDS QC), and each unknown sample.

3. Add 100 μL of urine to the ALS vial.

4. Add 50 μL of internal standard working solution to the ALS vial.

5. Add 850 μL of specimen preparation buffer, cap, and vortex.

6. Place vials in the ALS tray and program the batch using Cliquid® 1.0 Drug Screen and Quant software.

3.2. Instrument Operating Conditions

1. LC column used: XTerra® MS C18, 3.5 μm, 2.1 × 100 mm with cartridge guard column (XTerra® MS C18, 5 μm, 2.1 × 10 mm) and pre-column filter.

2. MS source used: Turbo V ion source in Positive ESI mode.

3. The LC and MS operating conditions, modified from those in the drug screening method supplied with the Cliquid® 1.0 software package are given in **Tables 19.2** and **19.3**.

4. Prior to sample injection perform equilibration procedure to insure reproducible Rt and peak areas for early eluting compounds (*see* **Note 3**).

5. Follow equilibration procedure with two "LC primer" injections that are directed to waste via the divert valve (instead of to the source). These improve the consistency of Rt of early eluting compounds (*see* **Note 4**).

6. Avoid exposing the XTerra® MS C18 stationary phase to low %B (e.g., 95:5 Mobile Phase A:Mobile Phase B) (*see* **Note 5**).

Table 19.2
LC Operating Conditions – Note that the 1200 system must be reconfigured for 2.1 mm i.d. columns from the default installation configuration to minimize extra-column volume affecting gradient delay, re-equilibration time, and peak width (consult Agilent Technical Support). The LC parameters shown are modified from the drug screening method parameters supplied with Cliquid® 1.0 Drug Screen and quant software

Component	Parameter	Details			
Binary pump	Gradient program	Time	Flow Rate (µL/min)		%B
		0	300		5
		11	300		100
		13	350		100
		15	350		5
		19	300		5
Autosampler	Temperature	4°C			
Autosampler	Volume injected	10 µL (use 1 µL to identify very large peaks [>1 e+6 area] or 30 µL for small peaks [<5 e+3 area])			
Autosampler	Wash program	DRAW default amount from sample, default speed, offset WASH NEEDLE in location 91, 2 times INJECT REMOTE start pulse, duration 40*12.5 msec WAIT 0.45 min VALVE bypass			
Thermal column compartment	Temperature	25°C			
Divert valve	Program	Time 0 = waste, Time 1 min = source, Time 15 min = waste			

7. Change the guard column every 500 injections or if changes are noted in Rts, peak areas, or peak shapes.

8. At the completion of each analytical run, wash the column/system to remove all traces of the acidic buffer. Place the column/system in standby only after a high-percent acetonitrile milieu is achieved. This protocol has been found to extend the lifetime of the column to several thousand injections (*see* **Note 6**).

9. Cliquid® 1.0 Drug Screen and Quant (with Analyst 1.4.2) software is required for the automated library search and report of the SRM peaks and EPI spectra generated with this method. User-defined parameters for this method that were modified from those in the drug screening method supplied with the Cliquid® Drug Screen and Quant software package were: "Report a Library match when Purity exceeds 50%."

Table 19.3
MS operating conditions – note that the MS parameters shown are modified from the drug screening method parameters supplied with Cliquid® Drug Screen and Quant software

Parameter	Setting	Details
Synchronization Mode & Duration	LC Sync 19 min	n/a
MRM experiment	Q1 mass, Q3 mass, CE & CEP for 264 transitions	Cliquid® users may consult the author re edits made to the Cliquid® Drug Screening MRMs (SRMs)
MRM experiment	Ionization mode & dwell time	Positive ESI and 5 msec
MRM experiment	Advanced MS	Unit resolution for Q1 and Q3
MRM experiment	Compound	DP 40, EP 10, CXP 5
MRM experiment	Source/Gas	Cur 20, CAD high, IS 4000, Temp 500, GS1 40, GS2 55, Ihe ON
IDA criteria	First level criteria	Select 1-3 most intense peaks which exceed 1,000 cps, exclude former target ions after 3 occurrences for 15 sec
3 EPI experiments	MS	Scan speed 4,000 da/s, mass range 50 to 700
3 EPI experiments	Advanced MS	Q0 Trap ON, Fixed LIT fill time of 50 msec, Q1 unit resolution
3 EPI experiments	Compound	CE 35, CE spread 15

3.3. Data Analysis

1. Analyze the data using Cliquid 1.0 and Analyst 1.4.2 software (Life Technologies/Applied Biosystems; Foster City, CA). Use both (1) a modified CES (collision energy spread) Best Candidate Detailed report and (2) a modified CES Confirmation Summary report. A portion (extracted ion chromatogram and library search summary) from the Best Candidate Detailed report for a patient sample analyzed with this method is shown in **Fig. 19.1**.

2. The run is considered acceptable if (1) for the negative control: No drugs are identified other than the internal standard, caffeine, theophylline (caffeine metabolite), and theobromine (caffeine metabolite) and (2) for the positive control: a) if peak areas are above threshold (established in-house), (b) if RRts are within +/−0.04 of the reference compound in the library, (c) if purity – from automated (Cliquid®) or manual (Analyst) library search – is ≥70, (d) if visual comparison of

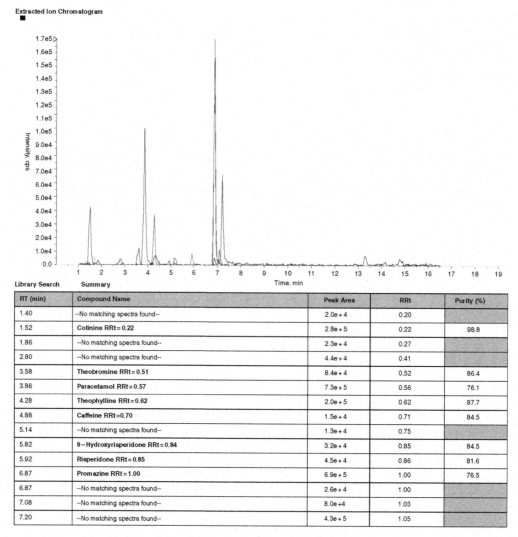

Fig. 19.1. Extracted ion chromatogram and library search summary from a Cliquid® 1.0 CES Best Candidate Detailed report (format modified in-house) for a patient specimen analyzed with this method.

RT (min)	Compound Name	Peak Area	RRt	Purity (%)
1.40	--No matching spectra found--	2.0e + 4	0.20	
1.52	Cotinine RRt = 0.22	2.8e + 5	0.22	98.8
1.86	--No matching spectra found--	2.3e + 4	0.27	
2.80	--No matching spectra found--	4.4e + 4	0.41	
3.58	Theobromine RRt = 0.51	8.4e + 4	0.52	86.4
3.86	Paracetamol RRt = 0.57	7.3e + 5	0.56	76.1
4.28	Theophylline RRt = 0.62	2.0e + 5	0.62	87.7
4.88	Caffeine RRt = 0.70	1.5e + 4	0.71	84.5
5.14	--No matching spectra found--	1.3e + 4	0.75	
5.82	9-Hydroxyrisperidone RRt = 0.84	3.2e + 4	0.85	84.5
5.92	Risperidone RRt = 0.85	4.5e + 4	0.86	81.6
6.87	Promazine RRt = 1.00	6.9e + 5	1.00	76.5
6.87	--No matching spectra found--	2.6e + 4	1.00	
7.08	--No matching spectra found--	8.0e +4	1.03	
7.20	--No matching spectra found--	4.3e + 5	1.05	

unknown and library EPI spectra passes reviewing criteria (**Note 6**), and (e) only expected compounds (11 spiked drugs and internal standard) are found.

3. Visual review of the match between each unknown spectra and library spectra using in-house developed criteria is necessary to avoid false positives (*see* **Note 7**) (7).

4. Manual library search (using Analyst Explore function) is necessary to avoid false negatives (*see* **Note 8**).

5. If spectra are not matched and appear to be of poor quality because a peak is too small (<5 e+3 area) or too large (>1 e+6 area), the sample can be reinjected with a larger (30 μL) or smaller (1 μL) injection volume than the original (10 μL).

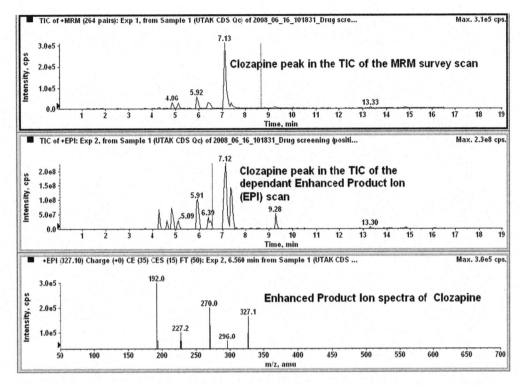

Fig. 19.2. Data acquired for the positive control using the SRM(MRM)-IDA-EPI acquisition method described and displayed with Analyst Explore function: *Top Pane* = Total ion chromatogram (TIC) of the SRM (MRM) acquisition; *Next Pane* = TIC of one EPI acquisition; *Next pane* = Clozapine spectrum manually selected at 6.56 min from the TIC of the EPI.

6. An example of an extracted ion chromatogram (XIC), an EPI total ion chromatogram (TIC of the EPI), and an EPI spectrum for one of the drugs (clozapine) in the positive control as displayed in Analyst Explore is shown in **Fig. 19.2**. An example of the library search results from Analyst Explore for the clozapine spectra is shown in **Fig. 19.3**.

7. Carryover was evaluated. Infrequent peaks with area >1 e+6 caused carryover to the next specimen. Carryover appeared to be drug dependent.

8. Results for limit of detection studies are shown in **Table 19.4**.

9. Variance of the RRts, purity, and peak areas for the positive control drugs are shown in **Fig. 19.4**.

10. Sixty-patient urine samples were tested with five broad spectrum screening methods: (1) this method, (2) an in-house electron ionization (EI) GCMS screening method, (3) the REMEDI HS®-LC-UV automated drug screening system from Bio-Rad Laboratories (Hercules, CA), (4) the Waters (Milford, MA) single quadrupole LC-MS (ZQ™ Mass Detector with MassLynx™ software), (5) the Thermo Scientific (Waltham, MA) Ion Trap LC-MSMS (Finnigan™

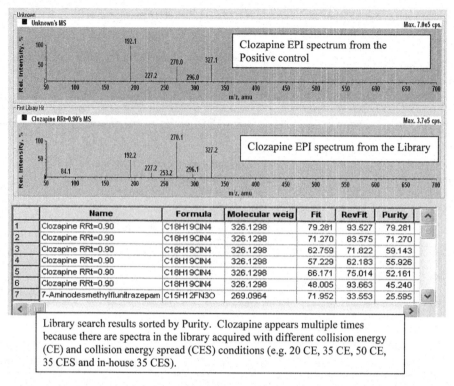

Fig. 19.3. Library search report (in Analyst) for the clozapine spectra in the positive control.

Table 19.4
Thirty-seven of the 264 compounds in the method were evaluated for limit of detection at 1,000 ng/mL and 250 ng/mL concentrations using dilutions of eight commercial drug mixes into drug-free urine. The mixed urine standards were analyzed with the procedure as described. (#) indicates the number of mixes containing the drug

Detection Verified to the Concentrations shown (ng/mL)

Drug	Drug mix	POS at	Drug mix	POS at	Drug mix	POS at
Alprazolam	1	250				
Amitriptyline (3)	2	250	6	250	4	250
Amphetamine (2)	2	250	8	250		
Caffeine (3)	3	>1000	7	>1000	8	>1000
Carbamazepine	3	250				
Clonazepam	1	250				
Cocaine (2)	2	250	3	250		

(continued)

Table 19.4 (continued)

Detection Verified to the Concentrations shown (ng/mL)

Drug	Drug mix	POS at	Drug mix	POS at	Drug mix	POS at
Codeine (2)	2	250	5	250		
Desipramine (3)	2	250	3	250	6	250
Dextromethorphan	4	250				
Diazepam	1	250				
Diphenhydramine	5	250				
Doxepin (2)	2	250	6	250		
EDDP	3	250				
Ephedrine	5	250				
Flunitrazepam	1	250				
Imipramine (3)	2	250	6	250	8	250
Lidocaine (2)	1	250	4	250		
Lorazepam	1	250				
Metamphetamine (2)	2	250	8	250		
Methadone (2)	2	250	3	250		
Methaqualone (3)	2	250	3	250	7	250
Methocarbamol	4	>1000				
Nitrazepam	1	250				
Nortriptyline (2)	6	250	5	250		
Oxazepam	1	250				
Oxycodone	2	250				
Paracetamol (acetaminophen)	7	250				
Pethidine (Meperidine)	2	250				
Phencyclidine (2)	2	250	8	250		
Phenylephrine	4	>1000				
Phenylpropanolamine	8	>1000				
Protriptyline	6	250				
Temazepam	1	250				
Triamterene	4	250				
Trimipramine	6	250				

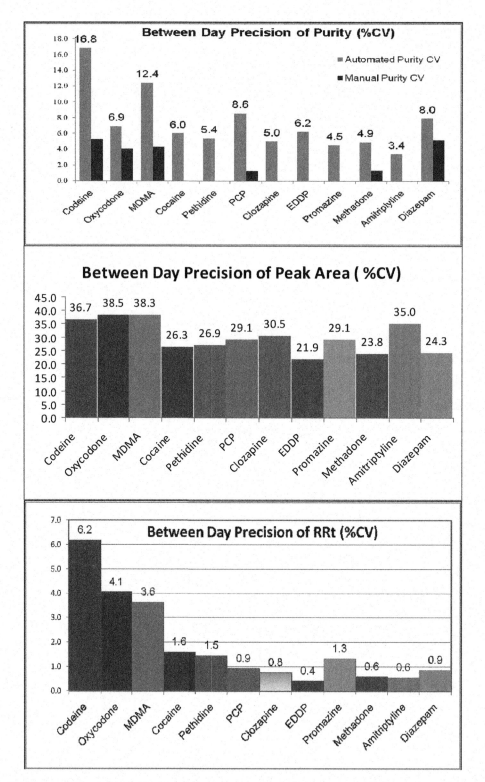

Fig. 19.4. Variance (%CV, *n*=37 days) for the positive control peak areas, RRts, and purity. The drugs in the positive control material are shown in Rt order (*left*, earliest to *right*, latest).

LXQ™ with ToxID™ software) and with 8 urine immunoassays for abused drugs (positives for six of the immunoassays – amphetamine, barbiturate, EDDP, opiate, oxycodone, and PCP – were tested with in-house-targeted LC-MSMS or GCMS confirmation methods). The results showed that the method described in this chapter detected almost twice the number of drugs that were found by Remedi® LC-UV, and within the subset of 60 drugs detectable by all three systems, this method detected more verified drugs than did any other method. Results of Patient Specimen Comparisons between five broad spectrum drug screening methods: LC-UV (Remedi®), GC-MS, one LC-MS, and two LC-MSMS. The Waters LC-ZQ-MS is "Single Quad." The LC-MSMS method described in this chapter is "QTRAP®" and the Thermo Scientific LXQ LC-MSMS is "Ion Trap." *See* **Tables 19.5**, **19.6**, and **19.7** for details.

Table 19.5
Drugs detected by each method and verified by another

Analyzer	Total number of drugs detected	Number of drugs detected in subset*	Percent detected by second method	Percent identified by second method or medical record review
LC-UV (Remedi®)	145	110	97.3	99.1
GC-MS	216	157	97.5	99.4
Single Quad (LC-MS)	193	157	90.4	92.4
QTrap® (LC-MSMS)	286	200	95.0	97.0
Ion trap (LC-MSMS)	331	217	81.1	85.3

*the subset included 60 drugs that all methods were capable of detecting.

Table 19.6
Percentage of Drugs detected by Remedi® and GC-MS that were also found by LC-MS or MSMS

Analyzer	Compared to Remedi (%)	Compared to GC-MS (%)
Single Quad (LC-MS)	75.5	65.0
QTrap® (LC-MSMS)	95.5	88.8
Ion trap (LC-MSMS)	93.6	77.7

Table 19.7
Comparison between LC-MS/MSMS systems only

Analyzer	Percent of Drugs not detected*
Single quad (LC-MS)	32.9
QTrap® (LC-MSMS)	9.4
Ion trap (LC-MSMS)	17.2

*compared to other LC-MS or MSMS systems only.

4. Notes

1. Avoid exposure of all organic solvents used in the method (in standards, internal standards, mobile phases, pump washes, etc.) to all forms of plastic (containers, vials, pipets, transfer pipets, parafilm, pipet tips, etc.). Phthalate contamination from plastics is common and can be significant. Use polytetrafluoroethylene (PTFE)-lined caps for containers and PTFE-lined septa for ALS vials.

2. This is a qualitative method, no calibrators are used and no concentrations are reported. Instead of a lower limit of quantitation, the criteria necessary for reporting a drug as "detected" or "positive" are (a) peak area must be ≥5 e+3, (b) purity (library match factor for EPI spectra) must be ≥70 and the spectral match passes visual review criteria, (c) relative retention time (RRt) must be within +/− 0.04 or 4%, whichever is larger of the library standard RRt, and (d) the drug must not be present in the blank (drug-free urine) control. RRts and retention times (Rts) were established by spiking drug standards into drug-free urine at a concentration of 5,000 ng/mL followed by analysis using the procedure described. From this same analyses of drug standards EPI spectra for drugs were reviewed and appended to existing library entries (drugs present in library included with Cliquid® 1.0 Drug Screen and Quant software from Life Technologies/Applied Biosystems, Santa Clara, CA) or new library entries were created (drugs not in Cliquid® 1.0 library).

3. LC equilibration from standby is performed as follows: with the MS off-line, program LC to run 50% mobile phase B at a flow rate of 300 μL/min for 10 min. Continue equilibrating for another 15 min using 10% B at a flow rate of 300 μL/min.

4. The "LC primer" injections are performed after LC equilibration as follows: 2 min gradient from 5 to 100% B at 350 μL/min; 2 min gradient from 100 to 5% B at 350 μL/min;

4 min re-equilibration at 5% B at 350 μL/min. A 20-min MS equilibration is programmed to insure thermal equilibrium in the ion source.

5. C18 stationary phases may "collapse" or "dewet" when exposed to a low percent organic component in the mobile phase (e.g., ~ <10%). This phenomenon can cause Rt shifts and bizarre peak shapes. This problem occurs if the column used for this method is exposed to <10% Mobile Phase B without flow or in isocratic conditions of <10% B. It does not occur with a continuously cycling gradient starting at 5% B. The equilibration and "primer" injection procedures described in **Notes 2 and 3** are successful at preventing this phenomenon.

6. The "column/system wash" is performed at the end of a run as follows: Switch solvent select valve to Channel 2 (Pump Wash A & B); 20 min at 50% Pump Wash B; 300 μL/min. Solvent conditions in standby (no flow): 25:75 HPLC water:acetonitrile.

7. In-house EPI Spectra Review Criteria (7): (a) The four most abundant ions in the library spectra must be present in the unknown spectra. (b) If there are ions in the unknown that are NOT present in the library spectra – they must be less abundant than the two most abundant ions in the unknown (and based on a. – those two most abundant ions will also be present in the library spectra), OR (c) if there are <5 ions in the library spectra then the three most abundant of these must be present in the unknown spectra, AND (d) any ions in the unknown that are NOT present in the library spectra must be of less abundant than the three matched ions in the unknown, AND (e) there must be other evidence that supports a positive result – such as a positive result with GCMS or immunoassay or for a metabolite of the drug, by any method.

8. Manual search is necessary for all spectra with purity between 50 and 70 and for all peaks with areas ≥1 e+6 and "– no matching spectra found– " on the Best Candidate Detailed Report. The Confirmation Summary Report lists the drug/ MRM transitions for every integrated peak, matched or unmatched, and can be useful to identify candidate peaks for manual searching. The most common cause of false negatives is from too many ions in the LIT causing space charging (high peak areas, e.g., ≥1 e+6). For such large peaks, a spectrum can usually be selected manually in Analyst from the tail of the peak (lower abundance of ions in the LIT) or from the second or third EPI experiment and matched with a purity ≥70.

References

1. Mueller CA, Weinmann W, Dresen S, Schreiber A, Gergov M. (2005) Development of a multi-target screening analysis for 301 drugs using a QTrap liquid chromatography/tandem mass spectrometry system and automated library searching. *Rapid Commun Mass Spectrom* **19,** 1332–1338.

2. Maurer HH. (2006) Hyphenated mass spectrometric techniques-indispensable tools in clinical and forensic toxicology and in doping control. *J Mass Spectrom* **4,** 1399–1413. Review.

3. Marquet P. (2002) Is LC-MS suitable for a comprehensive screening of drugs and poisons in clinical toxicology? *Ther Drug Monit* **24,** 125–133. Review.

4. Saint-Marcoux F, Lachâtre G, Marquet P. (2003) Evaluation of an improved general unknown screening procedure using liquid chromatography-electrospray-mass spectrometry by comparison with gas chromatography and high-performance liquid-chromatography–diode array detection. *J Am Soc Mass Spectrom* **14,** 14–22.

5. Sauvage FL, Saint-Marcoux F, Duretz B, Deporte D, Lachâtre G, Marquet P. (2006) Screening of drugs and toxic compounds with liquid chromatography-linear ion trap tandem mass spectrometry. *Clin Chem* **52,** 1735–1742.

6. Lynch K, et al. Performance evaluation of three liquid chromatograph-mass spectrometry methods for broad spectrun drug screening. *Clin Chem* Submitted (July 2009).

7. Chase D et al. (2007) Mass Spectrometry in the Clinical Laboratory: General Principles and Guidance; Approved Guideline **C50-A.** *Clinical and Laboratory Standards Institute*, Wayne, PA.

Chapter 20

High Sensitivity Measurement of Estrone and Estradiol in Serum and Plasma Using LC-MS/MS

Mark M. Kushnir, Alan L. Rockwood, Bingfang Yue, and A. Wayne Meikle

Abstract

Measurement of low concentrations of estrogens, encountered in pre-pubertal children, men, and post-menopausal women, is important for numerous clinical applications. We describe a method for high sensitivity analysis of estrogens that uses two-dimensional chromatographic separation and tandem mass spectrometry detection. Aliquots of serum or plasma samples are combined with stable isotope-labeled internal standard and estrogens are extracted with methyl t-butyl ether. The solvent is evaporated, estrogens derivatized to form dansyl derivatives, and the samples are analyzed. Quantitation is performed using triple quadrupole mass spectrometer equipped with electrospray ion source using positive ion mode ionization and multiple reaction monitoring acquisition.

Key words: Estrogens, estradiol, estrone, derivatization, liquid chromatography, mass spectrometry, tandem mass spectrometry.

1. Introduction

Estrogens (EST) are hormones responsible for the development and maintenance of female secondary gender characteristics, reproductive function, regulation of the menstrual cycle, and maintenance of pregnancy. Accurate methods with high sensitivity are necessary when EST measurements are used in the diagnosis of hormone-related disorders in both genders. Accurate measurements of low concentrations of estrogens in women are necessary for determining menopausal status, estrogen deficiency, and monitoring anti-estrogen treatment. Measurements are also useful for predicting risk of osteoporosis and cardiovascular disease and detecting of some forms of cancer (1–9). In pediatric endocrinology accurate measurement of estrogens is important for diagnosing

U. Garg, C.A. Hammett-Stabler (eds.), *Clinical Applications of Mass Spectrometry*, Methods in Molecular Biology 603,
DOI 10.1007/978-1-60761-459-3_20, © Humana Press, a part of Springer Science+Business Media, LLC 2010

delayed or precocious puberty, as well as disorders of growth and reproductive systems (1, 2). Analyzing samples from postmenopausal women, men, and children has been notoriously difficult because of very low endogenous concentrations of estrogens found in these populations. Some of the recently described liquid chromatography tandem mass spectrometry based methods use derivatization as a way of enhancing the sensitivity (10–14), but derivatization may also compromise the specificity. We describe an LC-MS/MS method (14) that exhibits both high specificity and sensitivity necessary for measurement of low endogenous concentrations of estrogens, characteristic of postmenopausal women, men, and prepubertal children.

2. Materials

2.1. Samples

Acceptable sample types are serum or EDTA, plasma. Samples can be stored at ambient temperature for 2 days, refrigerated (2–6°C) for 1 week, or frozen at −20°C or less for up to 1 month.

2.2. Reagents and Buffers

1. Estrone, estradiol, dansyl chloride, sodium carbonate, formic acid, trifluoroacetic acid, methanol, acetonitrile, isopropanol, methyl-tert-butyl ether (MTBE), and acetone (Sigma-Aldrich, St Louis, MO or VWR, West Chester, PA).

2. Estrone-d_4, and 17β-estradiol-d_3 (CDN Isotopes, Pointe-Claire, Quebec, Canada).

3. Phosphate buffered saline: 0.008 M dibasic sodium phosphate, 0.001 M monobasic sodium phosphate, 0.14 M sodium chloride adjusted to pH 7.4 with phosphoric acid.

4. Bovine serum albumin (Equitech-Bio Inc., Kerrville, TX) (*see* **Note 1**).

5. Estrogen-free matrix (referred in chapter as BSA): 0.05% BSA solution in phosphate buffered saline. Stable for 2 years at −70°C. (*see* **Note 1**)

6. 1 g/L dansyl chloride: Prepare in high purity acetonitrile. Stable for 6 month at −20°C.

7. 1 g/L sodium carbonate: Prepare in deionized water. Stable for 3 month at 4°C.

8. Derivatizing solution: mix equal volumes (50:50 v/v) 1 g/L dansyl chloride and 1 g/L sodium carbonate. Prepare immediately before use.

9. 10 mM formic acid in water.

10. 10 mM formic acid in methanol.

11. 10 mM formic acid in acetonitrile.

12. Autosampler syringe wash solution 1: 45% 2-propanol, 45% acetonitrile, 10% acetone, 0.3% trifluoroacetic acid, and

13. Autosampler syringe wash solution 2: 50% water, 50% methanol, 0.1% formic acid.

14. Autosampler (CTC PAL) injection valve cleaning solution 1: 45% 2-propanol, 45% acetonitrile, 10% acetone, 0.3% trifluoroacetic acid.

15. Autosampler (CTC PAL) injection valve cleaning solution 2: 50% water, 50% methanol, 0.1% formic acid.

2.3. Calibrators, Standards, and Quality Controls

1. Stock standards of estrone and estradiol: Prepare in methanol at concentration of 1 µg/µL. Stable for 3 years at −70°C.

2. Combined working calibration standard (0.4 ng/mL): Prepare from the stock standards using two consecutive dilutions, as follows:
   a. Dilution 1 (1 µg/mL). Transfer 25 µL of each of the stock standards (estrone and estradiol) to a 25 mL volumetric flask and dilute with methanol. Use solution freshly prepared; discard remaining solution after preparation of the working calibration standard (dilution 2).
   b. Dilution 2 (0.4 ng/mL). Transfer 40 µL of standard (dilution 1) to a 100 mL volumetric flask and dilute with a mix of water and methanol (1:1). Stable for 2 years at −70°C.

3. Stock internal standards of estrone-d_4 and estradiol-d_3. Prepare in methanol at concentration of 1 mg/mL. Stable for 5 years at −70°C.

4. Combined working internal standard (2 ng/mL). Pipet 1 µL of stock internal standards (estrone-d_4 and estradiol-d_3) to a 500 mL volumetric flask and bring to volume with a mixed solution of water and methanol (1:1). Stable for 2 years at −70°C.

5. Quality control samples:
   a. Negative control: 0.05% BSA in PBS. Stable for 2 years at −70°C.
   b. Control level 1. Prepare in BSA supplemented with estrone and estradiol to a concentration of 20–30 pg/mL. Transfer aliquots to microcentrifuge tubes for storage at −70°C. Control is stable for 6 months at −70°C.
   c. Control Level 2: Human serum spiked with estrone and estradiol to a concentration of 200–300 pg/mL. Transfer aliquots to microcentrifuge tubes for storage at −70°C. Control is stable for 6 months at −70°C.

2.4. Equipment

1. A triple quadrupole mass spectrometer API4000 with Tur-boV ion source (Applied Biosystems/Sciex, Foster City, CA). Software Analyst 1.4.2 or newer version (earlier versions do not support HPLC pumps Agilent 1200SL).

2. Two binary HPLC pumps series 1200 SL (Agilent Technologies, Santa Clara, CA), vacuum degasser, 6-port switching valve, autosampler CTC PAL (Carrboro, NC) equipped with fast wash station.

3. Vortex with adaptor for microcentrifuge tubes.

4. Evaporator for 96-well plates.

5. Centrifuge for 2 mL microcentrifuge tubes.

6. Centrifuge with buckets for 96-well plates.

2.5. Supplies

1. 2 mL microcentrifuge tubes.

2. 96-well plates (deep plates with square wells, 2 mL) and sealing mats for the plates.

3. Methods

3.1. Stepwise Procedure

1. Label 2.0 mL microcentrifuge tubes for each sample, calibrator, and control to be analyzed.

2. Pipet 200 μL of BSA into each calibrator and negative control tube.

3. Add the volume of combined working calibration standard described in **Table 20.1** to the appropriate calibrator tube.

4. Pipet 200 μL of each serum or plasma sample to be analyzed into the appropriately labeled tube.

Table 20.1
Preparation of calibration standards

Concentration of standard, pg/mL	Add working calibration standard, μL
5	2.5
20	10
50	25
80	40
120	60
200	100

5. To each tube add 20 μL of working internal standard.

6. Add 1.2 mL methyl-tert-butyl ether (MTBE).

7. Vortex tubes for 10 min (medium speed).

8. Centrifuge tubes at $14,000 \times g$ at 4°C for 5 min.

9. Label 96-well plate.

10. Add 20 μL of internal standard and 2.5 μL of the calibration standard into the 96th well of the plate. This will serve as the injection standard.

11. Transfer ~80% of the organic solution from each tube to its corresponding well of the 96-well plate.

12. Evaporate to dryness.

13. Prepare derivatizing solution for use.

14. Using 8-chanel pipet, add 50 μL of the derivatizing solution to each well.

15. Cover plate, vortex for 1 min, and incubate in an oven at 70°C for 10 min (*see* **Note 2**).

16. Carefully remove the cover from the plate. Using an 8-chanel pipet, add 50 μL of water/acetonitrile mix (ratio 1:1) to each well. Cover plate again and vortex for 1 min.

17. Centrifuge plate for 1 min at $4,000 \times g$.

3.2. Chromatographic Conditions

1. First dimension chromatographic separation: C1 cartridge 2 × 4 mm (Phenomenex, Torrance, CA).

2. Second dimension chromatographic separation: HPLC column Germini C6, 100 × 2 mm, 3 μm particles (Phenomenex, Torrance, CA), temperature 30°C.

3. Six-port/two-position switching valve (VICI VALCO Instruments Inc., Houston, TX) (**Fig. 20.1**).

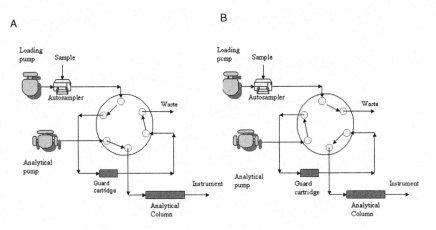

Fig. 20.1. Switching valves setup.

4. Mobile phase for the first dimension chromatographic separation: bottle A – water with 10 mM formic acid; bottle B – methanol with 10 mM formic acid.

5. Mobile phase for the second dimension chromatographic separation: bottle A – water with 10 mM formic acid, bottle B – acetonitrile with 10 mM formic acid.

6. Programs for the first and second dimension chromatographic separation are summarized in **Tables 20.2** and **20.3** (*see* **Note 3**).

7. Injection volume 60 μL.

8. Autosampler syringe cleaning, 10 washes with each: Solution #1 and Solution #2.

9. Autosampler (CTC PAL) injection valve cleaning, 1 ml each wash: Solution #1 and Solution #2.

Table 20.2
Mobile phase program for first dimension chromatographic separation (flow rate 1 ml/min)

Step	Time (min)	A (%)	B (%)
0	0.1	90	10.0
1	0.2	90	10.0
2	0.3	30	70.0
3	1.0	5	95
4	3.1	5	95
5	3.3	90	10
6	3.7	90	10
7	6.9	90	10

Table 20.3
Mobile phase program for the second dimension chromatographic separation (flow rate 0.6 ml/min)

Step	Time (min)	A (%)	B (%)
0	0.1	50	50.0
1	1.3	50	50.0
2	6.8	85	15.0
3	6.9	95	5.0

3.3. Mass Spectrometer Conditions

1. Mass transitions used for estrone, estradiol, and their internal standards are listed in **Table 20.4**.

2. Voltages and gases flow rates for the mass spectrometer, optimized for maximum sensitivity, are as follows:
   a. Ion spray voltage 5,000 V.

   b. Ion source temperature 650°C.

   c. Nebulizer gas: 50, heating gas: 50.

   d. Collision gas: 5.

   e. Declustering potential: 70 V.

   f. Entrance potential: 10 V.

   g. Collision cell exit potential: 14 V.

3. Collision energies (CE) for the mass transitions are compound-dependent (**Table 20.4**).

4. Mass analyzer Q_1 is tuned for unit resolution (0.7 Da at 50% height) and analyzer Q_3 is tuned for low resolution (1.0 Da at 50% height).

Table 20.4

Mass transitions* and corresponding collision energies for the dansyl derivatives of estrone, estradiol, and their internal standards

Compound	Primary transition, *m/z* (collision energy, V)	Secondary transition, *m/z* (collision energy, V)
Estrone	504.2–156.1 (40)	504.2–171.1 (50)
d_4- estrone	508.2–156.1 (40)	508.2–171.1 (50)
17β-estradiol	506.2–156.1 (40)	506.2–171.1 (50)
d_3-17β estradiol	509.2–156.1 (40)	509.2–171.1 (50)

* dwell time 40 ms.

3.4. Data Analysis

The data analysis is performed using software Analyst 1.4 (Applied Biosystems/Sciex, Foster City, CA). The quantitation of estrone and estradiol is performed using calibration curves generated with every batch of samples. Calculations are performed using peak area ratios of the primary mass transitions of the analytes and corresponding internal standards (**Fig. 20.2**). The test results are considered acceptable if correlation coefficient (r) for the calibration curve is greater than 0.99, calculated concentrations of estrone and estradiol in the controls are within 20% of target values, and the ratio of the concentrations

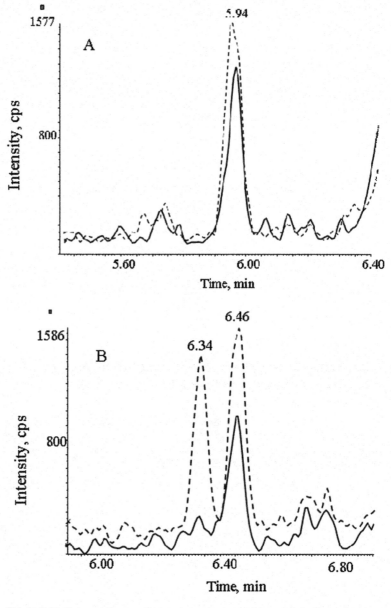

Fig. 20.2. Multiple reaction monitoring chromatograms of dansyl derivatives of estrone (**A**) and estradiol (**B**).

determined from two sets of mass transitions (primary and secondary, **Table 20.4**) is within 30% of each other (15). Concentration of estrogens in the negative control should be below the limit of quantitation for the method (*see* **Note 4**). Expected intra- and inter-assay variations for the method are below 10% (*see* **Note 5**).

4. Notes

1. Bovine serum albumin, used to prepare the estrogen-free matrix, may contain some amount of non-specifically bound estrogens; it is necessary to select a BSA lot with amount of estrogens below the limit of quantitation of the method.

2. The yield of the reaction between the estrogens and dansyl chloride is highly dependent on concentration of dansyl chloride, pH of solution, incubation temperature, and incubation time.

3. Among the advantages of using dansyl chloride as the derivatizing reagent for estrogens are significant gain in the sensitivity of detection and mild reaction conditions. The main disadvantage of the reagent is nonspecific fragmentation of the derivatives: all dansyl derivatives produce the same major product ions at m/z 156 and m/z 171. The ion m/z 171 originates from a cleavage of a C-S bond in the dansyl portion of the molecule, and the ion m/z 156 is produced by loss of the methyl group from the fragment m/z 171. Potential complication in the analysis is that the second isotopic ion (A+2) of the molecular ion of dansyl estrone is an isobar to the molecular ion of dansyl estradiol and first isotopic ion (A+1) of dansyl d_4-estrone is an isobar to the molecular ion of dansyl d_3-estradiol. Because of the above, extensive chromatographic separation utilized in this method plays an important role for eliminating potential interferences. The peaks of dansyl estrone and dansyl estradiol are resolved from each other in this method.

4. If a run contains sample with concentration of one of the estrogens above 2,000 pg/mL, the following sample should be evaluated for carryover. To evaluate for carryover the sample should be reinjected following acceptable solvent blank and the negative control reinjection.

5. Considering high sensitivity and high specificity of this method, it is suitable for measurement of endogenous concentrations of estrone and estradiol in samples from men, children, and post-menopausal women.

References

1. Haymond, S., Gronowski, A.M. (2005) Reproductive related disease. In: Ashwood, E.R., Burtis, C.A., Bruns, D.E. (eds.) Tietz Textbook of Clinical Chemistry and Molecular Diagnostics, 4th edition. New York: Saunders, pp. 2097–2152.

2. Grumbach, M.M., Hughes, I., Conte, FA. (2002) Disorders of sexual differentiation. In: Larsen, P.R., Kronenberg, H.M., Melmed, S., Polonsky, K.S. (eds.) Williams Textbook of Endocrionology, 10th edition. New York: Saunders, pp. 842–1002.

3. Kol, S. (2003) Hormonal therapy of the infertile women. In: Meikle, A.W. (ed.) Endocrine Replacement Therapy in Clinical Practice. New Jersey: Humana, pp. 525–537.

4. Napoli, N., Donepudi, S., Sheikh S., Rini, G.B., Armamento-Villareal, R. (2005) Increased 2-hydroxylation of estrogen in women with a family history of osteoporosis. *J Clin Endocrinol Metab* **90**, 2035–2041.

5. Singh, M., Dykens, J.A., Simpkins, J.W. (2006) Novel mechanisms for estrogen-induced neuroprotection. *Exp Biol Med* **231**, 514–521.

6. Green, P.S., Simpkins, J.W. (2000) Neuroprotective effects of estrogens: potential mechanisms of action. *Int J Dev Neurosci* **18**, 347–358.

7. Orwoll, E.S. (1998) Osteoporosis in men. *Endocrinol Metab Clin North Am* **27**, 349–367.

8. Meier, C., Nguyen, T.V., Handelsman, D.J., Schindler, C., Kushnir, M.M., Rockwood A.L., et al. (2008) Endogenous sex hormones and incident fracture risk in older men: The Dubbo osteoporosis epidemiology study. *Arch Intern Med* **168**, 147–154.

9. Adamski, J., Jakob F.J. (2001) A guide to 17β hydroxysteroid dehydrogenases. *Mol Cell Endocrinol* **171**, 1–4.

10. Kushnir, M.M., Rockwood, A.L., Roberts, W.L., Pattison, E.G., Bunker, A.M., Meikle, A.W. (2006) Performance characteristics of a novel tandem mass spectrometry assay for serum testosterone. *Clin Chem* **52**, 120–128.

11. Kushnir, M.M., Rockwood, A.L., Roberts, W.L., Owen, W.E., Bunker, A.M., Meikle, A.W. (2006) Development and performance evaluation of a tandem mass spectrometry assay for four adrenal steroids. *Clin Chem* **52**, 1559–1567.

12. Anari, M.R., Bakhtiar, R., Zhu, B., Huskey, S., Franklin, R.B., Evans, D.C. (2002) Derivatization of ethinylestradiol with dansyl chloride to enhance electrospray ionization: application in trace analysis of ethinylestradiol in rhesus monkey plasma. *Anal Chem* **74**, 4136–4144.

13. Nelson, R.E., Grebe, S.K., OKane, D.J., Singh, R.J. (2004) Liquid chromatography-tandem mass spectrometry assay for simultaneous measurement of estradiol and estrone in human plasma. *Clin Chem* **50**, 373–384.

14. Kushnir, M.M., Rockwood, A.L., Bergquist, J., Varshavsky, M., Roberts, W.L., Yue, B., et al. (2008) A tandem mass spectrometry assay for estrone and estradiol in serum of postmenopausal women, men and children. *Am J Clin Pathol* **129**, 530–539.

15. Kushnir, M.M., Rockwood, A.L., Nelson, G.J., Yue, B., Urry, F.M. (2005) Assessing analytical specificity in quantitative analysis using tandem mass spectrometry. *Clin Biochem* **38**, 319–327.

Chapter 21

3-Hydroxy-Fatty Acid Analysis by Gas Chromatography-Mass Spectrometry

Patricia M. Jones and Michael J. Bennett

Abstract

The mitochondrial fatty acid β-oxidation is integral to normal cellular metabolism and maintenance of cellular energy supplies. Disorders of this pathway interrupt the body's ability to deal with fasting states, as well as compromising the functioning of organs and systems whose high-energy requirements utilize fats for a continuous energy source, such as heart and skeletal muscle. This method quantitatively measures intermediate metabolites of fatty acid β-oxidation, specifically the 3-hydroxy-fatty acids produced by the third step in the pathway. The method is useful for helping to diagnose disorders of the pathway, especially defects in the L-3-hydroxyacyl CoA dehydrogenases. Serum or plasma samples are used for routine clinical evaluation; however, measurement of 3-hydroxy-fatty acid intermediates in fibroblast cell culture media and in samples from mice also allows the method to be used for research into fatty acid oxidation and interconnected pathways. The method is a stable isotope dilution, electron impact ionization gas chromatography/mass spectrometry (GC/MS) procedure.

Key words: Stable isotope dilution, fatty acid oxidation, 3-hydroxy-fatty acids, L-3-hydroxyacyl CoA dehydrogenase.

1. Introduction

Fatty acid β-oxidation (FAO) is comprised of four enzymatic steps that break down fatty acids by consecutively removing two carbon units in the form of acetyl CoA from the fatty acid chain (1, 2). The basic process is depicted in **Fig. 21.1**. As shown in **Fig. 21.1**, each of the four enzymatic steps has chain-length specific enzymes. Although in general the short chain enzymes are believed to oxidize fatty acids with chain lengths of 4–6 carbons, the medium-chain enzymes function best with carbon chains of 8–12 carbons and the long-chain enzymes utilize carbon chain length fatty acids of 14–18

U. Garg, C.A. Hammett-Stabler (eds.), *Clinical Applications of Mass Spectrometry*, Methods in Molecular Biology 603, DOI 10.1007/978-1-60761-459-3_21, © Humana Press, a part of Springer Science+Business Media, LLC 2010

FATTY ACID β-OXIDATION

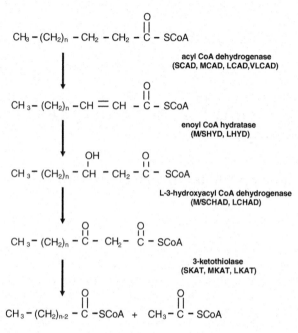

Fig. 21.1. The four enzymatic steps involved in mitochondrial fatty acid β-oxidation (FAO) with the different chain-length enzymes which occur at each step.

carbons; in vitro systems for measuring the enzymes show some ability of the enzymes to utilize other chain lengths. The third step in the FAO pathway utilizes L-3-hydroxyacyl CoA dehydrogenases and converts the fatty acids from 3-hydroxy-fatty acid (3-OHFA) to 3-keto-intermediates. These intermediates are quantified in serum or plasma samples and in media from cells in culture. The concentrations of 3-OHFA present in serum and plasma samples can be used to assist in the diagnosis of disorders of the FAO pathway. Elevations in the concentration of different chain lengths of the 3-OHFAs in blood are indicative of defects in those specific enzymes (3, 4). For example, individuals with long-chain L-3-hydroxyacyl CoA dehydrogenase (LCHAD) deficiency have elevated concentrations of 3-OHFAs with chain lengths of 14–18 carbons. The cell culture adaptation has allowed the study of various aspects of disorders of FAO and possibly other interconnected pathways using fibroblasts in cell culture (5–8). The utilization of samples from mice will also feasibly allow study of these pathways in mice as well.

The method utilizes stable-isotope dilution electron-impact ionization gas chromatography/mass spectrometry. Stable isotope standards of the 3-OHFA species from 3-hydroxy-hexanoic acid (C6) to 3-hydroxy-octadecanoic acid (C18) (9) are added to the samples and then the samples are extracted and derivatized. The treated samples are separated via gas chromatography and then introduced into the source of a mass spectrometer where they are fragmented via electron impact ionization. The mass spectra produced from the fragmentation of each compound are analyzed and the 3-OHFAs are quantified.

2. Materials

2.1. Samples

1. Serum, heparinized plasma, or EDTA anticoagulated plasma are acceptable (*see* **Notes 10** and **11**).

2. Samples of cell culture media and mouse tissue are also acceptable.

3. No special handling is required for the samples. All sample types are stable at room temperature for 8 h, refrigerated at 2–8°C for 6 months, and frozen at −20°C for more than 2 years. Generally samples are shipped frozen.

2.2. Native and Stable-Isotope 3-Hydroxy-Fatty Acids

1. All native and stable-isotope 3-hydroxy-fatty acids were synthesized by the Pediatric Mass Spectrometry facility, University of Colorado Health Science Center at Denver (3). The stable isotope form was synthesized with isotopic carbon (^{13}C) in place of regular carbon at positions 1 and 2 of the fatty acid carbon chain, as shown in **Fig. 21.2**.

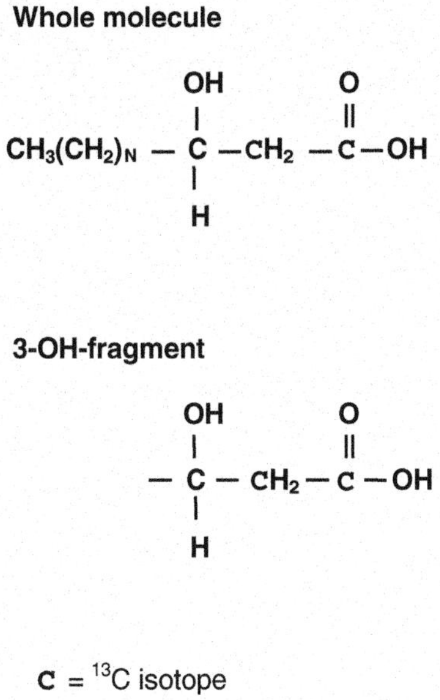

Whole molecule

$$CH_3(CH_2)_N - \underset{\underset{H}{|}}{\overset{\overset{OH}{|}}{C}} - CH_2 - \overset{\overset{O}{\|}}{C} - OH$$

3-OH-fragment

$$- \underset{\underset{H}{|}}{\overset{\overset{OH}{|}}{C}} - CH_2 - \overset{\overset{O}{\|}}{C} - OH$$

C = ^{13}C isotope

Fig. 21.2. Isotope labeling of the 3-hydroxy-fatty acids and the common 3-hydroxy-fragment resulting from fragmentation of each of them.

2. Each 3-OHFA species was prepared for use by weighing the solid compound and diluting to a final concentration of 500 μM. The solutions were then aliquotted into 1.5 ml eppendorf tubes and frozen at −70°C until use.

    a. 3-hydroxy-hexanoic acid (3OHC6) and its stable-isotope 1,2-13C$_2$ - 3-hydroxy-hexanoic acid (1,2-13C$_2$ -3OHC6) were dissolved in anhydrous ethyl alcohol (Sigma-Aldrich, St. Louis. MO) (*see* **Note 1**).

    b. 3-hydroxy-octanoic acid (3OHC8), 3-hydroxy-decanoic acid (3OHC10), 3-hydroxy-dodecanoic acid (3OHC12), 3-hydroxy-tetradecanoic acid (3OHC14), and their respective stable-isotopes were each dissolved in chloroform (Sigma-Aldrich, St. Louis, MO).

    c. 3-hydroxy-hexadecanoic acid (3OHC16), 3-hydroxy-octadecanoic acid (3OHC18) and their respective stable isotopes were dissolved in dichloromethane (Sigma-Aldrich, St. Louis, MO) (*see* **Note 2**).

3. Three concentrations of controls that are assayed along with clinical patient samples are made from the native 3-hydroxy-fatty acid species. These controls are made by first weighing each native species into a single container so that they can be diluted in chloroform to a final concentration of 1000 μM. (*see* **Note 3**). This is the control stock solution. The three concentrations of controls are made from this control stock solution as follows:

    a. High control (~10 μM) (*see* **Note 4**): 1 mL of control stock solution into 99 mL of 4 g/dL bovine serum albumin (BSA) (Sigma-Aldrich, St. Louis. MO) made in phosphate buffered saline (PBS), pH 7.4.

    b. Medium control (~5 μM): 0.5 mL of control stock into 99.5 mL of 4 g/dL BSA in PBS.

    c. Low control (~0.5 μM): 50 μL of control stock into 99.95 mL of 4 g/dL BSA in PBS.

Mix the controls thoroughly, aliquot, and freeze at −70°C until use. Stable for 5 years.

2.3. Extraction and Derivatization Reagents

1. 6.0 M Hydrochloric Acid (HCl) (Ricca Chemical, Arlington, TX)

2. Ethyl Acetate, HPLC grade (Mallinckrodt, St. Louis. MO)

3. Sodium sulfate, anhydrous (Sigma, St. Louis, MO)

4. Derivatizing reagent, 99:1: N,O-bis(trimethylsilyl) trifluoroacetamide and trimethylchlorosilane (BTSFA+TMCS), 99:1 (Supelco, Bellefonte, PA)

5. 10 M NaOH: Prepare by adding 40 g of NaOH pellets (J.T.Baker/Mallinkrodt) to 100 mL de-ionized water.

6. Phosphate buffered saline (PBS): Prepare by dissolving one pouch of commercially available PBS (Sigma, St. Louis) in one liter of deionized water.

2.4. Cell Culture Media

1. Dulbecco's Modified Eagle Medium (DMEM) (Gibco/BRL, Bethesda, MD) is used for all cell culture with 10% fetal bovine serum (FBS) (Gibco/BRL, Bethesda, MD) added.

3. Methods

The method is calibrated and the linearity, precision, and lower limit of detection are established using the prepared native compounds. The assay is linear from 0.2 to 50 μM and has a lower limit of detection of 0.2 μM. Precision ranged from 1 to 15% across the various chain-length species and concentrations (*see* **Notes 5, 6 and 7**). Once validated, the assay can be used for testing various sample types including serum, plasma, bile, and culture media from cell culture. Samples from mice have also been used.

Clinical patient samples are run in duplicate: one sample is left unhydrolyzed to give the free 3-OHFAs and the second sample is hydrolyzed using sodium hydroxide to yield the total 3-OHFAs. Samples are pipetted into test tubes, the second tube hydrolyzed and then the stable isotope standards are added. All samples are then extracted twice with ethyl acetate to retrieve the organic/hydrophobic 3-OHFAs in the top ethyl acetate layer. Sodium sulfate is added to the ethyl acetate portions containing the 3-OHFAs to remove all water and then evaporated to dryness down under nitrogen. The dried samples are reconstituted in BSTFA+TMCS, derivatized, and analyzed using GC/MS.

The GC/MS method is a single ion monitoring (SIM) method which monitors the native and isotope molecular weight minus a methyl group $[M-CH_3]^+$ *m/z* ratios specific for the seven 3-OHFAs, as well as the 3-OH fragment ion. **Figure 21.2** demonstrates the isotope labeling of the 3-OHFAs and the 3-OH fragment that is released upon fragmentation. The $[M-CH_3]^+$ *m/z* for each of the native and stable-isotope compounds is given in **Table 21.1**. Quantification of the amount of native compound is achieved by calculating the ratio of native to isotope and multiplying by the known concentration of the isotope in the sample. Alternatively, several calibration curves can be run for each of the seven chain-length species and a linear regression equation of the line determined for each species and used to calculate the concentrations (*see* **Note 8**).

3.1. Serum/Plasma Sample Extraction and Derivatization

1. Samples are run in duplicate. The first should be identified as "unhydrolyzed" and the second as "hydrolyzed". For *each* duplicate, label three 16 × 100 mm (10 mL) glass screw cap tubes (i.e., there will be six tubes per patient sample – 3 tubes for the unhydrolyzed sample and three for the hydrolyzed) (*see* **Note 9**).

Table 21.1
Molecular weight *m/z* ion fragments for the 3-OHFA species

	[M-CH₃]⁺ *m/z* ion		3-OH fragment *m/z* ion	
3-OH-fatty acid	Native	Isotope	Native	Isotope
3-OH-hexanoic (C6)	261	263	233	235
3-OH-octanoic (C8)	289	291	233	235
3-OH-decanoic (C10)	317	319	233	235
3-OH-dodecanoic (C12)	345	347	233	235
3-OH-tetradecanoic (C14)	373	375	233	235
3-OH-hexadecanoic (C16)	401	403	233	235
3-OH-octadecanoic (C18)	429	431	233	235

2. Pipet 0.5 mL of serum or plasma (*see* **Note 10**) into each tube (*see* **Note 11**).

3. Into the "hydrolyzed" set of tubes, pipet an equal volume (0.5 mL if 0.5 mL of sample is used) of 10 M NaOH. Cap tightly and vortex 30 sec.

4. Place the capped tubes containing NaOH in a 37°C heat block for 30 min to hydrolyze. After hydrolysis, uncap and return tubes to rack for next step.

5. Pipet 10 μL of each of the seven 500 μM stable-isotope standards into each tube containing a sample. Vortex tubes for 15 sec.

6. Add 125 μL of 6 M HCl to all unhydrolyzed samples and 2 mL of 6 M HCl to all hydrolyzed samples that used 0.5 mL of sample and NaOH. For other sample volumes, *see* HCl calculation in **Note 12** below.

7. Vortex all tubes for 15 sec.

8. Add 3 mL of ethyl acetate to each sample to extract the fatty acids. Cap tightly and vortex for 30 sec.

9. Centrifuge at 2,000 × *g* for 2 min to separate the phases.

10. Remove the top ethyl acetate later to the second labeled tube and repeat Steps 8–9.

11. Remove the ethyl acetate layer from the second extraction and combine it with the first extracted layer in the second test tube. Discard the bottom layer and the first test tube.

12. Add approximately 0.5–1.0 g anhydrous sodium sulfate to the combined ethyl acetate layers in each tube (*see* **Note 13**). Cap and vortex for 15 sec.

13. Centrifuge at $3,000 \times g$ for 2 min. Pour off the ethyl acetate into the clean third tube (*see* **Note 14**). Discard second tube.

14. Dry the samples down under a stream of nitrogen in a 37°C heat block.

15. Add 65 µL of BSTFA+TMCS to each tube. Cap tightly and derivatize at least 30 min (up to 2 h) at 70–80°C.

16. Transfer each sample to a 0.2 mL conical insert inside a labeled 1.5 mL autosampler vial. Cap tightly.

17. Load on a GC/MS autosampler, inject, and analyze.

3.2. Samples Other Than Serum or Plasma

3.2.1. Media Removed from Cell Culture

1. Label three 16 × 100 mm glass screw cap tubes for each sample.

2. Pipet 4 mL of media into its labeled tube (*see* **Notes 15** and **16**).

3. Pipet 10 µL of each of the seven 500 µM stable-isotope standards into each tube containing a sample. Cap and vortex tubes for 15 sec.

4. Add 0.25 mL of 6 M HCl to acidify sample, cap, and vortex for 15 sec.

5. Follow the regular procedure from step #8 onward with the following exceptions:
   a. DO NOT vortex the samples after adding ethyl acetate. Shake or rotate the tubes gently for 30 sec to accomplish the extraction (*see* **Note 17**).
   b. Add about twice the normal amount of sodium sulfate.
   c. Derivatize with 50 µL of BSTFA and after derivatizing carefully transfer as much as possible to the conical insert for analysis.

3.2.2. Samples from Mice

1. Make 100 µM stable-isotope standards from the 500 µM stable-isotope standards by diluting each standard with its appropriate diluent, i.e., absolute ethanol for $1,2–13C_2$ – 3OHC6, chloroform for C8–C14, etc.

2. Label three 16 × 100 mm glass screw cap tubes for each sample.

3. Pipet 50 µL of mouse serum or plasma into the labeled tube (*see* **Note 18**).

4. Hydrolyze the sample with an equal volume of 10 M NaOH for 30 min at 37°C.

5. Pipet 10 µL of each of the seven 100 µmo/L stable-isotope standards into each tube containing a sample. Vortex tubes for 15 sec.

6. Add 125 µL of 6 M HCl, cap and vortex for 15 sec.

7. Follow the regular procedure from step #8 onward with the following exceptions:
   a. Derivatize with 30 μL of BSTFA and after derivatizing carefully transfer as much as possible to the conical insert for analysis.
   b. Inject 2 μL into the GC/MS instead of the 1 μL injection used for the other sample types (*see* **Note 19**).

3.3. GC/MS Analysis

1. Perform gas chromatography-mass spectrometry analysis on a gas chromatograph with a quadrupole mass spectrometer using helium as the carrier gas.
2. For the GC/MS method, utilize a split/splitless injector in splitless mode at 270°C to introduce a 1 μL sample onto a Hewlett-Packard HP-5MS capillary column [30 m × 0.25 mm(i.d.)] coated with a 0.25 μm film of cross-linked 5% PH ME Siloxane. After passing through the column, the sample is introduced directly into the mass spectrometer source, subjected to electron impact ionization and fragmentation, and analyzed in the mass spectrometer.
3. The GC temperature program for the method is as follows: 80°C for 5 min; ramp 3.8°C/min to 140°C; ramp 2.3°C/min to 200°C; ramp 15°C/min to 290°C, then hold for 6 min. This results in a 59-min program.
4. The SIM settings monitor four ion fragments at each time setting with a 50-msec dwell time per ion. **Table 21.2** shows the window start times, the times when the peak occurs, and the fragments scanned for at each window (*see* **Note 20**).

Table 21.2
SIM program parameters and peak time windows

SIM start time (min)	Chain length	Peak time window	Ions to scan for			
			$[M-CH_3]^+$		3-OH fragment	
0.0	C6	16.0–19.0	261	263	233	235
19.0	C8	22.0–25.0	289	291	233	235
25.0	C10	28.0–32.0	317	319	233	235
33.0	C12	36.0–39.0	345	347	233	235
40.0	C14	43.0–47.0	373	375	233	235
47.0	C16	48.0–51.0	401	403	233	235
51.0	C18	51.0–53.0	429	431	233	235

5. After the GC/MS has finished analyzing a sample, retrieve the abundances of each of the monitored ions from the computer, along with the background abundances (*see* **Note 21**).

6. Use the $[M\text{-}CH_3]^+$ (molecular weight) *m/z* abundances to calculate all compounds except 3-OHC12 and 3-OHC18. For quantification of these two compounds use the 3-OH fragments (233, 235) due to interfering fragments from other compounds (*see* **Note 22**).

7. Quantify the amount of each 3-OHFA in the sample by calculating the ratio of native to isotope for each species and multiplying the ratio times the concentration of the isotope. The following formula is used:

$$\frac{\text{Native compound abundance } - \text{ background abundance}}{\text{Isotope compound abundance } - \text{ background abundance}}$$

$$\times \ 10 \ \mu M = 3\text{-OHFA concentration}$$

8. Calculations are performed in an excel program set up with these calculations for this purpose (*see* **Note 23**).

9. Report a quantitative result for each of the seven chain-length 3-OHFAs. Normal reference intervals for serum/plasma samples can be found in **Table 21.3** (*see* **Note 24**).

Table 21.3
Serum/plasma reference intervals in μM

3-OHFA	Free	Total	MCT in diet
3-OHC6	0.4–2.2	0.4–2.2	1.6–18.5
3-OHC8	0.2–1.0	0.2–1.0	0.4–8.5
3-OHC10	0.12–0.6	0.12–0.6	0.2–2.3
3-OHC12	0.1–0.5	0.1–0.6	0.1–1.4
3-OHC14	< 0.4	< 0.7	< 0.7
3-OHC16	< 0.5	< 0.9	< 0.9
3-OHC18	< 0.5	< 0.9	< 0.9

4. Notes

1. 3-OHC6 is an oil at room temperature. The easiest way to mix it is to first weigh the container that will be used for the solution and then carefully add the oil to the container until the correct weight is reached.

2. The long-chain (C16 and C18) 3-OHFAs are not very soluble. The solutions were vortexed frequently and heated mildly (up to 50°C) for about an hour prior to aliquotting and freezing. Once frozen at −70°C all the various chain-length aliquots are stable for several years.

3. Native 3-OHFAs were used to design and validate the method and to spike control samples for use with clinical patient samples. They are not used in the daily performance of the assay. For the Control Stock, the following weights of each native compound were placed into a single vial and the whole was diluted to 10 mL with chloroform to give an approximate concentration of 1,000 μM for each compound.

Native compound	weight
3-OH-C6	~1.2 μl (oil)
3-OH-C8	1.602 mg
3-OH-C10	1.883 mg
3-OH-C12	2.163 mg
3-OH-C14	2.444 mg
3-OH-C16	2.724 mg
3-OH-C18	3.376 mg

Obviously, weighing these small numbers is problematic, but they do not have to be exact to three decimal places. Like any unassayed control, the controls will be repeatedly assayed to determine the acceptable range of concentration for each control.

4. The concentrations are approximate because this is what the concentration would be if a concentration of 1,000 μM is achieved in the control stock solution. Actual concentration is determined by repeated assays. 4 g/dL BSA in PBS is used to approximate a serum sample matrix.

5. Linearity and lower limit of detection were determined by repeatedly assaying the seven 3-OHFA native species at concentrations ranging from 0.001 to 250 μM. The ratio of native to isotopic compound was obtained for each measurement and the ratios were then plotted against known concentration. It was determined that best linearity was achieved between 0.2 and 50 μM, with 0.2 μM being the lowest concentration that could be reliably reproduced.

6. Calibration curves using native compounds in the range between 0.2 and 50 μM were assayed repeatedly (4–6 runs) and a linear regression equation was derived for each chain-length species using the mean slope and intercept from the repeated runs.

7. Precisions were determined from the repeated analyses at the bottom, middle and top of the range and coefficient of variation (CV) were found to range from 1 to 15% across the range for all species, with CVs between 1 and 10% for all species at the high end of the range and CVs of 12–15% for all species at the low end of the concentration range.

8. The results obtained using the two different means of calculation were so similar for most of the chain-length species that the ratio × concentration method was used for all species except the C18 because it was easier. For the 3-OHC18, the concentrations derived from the linear regression equation of the line were more accurate, so this method was exclusively used to calculate 3-OHC18 in patient samples.

9. Controls are run unhydrolyzed since the same result is obtained whether the control is hydrolyzed or not.

10. Citrated plasma is an unacceptable specimen. Citrate has an *m/z* fragment of 247 that interferes with the quantification of 3-OHC12.

11. For blood samples, 0.5 mL is the optimum sample volume. Volumes of serum or plasma down to 0.25 mL will give acceptable results. Volumes below 0.25 mL will give significantly falsely elevated results.

12. In order to acidify a sample with a volume other than 500 µL that has had 10 M NaOH added to it, we calculated the amount of 6 M HCl to use by first calculating the amount needed to neutralize the NaOH, using the following formula:

$$(\text{sample volume}\,(\mu L)) \times (10\,M\,NaOH) = (?\,\mu L) \times (6\,M\,HCl)$$

Solving this formula gives the amount of acid needed to neutralize the NaOH. To acidify the sample, we then multiply the answer times 2 and round up to the nearest half-milliliter for the volume of 6 M HCl to add. For example, solving this equation for a 500 µL sample gives a result of 833 µL of HCl. 833 µL × 2 = 1666 µL = 1.666 mL, rounded up to 2.0 mL. 2.0 mL is used for 0.5 mL samples.

13. This does not have to be an exact amount. We use a small mound on the end of a reagent scoop/spatula. The purpose is to bind any water that may still be in the sample.

14. Occasionally it may be necessary to remove the ethyl acetate layer to the third tube with a Pasteur pipet; however, usually the sodium sulfate stays packed in the tube bottom and the ethyl acetate can be poured off.

15. For acceptable results in media samples, media volumes can be used down to 2 mL.

16. For media samples there is essentially no difference in hydrolyzed and unhydrolyzed results, suggesting that 3-OHFAs in media samples are not bound to other substances. However, if hydrolysis of the samples is desired, hydrolyze by,

    a. Add 0.5 mL of 10 M NaOH to each sample. Cap tightly and vortex for 15 sec. Hydrolyze for 30 min at 37°C.

    b. To acidify the sample after adding the stable-isotope standards, add 1.5 mL of 6 M HCl.

17. The amount of water in media samples causes a semi-solid gel to form with the ethyl acetate if the samples are mixed too vigorously. If this plug forms after mixture or after centrifugation, break the plug-up by shaking and re-centrifuge as many times as necessary to allow the ethyl acetate to be pulled off the top.

18. For mice, volumes less than about 15 μL gave undetectable results. Samples are usually assayed hydrolyzed for total 3-OHFA since there is rarely enough sample to run in duplicate.

19. The 2 μL injection may need to be done manually rather than by the autosampler, depending on the autosampler's capabilities in dealing with an extremely low volume of sample.

20. SIM time window start times may occasionally need to be adjusted, depending on column conditions. **Figure 21.3** demonstrates representative chromatographs generated from looking at the specific ions in their individual time windows for patient samples.

21. Abundances are retrieved manually by individually pulling up the peak for each chain length m/z ion, clicking on the apex of the native and then the isotopic peak, and printing the highest abundance for each one. The peaks from the same chain length migrate in the same time window. Background abundances are obtained from the baseline near the peaks in that time window. These abundances are used to calculate the concentration of each native compound.

22. Abundances must be at least three times background abundance to be used for calculation. If the abundances are not three times background for either the $[M-CH_3]^+$ ion pair or the 3-OH fragment ion pair, the compound is reported as less than lower limit of detection.

23. For sample volumes other than 0.5 mL, the calculations of native compound final concentration must be adjusted by multiplying the result by the volume factor (0.5 mL/actual volume used in mL). This calculation is included in the excel file. *See* **Fig. 21.4** for an example excel file for 3-OH-C6, calculated using the ratio x concentration, the regression equation, and the 3-OH fragment by ratio × concentration. The file would be set up to calculate each chain-length species in this manner.

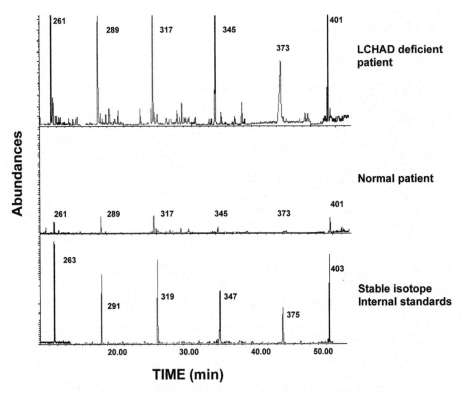

Fig. 21.3. Representative chromatographs from an LCHAD-deficient individual (**top panel**), a normal individual (**middle panel**), and the stable isotope [M-CH$_3$]$^+$ *m/z* ions (**bottom panel**). All three panels are scaled to the same abundance.

24. Elevated concentrations of 3-OHC6 to 3-OHC10 can be seen in ketotic individuals, in individuals receiving medium-chain triglycerides in their diet, and in individuals who are deficient in short-/medium-chain L-3-hydroxyacyl CoA dehydrogenase (S/MCHAD). Elevations in concentrations of 3-OHC14 through 3-OHC18 have only been seen in individuals deficient in long-chain L-3-hydroxyacyl CoA dehydrogenase (LCHAD). Along with the elevated concentrations of 3-OHC14 through -C18, individuals who are LCHAD deficient will also have elevated concentrations of the mono- and di-unsaturated forms of -C16 and -C18. These unsaturated forms are undetectable in normal individuals. The unsaturated forms can be calculated by adding two more ions for each species in the SIM program. For 3-OHC16:1 and -16:2, add the ions 399 and 397, respectively to the C16 time window in the SIM program. Calculate the native concentration using the C16 isotope standard ion of 403. For 3-OHC18:1 and 18:2, add the ions 427 and 425, respectively to the C18 time window and calculate using the 431 ion from the C18 isotope standard.

Set-up example

	A	B
1	**3-OH-C6**	
2	261 abundance:	
3	261 background:	
4	263 abundance:	
5	263 background:	
6	Sample volume (µL):	
7	Ratio:	
8	RESULT (µmol/L):	
9	Equation result (µmol/L):	

B7 field is a calculation:
   =(B2-B3)/(B5-B5)
B8 field is a calculation:
   =B7*(500/B6)*10
B9 field is a calculation using the linear regression equation for the C6 species:
   =(B7/0.098+0.0054)*(500/B6)

	A	B
11	**3-OH-C6 (233/235)**	
12	233 abundance:	
13	233 background:	
14	235 abundance:	
15	235 background:	
16	Sample volume (µL):	
17	Ratio:	
18	RESULT (µmol/L):	

B16 field: **=B6** (typing in sample volume in B6 copies it to B16)
B17 field is a calculation:
   =(B12-B13)/(B14-B15)
B18 field is a calculation:
   =B17*(500/B16)*10

Calculation example

	A	B
1	**3-OH-C6**	
2	261 abundance:	595
3	261 background:	98
4	263 abundance:	9015
5	263 background:	201
6	Sample volume (µL):	400
7	Ratio:	0.05639
8	RESULT (µmol/L):	0.56
9	Equation result (µmol/L):	0.58

Type in values for fields B2 through B6.

B7 field calculates:
   =(595-98)/(9015-201)   = 0.05639
B8 field calculates:
   =0.05639*(500/400)*10   = 0.56
B9 field calculates using the linear regression equation:
   =(0.05639/0.098 + 0.0054)
   *(500/400)   = 0.58

	A	B
11	**3-OH-C6 (233/235)**	
12	233 abundance:	402
13	233 background:	78
14	235 abundance:	6012
15	235 background:	61
16	Sample volume (µL):	400
17	Ratio:	0.054445
18	RESULT (µmol/L):	0.54

Type in B12 through B15.

B16 field is populated by typing in volume in B6
B17 field calculates:
   =(402-78)/(6012-61)   = 0.054445
B18 field calculates:
   =0.05445*(500/400)*10 = 0.54

Fig. 21.4. Example set-up and calculations for quantifying each 3OHFA species.

References

1. Eaton, S., Bartlett, K., Pourfarzam, M. (1996) Mammalian mitochondrial β-oxidation. *Biochem J*, **320**, 345–357.

2. Schultz, H. (1990) Mitochondrial β-oxidation, in: *Fatty acid Oxidation: Clinical, Biochemical and Molecular Aspects* (Tanaka, K., Coates, P.M., eds.), Alan R. Liss, New York, pp. 23–36.

3. Jones, P.M., Quinn, R., Fennessey, P.V., Tjoa, S., Goodman, S.I., Fiore, S., Burlina, A.B., Rinaldo, P., Boriack, R.L., Bennett, M.J. (2000) Improved stable isotope dilution-gas chromatography-mass spectrometry method for serum or plasma free 3-hydroxy-fatty acids and its utility for the study of disorders of mitochondrial fatty acid β-oxidation. *Clin Chem*, **46(2)**, 149–155.

4. Jones, P.M., Burlina, A.B., Bennett, M.J. (2000) Quantitative measurement of total and free 3-hydroxy-fatty acids in serum or plasma samples: Short-chain 3-hydroxy-fatty acids are not esterified. *J Inher Metab Dis*, **23(7)**, 745–750.

5. Jones, P.M., Moffitt, M., Joseph, D., Harthcock, P.A., Boriack, R.L., Ibdah, J.A., Strauss, A.W., Bennett, M.J. (2001) Accumulation of free 3-hydroxy-fatty acids in the culture media of fibroblasts from patients deficient in long-chain L-3-hydroxyacyl-CoA dehydrogenase: a useful diagnostic aid. *Clin Chem*, **47(7)**, 1190–1194.

6. Jones, P.M., Butt, Y., Bennett, M.J. (2003) Accumulation of 3-hydroxy-fatty acids in the culture medium of long-chain L-3-hydroxyacyl CoA dehydrogenase and mitochondrial trifunctional protein deficient skin fibroblasts: Implications for medium chain triglyceride dietary treatment of LCHAD deficiency. *Pediatr Res*, **53(5)**, 783–787, doi:10.1203/01.PDR.0000059748.67987.1F.

7. Jones, P.M., Butt, Y.M., Bennett, M.J. (2004) Effects of odd-numbered medium-chain fatty acids on the accumulation of long-chain 3-hydroxy-fatty acids in long-chain L-3-hydroxyacyl CoA dehydrogenase (LCHAD) and mitochondrial trifunctional protein (MTFP) deficient skin fibroblasts. *Mol Genet Metabol*, **81(2)**, 96–99.

8. Jones, P.M., Butt, Y.M., Messmer, B., Boriack, R., Bennett, M.J. (2006) Medium-chain fatty acids undergo elongation before β-oxidation in fibroblasts. *Biochem Biophys Res Comm*, **346**, 193–197.

9. Jones, P.M., Tjoa, S., Fennessey, P.V., Goodman, S.I., Bennett, M.J. (2002) Addition of quantitative 3-hydroxy-octadecanoic acid (3-OH-C18) to the stable isotope gas chromatography-mass spectrometry method for measuring 3-hydroxy-fatty acids. *Clin Chem*, **48(1)**, 176–179.

Chapter 22

Quantitation of Fentanyl in Blood and Urine Using Gas Chromatography-Mass Spectrometry (GC-MS)

Bruce A. Goldberger, Chris W. Chronister, and Michele L. Merves

Abstract

Fentanyl is a potent, short-acting synthetic opioid analgesic. Fentanyl is measured in blood and urine following mixed-mode solid phase extraction. The specimens are fortified with deuterated internal standard and a five-point calibration curve is constructed. The final extracts are reconstituted in methanol and analyzed using selected ion monitoring gas chromatography-mass spectrometry.

Key words: Opiate, opioid, fentanyl, gas chromatography, mass spectrometry, solid phase extraction.

1. Introduction

Fentanyl is a synthetic opioid analgesic medication that was originally introduced in 1963 as an intravenous anesthetic supplement. Fentanyl is available as an injectable solution, buccal tablets, lozenges, and transdermal patches. Fentanyl is fast-acting, has a short duration of action, and is approximately 100 times more potent than morphine. Despite increased analgesic potency relative to morphine, it has comparable tolerance and physical dependence liability. Due to its lipophilicity, fentanyl quickly crosses the blood-brain barrier where it has a pronounced effect on the central nervous system with a potential for heightened euphoria and respiratory depression. Fentanyl is rapidly metabolized to norfentanyl, hydroxyfentanyl, and despropionylfentanyl (1–3).

Transdermal fentanyl is utilized for the long-term management of moderate to severe chronic pain. Upon initiation of transdermal administration, therapeutic blood concentrations will not be reached for a period of 12–24 h. Abuse of fentanyl

U. Garg, C.A. Hammett-Stabler (eds.), *Clinical Applications of Mass Spectrometry*, Methods in Molecular Biology 603, DOI 10.1007/978-1-60761-459-3_22, © Humana Press, a part of Springer Science+Business Media, LLC 2010

has become popular among healthcare workers due to its euphoric effect and availability. Transdermal fentanyl patches are also abused (1–3).

The analysis of fentanyl includes immunoassay, as well as confirmation and quantitation by gas chromatography-mass spectrometry (GC-MS). Solid phase extraction methods have been reported for the isolation of drug from blood and urine matrices (4–7).

The method described below is validated for the analysis of fentanyl in blood and urine (*see* **Note 1**).

2. Materials

2.1. Chemical, Reagents, and Buffers

1. Blood, drug-free prepared from human whole blood purchased from a blood bank and pretested to confirm the absence of analyte or interfering substance.

2. 1.0 M Acetic Acid: Add 28.6 mL acetic acid to 400 mL deionized water. Dilute to 500 mL with water. Stable for 1 year at room temperature.

3. Methylene Chloride: Isopropanol:Ammonium Hydroxide Solution (78:20:2; v:v:v): To 20 mL of isopropanol, add 2 mL of ammonium hydroxide. Add 78 mL of methylene chloride. Mix well. Prepare fresh.

4. 0.1 M Phosphate Buffer, pH 6: Dissolve 13.61 g of potassium phosphate monobasic in 900 mL of water. Adjust the pH to 6.0 with 5.0 M potassium hydroxide. Q.S. to 1,000 mL with water. Stable for 1 year at 0–8°C.

5. 5.0 M Potassium Hydroxide: Dissolve 70.13 g of potassium hydroxide in 100 mL of water in a 250 mL volumetric flask. Mix well and Q.S. to 250 mL with water. Stable for 1 year at room temperature.

2.2. Preparation of Fentanyl Calibrators (see Note 2)

1. Fentanyl Internal Standard Stock Solution (10 µg/mL): Add the contents of 100 µg/mL D5-fentanyl vial (Cerilliant Corporation) to a 10 mL volumetric flask and bring to volume with methanol. Store at ≤–20°C.

2. Fentanyl Internal Standard Solution (1.0 µg/mL): Dilute 1.0 mL of the 10 µg/mL D5-fentanyl standard stock solution to 10 mL with methanol in a 10 mL volumetric flask. Store at ≤–20°C.

3. Fentanyl Standard Stock Solution (10 µg/mL): Add the contents of 100 µg/mL fentanyl vial (Cerilliant Corporation) to a 10 mL volumetric flask and bring to volume with methanol. Store at ≤–20°C.

Table 22.1

Preparation of fentanyl calibration curve

Calibrator	Calibrator concentration (ng/mL)	Volume of 0.1 µg/mL standard solution	Volume of 1.0 µg/mL standard solution
1	2.5	25 µL	–
2	5	50 µL	–
3	10	100 µL	–
4	25	–	25 µL
5	50	–	50 µL

4. Fentanyl Standard Solution (1.0 µg/mL): Dilute 1.0 mL of the 10 µg/mL fentanyl standard stock solution to 10 mL with methanol in a 10 mL volumetric flask. Store at ≤–20°C.

5. Fentanyl Standard Solution (0.1 µg/mL): Dilute 1.0 mL of the 1.0 µg/mL fentanyl standard solution to 10 mL with methanol in a 10 mL volumetric flask. Store at ≤–20°C.

6. Aqueous Fentanyl Calibrators: Add standard solutions to 1.0 mL of drug-free blood according to **Table 22.1**.

2.3. Preparation of Control Samples (see Note 2)

1. Negative (drug-free) Control: Prepare using human whole blood. Blood is pretested to confirm the absence of an opioid or interfering substance.

2. Fentanyl Control Stock Solution (10 µg/mL): Add the contents of 100 µg/mL fentanyl vial (Cerilliant Corporation) to a 10 mL volumetric flask and bring to volume with methanol. Store at ≤–20°C.

3. Fentanyl Control Solution (1.0 µg/mL): Dilute 1.0 mL of the 10 µg/mL fentanyl control stock solution to 10 mL with methanol in a 10 mL volumetric flask. Store at ≤–20°C.

4. Fentanyl Control Solution (0.1 µg/mL): Dilute 1.0 mL of the 1.0 µg/mL fentanyl control solution to 10 mL with methanol in a 10 mL volumetric flask. Store at ≤–20°C.

5. Aqueous Fentanyl Control: Add fentanyl control solution to 1.0 mL of drug-free blood according to **Table 22.2**.

2.4. Supplies

1. Clean Screen Extraction Columns (United Chemical Technologies)

2. Autosampler vials

3. Disposable glass culture tubes

4. Volumetric pipet with disposable tips

Table 22.2
Preparation of fentanyl controls

Control	Quality control concentration (ng/mL)	Volume of 1.0 µg/mL control solution
Low	10	10 µL
High	25	25 µL

2.5. Equipment

1. 6890 Gas Chromatograph (Agilent Technologies, Inc.) or equivalent

2. 5973 Mass Selective Detector (Agilent Technologies, Inc.) or equivalent

3. 7673 Automatic Liquid Sampler (Agilent Technologies, Inc.) or equivalent

4. ChemStation with DrugQuant Software (Agilent Technologies, Inc.) or equivalent

5. Extraction manifold

6. Vortex mixer

7. Centrifuge

8. Caliper Life Sciences TurboVap connected to nitrogen gas

3. Methods

3.1. Stepwise Procedure for Fentanyl Assay

1. Label culture tubes for each calibrator, control, and specimen to be analyzed and add 1.0 mL of the appropriate specimen to each corresponding tube.

2. Add 25 µL of the 1.0 µg/mL fentanyl internal standard solution to all culture tubes except the negative control.

3. Add 2.0 mL of 0.1 M phosphate buffer (pH 6) and vortex well.

4. Centrifuge at ~1,500 × g for 5 min.

5. Place columns into the extraction manifold.

6. Prewash the columns with 3.0 mL of the methylene chloride:isopropanol:ammonium hydroxide solution.

7. Pass 3.0 mL of methanol through each column. Do not permit the column to dry.

8. Pass 3.0 mL of water through each column. Do not permit the column to dry.

9. Pass 1.0 mL of 0.1 M phosphate buffer (pH 6) through each column. Do not permit the column to dry.

10. Pour specimen into column. Slowly draw specimen through column (at least 2 min) under low vacuum.

11. Pass 3.0 mL of water through each column.

12. Pass 1.0 mL of 1.0 M acetic acid through each column.

13. Dry column with full vacuum.

14. Pass 2.0 mL of hexane through each column.

15. Pass 3.0 mL of methanol through each column.

16. Dry column with full vacuum.

17. Turn off vacuum. Dry tips. Place labeled disposable culture tubes into column reservoir.

18. Add 3.0 mL of the methylene chloride:isopropanol:ammonium hydroxide solution to each column and collect in disposable culture tubes.

19. Evaporate to dryness at 40°C ± 5°C with a stream of nitrogen.

20. Add 50 μL of methanol to each tube and vortex well.

21. Transfer extract to autosampler vial and submit for GC-MS analysis.

3.2. Instrument Operating Conditions

1. The GC-MS operating parameters are presented in **Tables 22.3** and **22.4**.

2. The capillary column phase used for the fentanyl assay is either 100% methylsiloxane or 5% phenyl-95% methylsiloxane.

3. Set-up of the autosampler should include the exchange of solvent in both autosampler wash bottles with fresh solvent (bottle 1 – methanol; bottle 2 – ethyl acetate).

Table 22.3
GC operating conditions for fentanyl assay

Initial oven temp.	120°C
Initial time	0.5 min
Ramp 1	30°C/min
Final temp.	320°C
Final time	5.00 min
Total run time	12.17 min
Injector temp.	250°C
Detector temp.	300°C
Purge time	0.5 min
Column flow	1.0 mL/min

Table 22.4
Quantitative and qualifier ions for the analysis of fentanyl

Analyte	Quantitative ion (m/z)	Qualifier ions (m/z)
Fentanyl	245	189, 146
D5- Fentanyl	250	194

4. A daily autotune must be performed with perfluorotributylamine (PFTBA) as the tuning compound prior to each GC-MS run.

3.3. Data Analysis

1. The review of the data requires the following information: retention times obtained from the selected ion chromatograms; ion abundance and ion peak ratios obtained from the selected ion chromatograms; and quantitative ion ratios between the native drug and its corresponding deuterated internal standard.

2. Typical Agilent ChemStation reports for the fentanyl assay is illustrated in **Fig. 22.1**.

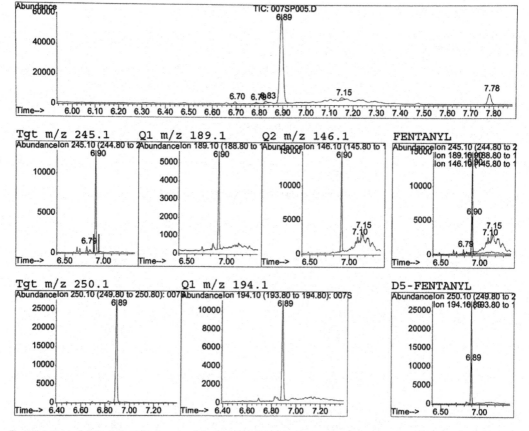

Fig. 22.1. Typical Agilent ChemStation report for fentanyl. **Top panel** is the total ion chromatogram; **middle and bottom panels** are selected ion chromatograms for fentanyl (retention time 6.90 min) and deuterated fentanyl (retention time 6.89 min), respectively.

3. In order for a result to be reported as "Positive", the following criteria must be satisfied: (1) the retention times of the ion peaks must be within ± 1% of the corresponding ions of the intermediate calibrator; (2) the ion peak ratio for the specimen must be within ± 20% of corresponding ion peak ratio of the intermediate calibrator; and (3) the correlation coefficient for the calibration curve must be 0.99 or greater if linear regression is used. Failure to meet one of the above criteria requires that the specimen be reported as "None Detected."

4. The limit of detection for the fentanyl assay is 0.62 ng/mL and the range of linearity is 2.5–50 ng/mL. The reporting criteria for blood are presented in **Table 22.5**. Urine results are qualitative only and are reported as "Positive" when the concentration is greater than the limit of detection.

5. The intra- and inter-assay variability (%CV) is less than <10%.

Table 22.5
Reporting criteria for fentanyl in blood

Concentration (ng/mL)	Reported result
0.62–<1.25	Trace
1.25–<2.5	<2.5 ng/mL
2.5–60	report concentration
>60	reanalyze "diluted" specimen

4. Notes

1. The assay can be applied to other specimens, including those obtained at autopsy.

2. Separate sources of the fentanyl analyte must be used when preparing standard and control solutions.

References

1. Stout, P.R. and Farrell, L.J. (2003) Opioids: Effects on human performance and behavior. *Forensic Sci Rev*, **15**, 29–59. Review.

2. Gutstein, H.B. and Akil, H. (2006) Opioid Analgesics. In Brunton, L.L., Lazo, J.S. and Parker, K.L. (eds), *Goodman & Gilman's the Pharmacological Basis of Therapeutics*, 11th edition. McGraw-Hill, New York, pp. 547–590.

3. Lewis, L.S. (2006) Opioids. In Flomenbaum, N.E., Goldfrank, L.R., Hoffman, R.S., Howland, M.A., Lewin, N.A. and Nelson, L.S. (eds), *Goldfrank's Toxicologic*

Emergencies, 8th edition. McGraw-Hill, New York, pp. 590–613.

4. Clean Screen Extraction Column Application Manual, United Chemical Technologies (2009).

5. Anderson, D.T. and Muto, J.J. (2000) Duragesic transdermal patch: postmortem tissue distribution of fentanyl in 25 cases. *J Anal Toxicol*, **24**, 627–634.

6. Poklis, A. and Backer, R. (2004) Urine concentrations of fentanyl and norfentanyl during application of Duragesic transdermal patches. *J Anal Toxicol*, **28**, 422–425.

7. Thompson, J.G., Baker, A.M., Bracey, A.H., Seningen, J., Kloss, J.S., Strobl, A.Q. and Apple, F.S. (2007) Fentanyl concentrations in 23 postmortem cases from the hennepin county medical examiner's office. *J Forensic Sci*, **52**, 978–981.

Chapter 23

Determination of Total Homocysteine in Plasma Using Liquid Chromatography Coupled to Tandem Mass Spectrometry (LC/MS/MS)

Bruno Casetta

Abstract

A detailed protocol using liquid chromatography coupled to tandem mass spectrometry (LC/MS/MS) is described for the determination of the total homocysteine (HCY) in plasma. Sample preparation is simple and relies on the complete reduction of any homocystine ($(HCY)_2$) or protein-bound homocysteine, followed by the plasma protein precipitation. Any factor influencing the measurement is addressed. The protocol is robust and suitable for large-scale routine determinations.

Key words: Homocysteine, electrospray, liquid chromatography, tandem mass spectrometry.

1. Introduction

First described in 1932 by Butz and Du Vigneaud, homocysteine (HCY) is a sulfur-containing amino acid derived completely through the metabolism of methionine. Under normal physiological conditions, the concentration of HCY is relatively low, less than $10\,\mu moles/L$, due to its enzymatic back conversion to methionine. A deficiency in any of these enzymes leads to increased concentration of HCY with serious implications due to its toxicity. In the 1960s, Carson and Neil first described an association between increased urinary excretion of homocysteine and mental retardation. Values around or above $100\,\mu moles/L$ are considered very dangerous. Since then, homocysteine has been implicated in several other diseases such as atherosclerosis, osteoporosis, diabetes, and some forms of dementia.

U. Garg, C.A. Hammett-Stabler (eds.), *Clinical Applications of Mass Spectrometry*, Methods in Molecular Biology 603, DOI 10.1007/978-1-60761-459-3_23, © Humana Press, a part of Springer Science+Business Media, LLC 2010

Most clinical laboratories employ commercial immunoassay methods for the determination of HCY concentrations; however, both gas and liquid chromatography-based methods have been described for HCY (1). Chromatographic approaches have advantages of being more specific and relatively inexpensive compared to immunoassay, but they are quite demanding in terms of sample preparation and instrument maintenance. In the late nineties, tandem mass-spectrometry coupled to liquid chromatography (LC/MS/MS) was proposed as an alternate to the immunoassay (2–6). The protocol described here is a modification of one of the first LC/MS/MS methods described (2). Although the procedure is centered on a commercial kit specifically designed for LC/MS/MS containing the internal standard, the calibrators, and quality controls, the analytical laboratory may choose to prepare these reagents in-house.

Because of the SH moiety in its structure, HCY has the propensity to be either oxidized to homocystine ["(HCY)$_2$," the dimer generated by bridging two sulfhydryl groups] or to be bound by disulfide bridges to circulating proteins such as albumin and hemoglobin. The principle of the method is centered on "reducing" any homocystine and bound-homocysteine back to the free-homocysteine. The presence of a reducing agent, such as dithiothreitol (DTT) helps maintain the HCY in an un-oxidized state during the electrospray ionization process (in positive mode, electrospray provides an oxidizing environment). The internal standard, d_8-homocystine (d_8-(HCY)$_2$) is subjected to the same reduction reaction to yield d_4-homocysteine d_4-HCY. Any problem in the reduction reaction affects both endogenous HCY and the d_4-HCY, by keeping their ratio unaltered.

2. Materials

2.1. Specimens

Blood samples are drawn into 4.5 mL evacuated glass tubes containing EDTA (BD Vacutainer Systems), placed on ice immediately, and centrifuged at $3,500 \times g$ for 5 min. The plasma is separated and stored at $-20°C$ until analysis.

2.2. Reagents

1. HCY-kit from Recipe-GmbH (Sandstraße 37–39, D-80335 Munich / Germany) containing the following individual components:

    a. Internal standard, 5 μg d8-Homocystine: Reconstitute with 1 mL deionized water for a final concentration of 5 μg/mL.

    b. HCY calibration material: Add 3 mL deionized water, gently mix, and allow to sit for 15 min for the complete dissolution. After reconstitution, store calibrator at $-20°C$ and discard after 1 week.

c. Two levels of quality control materials: Add 3 mL deionized water, gently mix, and allow to sit for 15 min for the complete dissolution. After reconstitution, store calibrator at −20°C and discard after 1 week.

2. Reducing reagent, 1,4-Dithiothreitol (DTT): Store desiccated in refrigerator. Dissolve 77 mg 1,4-dithiothreitol into 1 mL deionized water. Make fresh daily. (*see* **Note 2**)

3. Precipitation reagent: Prepare in fume-hood by adding 25 μL of liquid-chromatography grade trifluoroacetic acid and 50 μL of liquid-chromatography grade formic acid to 50 mL of chromatography-grade acetonitrile. Mix well. Stable for 1 week.

4. Mobile phase A, HPLC-grade water containing 0.1 % v/v formic acid: Mix 1 mL formic acid to 999 mL water. Prepare fresh weekly.

5. Mobile phase B, HPLC-grade acetonitrile containing 0.1 % v/v formic acid: Mix 1 mL formic acid with 999 mL acetonitrile. Prepare fresh weekly.

2.3. Instruments

1. Applied Biosystems/MDS Analytical Technologies API 3200 Tandem Mass Spectrometer.

2. Shimadzu Prominence Liquid chromatography system with autosampler and column oven.

3. LC-column: CYANO 3 um, 4.6 × 33 mm (Supelco Inc).

3. Methods

3.1. Instrument Conditions

The following parameters are used for the Acquisition Method:
Source parameters:
Curtain gas: 20 psi

CAD setting value: 3

Ionspray voltage: 5,200 V

Temperature: 650°C

Gas source 1: 55 psi

Gas source 2: 65 psi

Compound parameters:
Declustering potential: 20 V

Entrance potential: 6 V

Collision energy: 15 eV

Collision exit potential: 2 V (indicative value)

MS section:

Scan type	MRM	
Polarity	Positive	
Q1 mass (amu)	Q3 mass (amu)	Time (msec)
136.1	90.0	250
140.1	94.0	150
Run time:	1.5 min	

LC-module section:
Pumping mode: Binary flow

Total flow: 1 mL/min

B concentration: 40%

B curve: 0

Oven temperature: 30°C

Injection volume: 1 μL

Running time: 1.5 min.

3.2. Procedure

1. Place 20 μL water (blank), calibrator, quality controls, or plasma samples into an appropriated labeled 1.5 mL Microfuge tube.

2. Add 20 μL of the internal standard solution.

3. After vortex mixing, centrifuge $300 \times g$ for 5 min.

4. Add 20 μL of the reduction reagent. (*see* **Note 1**)

5. Cap the tube, vortex-mix the content for 10 sec, and incubate at room temperature for 15 min. (*see* **Note 4**)

6. Add 100 μL of the protein-precipitation reagent, vortex-mix for 10 sec, centrifuge $300 \times g$ for 5 min.

7. Incubate at 4°C for 10 min.

8. Vortex-mix and centrifuge at $13,000 \times g$ for 5 min.

9. Transfer the supernatants to labeled autosampler vials, cap the vials carefully, and place in autosampler rack. (*see* **Note 3**)

10. Start the instrument for samples analysis. (*see* **Note 5**)

11. Acquire and process data by defining the blank and entering the nominal concentration of the calibrator (single-point calibration). Homocysteine data (transition $136 > 90$) are integrated and results are calculated by using the d_4-homocysteine signal (transition $140 > 94$) as for internal standard. **Figure 23.1** shows an example of the obtained tracings. (*see* **Notes 6, 7, 8**)

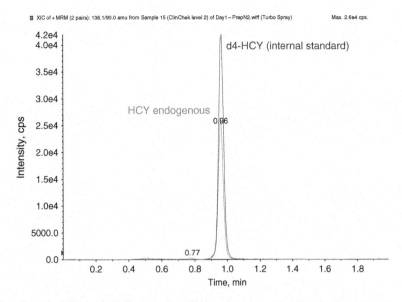

Fig. 23.1. Tracings obtained by the measurement of a QC control plasma. Traces represent the multiple reaction monitoring (MRM) of either the endogenous homocystine, and the internal standard. Measurement by MRM is typical for tandem mass-spectrometry coupled to liquid chromatography (LC/MS/MS).

4. Notes

1. The methodology is centered on the "reductive" media provided by the DTT reagent. Most of the problems are related to this step.

2. Too little DTT, either because not freshly prepared or because an early evaporation/oxidation, results in formation of homocystine dimer $[(HCY)_2]$. This can be seen by the appearance of a second LC-peak at a retention time longer than the HCY itself (*see* **Fig. 23.2**) (7).

3. Microtiter plates can be used as an alternative to sample vials as an economical approach to sample injection. The microtiter plate must be covered to minimize evaporation to prevent early DTT reagent depletion with the subsequent samples exposure to oxidation.

4. The incubation of the samples at room temperature ($\sim$24°C) after the DTT addition is pivotal for completing the reduction process. If the room is cold (<20°C), it could be sensible to incubate the samples in an 37°C oven. Some authors suggest addition of sodium hydroxide to facilitate reduction reaction (6).

5. The protocol presented here has been validated to be robust. **Figure 23.3** shows that even after some 1,000 injections, the proposed configuration does not show any degradation of the analytical performances.

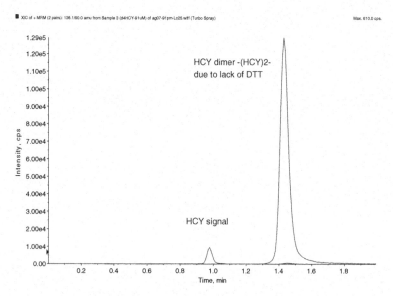

Fig. 23.2. Tracing for the HCY obtained on a sample altercated by a deficiency of DTT. The appearance of a second LC-peak at a retention time longer than the HCY itself denotes the strong presence of homocystine (the "dimerized" homocysteine).

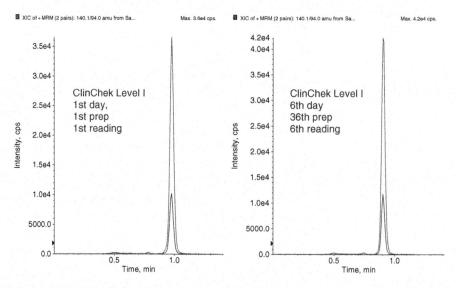

Fig. 23.3. Comparison between the analytical performances at the first injection and after roughly 1,000 injections during a validation survey.

6. **Table 23.1** shows the expected performances on an intra-day precision assay.

7. **Figure 23.4** shows the expected tracings for concentrations approaching the limit of quantitation, estimated in between 0.1 and 0.2 μmol/L.

Table 23.1
Typical performances on an intra-day precision assay during a validation survey

Day 5			
Prep. N25			
Sample Name	Readings Average (uM)	Readings SD (uM)	Readings Precision %
Blank			
ClinCal	14.42	0.22	1.55
ClinCheck level 1	9.87	0.13	1.27
ClinCheck level 2	22.45	0.21	0.92
Prep. N26			
Sample Name	Readings Average (uM)	Readings SD (uM)	Readings Precision %
Blank			
ClinCal	14.40	0.06	0.44
ClinCheck level 1	9.68	0.16	1.68
ClinCheck level 2	22.28	0.38	1.69
Prep. N27			
Sample Name	Readings Average (uM)	Readings SD (uM)	Readings Precision %
Blank			
ClinCal	14.42	0.15	1.02
ClinCheck level 1	9.74	0.13	1.31
ClinCheck level 2	22.83	0.38	1.68
Prep. N28			
Sample Name	Readings Average (uM)	Readings SD (uM)	Readings Precision %
Blank			
ClinCal	14.42	0.08	0.52
ClinCheck level 1	9.25	0.12	1.33
ClinCheck level 2	22.35	0.14	0.68
Prep. N29			
Sample Name	Readings Average (uM)	Readings SD (uM)	Readings Precision %
Blank			
ClinCal	14.40	0.11	0.76
ClinCheck level 1	9.85	0.12	1.25
ClinCheck level 2	22.82	0.38	1.65
Prep. N30			
Sample Name	Readings Average (uM)	Readings SD (uM)	Readings Precision %
Blank			
ClinCal	14.42	0.21	1.48
ClinCheck level 1	9.51	0.10	1.06
ClinCheck level 2	21.50	0.32	1.47
Intra-day results	Average (uM)	SD (uM)	Precision %
ClinCheck level 1	9.65	0.24	2.45
ClinCheck level 2	22.04	0.96	4.36

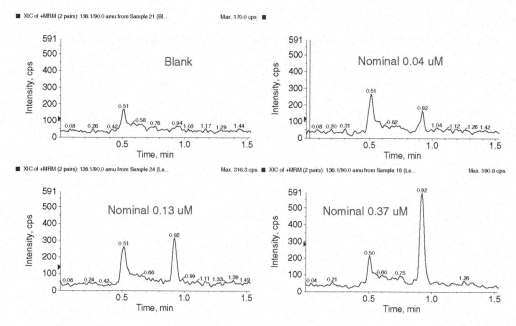

Fig. 23.4. Tracings for concentrations approaching the limit of quantitation, estimated in between 0.1 and 0.2 μmol/L.

8. The inter-day precision is 3.2% for the level I quality control sample with an average value of 9.51 μmol/L (reference value 9.41 μmol/L by LC-UV) and 5.1% for the level II sample with an average value of 21.91 μmol/L. (reference value 21.7 μmol/L by LC-UV).

References

1. Hanson NQ, Eckfeldt JH, Schwichtenberg K, Aras O, Tsai MY. Interlaboratory variation of plasma total homocysteine measurements: results of three successive homocysteine proficiency testing surveys. Clin Chem. 2002; 48(9):1539–1545.

2. Magera MJ, Lacey JM, Casetta B, Rinaldo P. Method for the determination of total homocysteine in plasma and urine by stable isotope dilution and electrospray tandem mass spectrometry. Clin Chem. 1999; 45(9):1517–1522.

3. Gempel K, Gerbitz KD, Casetta B, Bauer MF. Rapid determination of total homocysteine in blood spots by liquid chromatography-electrospray ionization-tandem mass spectrometry. Clin Chem. 2000; 46(1):122–123.

4. Nelson BC, Pfeiffer CM, Sniegoski LT, Satterfield MB. Development and evaluation of an isotope dilution LC/MS method for the determination of total homocystine in human plasma. Anal Chem. 2003; 75:775–784.

5. V. Ducros, Belva-Besnet H, Casetta B, Favier A. A robust LC/MS/MS method for total plasma homocysteine determination in clinical practice. Clin Chem Lab Med. 2006; 44:987–990.

6. Satterfield MB, Sniegoski LT, Welch MJ, Nelson BC. Comparison of isotope dilution mass spectrometry methods for the determination of total homocysteine in plasma and serum. Anal Chem. 2003; 75:4631–4638.

7. Tomaiuolo M, Vecchione G, Grandone E, Cocomazzi N, Casetta B, Di Minno G, Margaglione M. A new method for determination of plasma homocystine by isotope dilution and electrospray tandem mass spectrometry. J Chromatogr B, 2006; 842:64–69.

Chapter 24

Quantitation of Homovanillic Acid (HVA) and Vanillylmandelic Acid (VMA) in Urine Using Gas Chromatography-Mass Spectrometry (GC/MS)

Ryan Allenbrand and Uttam Garg

Abstract

Neuroblastoma, in most cases, is characterized by increased production of catecholamines and their metabolites. Laboratory diagnosis and clinical follow-up include the measurement of urinary homovanillic acid (HVA) and vanillylmandelic acid (VMA). In the following procedure, urine samples are diluted to give a creatinine concentration of 2 mg/dL. Deuterated internal standards are added to the diluted urine samples followed by acidification using HCl. Ethyl acetate is used to extract HVA and VMA from the acidified samples, and the extract is dried. The residue is treated with bis-(trimethylsilyl)trifluoroacetamide (BSTFA), 1% trimethylchlorosilane (TMCS), and pyridine to prepare trimethylsilyl derivatives of HVA and VMA. The derivatized samples are injected to into gas-chromatograph mass spectrometer. The concentration of HVA and VMA is determined by comparing responses of unknown sample to the responses of calibrators using selected ion monitoring.

Key words: Homovanillic acid, vanillylmandelic acid, neuroblastomas, epinephrine, norepinephrine, catecholamines, dopamine.

1. Introduction

Tumors of neural crest origin, such as neuroblastomas and pheochromocytomas, are characterized by their secretion of large amounts of catecholamines. Which catecholamine, dopamine, epinephrine, or norepinephrine is produced in excess varies. Therefore, the measurement of these compounds and of their metabolites homovanillic acid (HVA) and vanillylmandelic acid (VMA) is important in the diagnosis and monitoring of patients found to have such tumors (1–3).

U. Garg, C.A. Hammett-Stabler (eds.), *Clinical Applications of Mass Spectrometry*, Methods in Molecular Biology 603, DOI 10.1007/978-1-60761-459-3_24, © Humana Press, a part of Springer Science+Business Media, LLC 2010

The methods for analysis of HVA and VMA include spectrophotometric techniques and gas or liquid chromatography. Unfortunately, most spectrophotometric methods have poor specificity and, for this reason, have been replaced by chromatography methods. Liquid chromatography-based methods are the most frequently used. Many of these methods employ fluorescence, electrochemistry, or mass spectrometry to permit simultaneous measurement of both HVA and VMA (4–6). Gas chromatography-mass spectrometry (GC-MS) methods are also quite sensitive and specific and can detect both HVA and VMA simultaneously (7–9). In this chapter, we describe a robust GC-MS method involving acidic extraction and simultaneous measurement of HVA and VMA.

2. Materials

2.1. Sample

Randomly collected urine is acceptable sample for this procedure. Freeze urine after collection. Sample is stable for at least 1 month when frozen at –20°C.

2.2. Reagents

1. Bis-(Trimethylsilyl)trifluoroacetamide (BSTFA) with 1% TMCS (trimethylchlorosilane) (United Chemical Technologies, Bristol, PA).

2. d3-homovanillic acid and d3- vanillylmandelic acid (Cambridge Isotopes, Andover, MA)

3. 0.1 N HCl: Add approximately 50 mL of deionized water into a 100 mL volumetric flask. Add 0.83 mL of concentrated HCL to the flask and bring the volume to 100 mL with deionized water (solution stable for 1 year at room temperature).

4. 1 N HCl: Add approximately 50 mL of deionized water into a 100 mL volumetric flask. Add 8.3 mL of concentrated HCL to the flask and bring the volume to 100 mL with deionized water (solution stable for 1 year at room temperature).

5. 6 N HCl: Add 50 mL of DI water into a 100 mL volumetric flask. Bring the volume to 100 mL with concentrated HCL (solution stable for 1 year at room temperature).

2.3. Standards and Calibrators

1. HVA/VMA primary stock standard (20 mg/dL): Weigh out 20 mg of HVA (Sigma Chemical Co., St. Louis, MO) and 20 mg of VMA (Sigma Chemical Co., St. Louis, MO) into a 100 mL volumetric flask and bring the volume to 100 mL with 0.1 N HCl. The standard solution is stable for 1 year at –20°C.

Table 24.1
Preparation of secondary standards and calibrators

20 mg/dL Primary standard (mL)	0.1 N HCL (mL)	Concentration of secondary standard (mg/dL)	Volume (mL) of secondary standard	Volume (mL) of deionized water	Concentration of calibrators (mg/dL)
0.050	9.950	0.1	0.4	9.6	0.004
0.200	9.800	0.4	0.4	9.6	0.016
1.000	9.000	2.0	0.4	9.6	0.080
2.500	7.500	5.0	0.4	9.6	0.200

Table 24. 2
GC/MS operating conditions

Column pressure	5 psi
Injector temp.	250°C
Purge time on	0.5 min
Detector temp.	280°C
Initial oven temp.	70°C
Initial time	1.0 min
Temperature ramp	15°C/min
Final oven temp.	270°C
Final time	4.0 min
MS source temp.	230°C
MS mode	Electron impact at 70 eV, Selected ion monitoring
MS tune	Autotune

2. HVA/VMA secondary stock standards: Prepare in 10 mL volumetric flasks as per **Table 24.1**. The standard solution is stable for 1 year at −20°C.

3. Calibrators: Prepare in 10 mL volumetric flasks as per **Table 24.2**.

2.4. Internal Standard and Quality Control Samples

1. Working internal standard (1 mg/dL): Add 5 mg of d,l-HVA-d3 and 5 mg of d,l-VMA d3 into a 500 mL volumetric flask. Add 250 mL of deionized water and mix the contents.

Add 100 µL of concentrated HCl and bring the volume to 500 mL with deionized water. Internal standard is stable for 2 years at −20°C, or 1 year at 0°C.

2. Quality control samples (Bio-Rad Laboratories, Sunnydale, CA): These are lypholized controls. Reconstitute each normal and abnormal control in 10 mL of 0.01 N HCl (*see* **Note 1**).

2.5. Supplies

1. 13 × 100 mm test tubes and Teflon-lined caps (Fisher Scientific, Fair Lawn, NJ).

2. Autosampler vials (12 × 32 mm; crimp caps) with 0.3 mL limited volume inserts (P.J. Cobert Associates, Inc., St. Louis, MO).

3. GC column: Zebron ZB-1 with dimensions of 30 m × 0.25 mm × 0.25 µm (Phenomenex, Torrance, California).

2.6. Equipment

1. TurboVap®1 V Evaporator (Zymark Corporation, Hopkinton, MA, USA)

2. A gas chromatograph/mass spectrometer (GC-MS), model 6890 N/5973 operated in electron impact mode (Agilent Technologies, Santa Clara, CA)

---

3. Methods

3.1. Stepwise Procedure

1. Determine creatinine concentration on urine samples to be analyzed (*see* **Note 2**).

2. For each sample, add volume of urine containing 200 µg of creatinine to 10 mL volumetric flasks and bring the volume to mark using deionized water. This provides a creatinine concentration of 2 mg/dL (*see* **Note 3**).

3. Add 1 mL of diluted sample or calibrator or control to appropriately labeled 13 × 100 mm tubes.

4. Add 1 mL of 1.0 N HCL to each 13 × 100 tube.

5. Add 0.100 mL of working internal standard to each tube.

6. Add 4 mL of ethyl acetate to each tube.

7. Cap the tubes and vortex gently to mix.

8. Rock the tubes for 10 min.

9. Centrifuge the tubes at ∼1,600 × g for 10 min.

10. Transfer the upper organic layer to appropriately labeled 13 × 100 mm concentration tubes (*see* **Note 4**).

11. Evaporate the organic layer to dryness under nitrogen in a water bath at 37°C (*see* **Note 5**).

12. To each tube add 80 μL of BSTFA containing 1% TMCS and 40 μL pyridine.

13. Place the tubes in dry heat block 65°C for 15 min (*see* **Note 6**).

14. Allow tubes to cool at room temperature and transfer the contents to appropriately labeled autosampler vials containing tapered Inserts.

15. Position and crimp vial caps. Place into the autosample and inject 1 μL into GC-MS for analysis.

3.2. Instrument Operating Conditions

The instrument's operating conditions are given in **Table 24.2**.

3.3. Data analysis

1. Representative GC-MS chromatogram of derivatized HVA and VMA is shown in **Fig. 24.1**. GC-MS selected ion chromatograms of labeled and unlabeled HVA and VMA are shown in **Figs. 24.2 and 24.3**, respectively. Electron impact ionization mass spectra of these compounds are shown in **Figs. 24.4 and 24.5**, respectively (*see* **Note 7**). Ions used for identification and quantification is listed in **Table 24.3**.

2. Analyze data using Target Software (Thru-Put Systems, Orlando, FL) or similar software. The quantifying ions (**Table 24.3**) are used to construct standard curves of the peak area ratios (calibrator/internal standard pair) vs. concentration. These curves are then used to determine the concentrations of the controls and unknown samples.

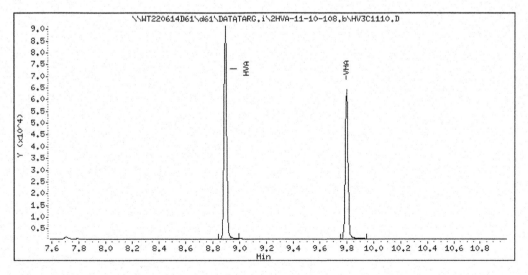

Fig. 24.1. GC-MS chromatogram of TMS-derivatives of HVA and VMA for 0.08 mg/dL calibrator.

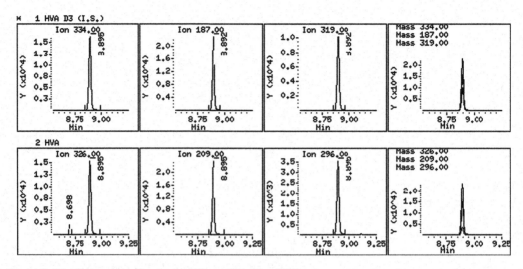

Fig. 24.2. Selected ion chromatograms of HVA and HVA-d3 – diTMS

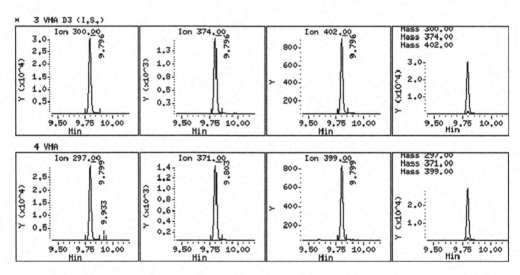

Fig. 24.3. Selected ion chromatograms of VMA and VMA-d3 – tri-TMS

3. The linearity/limit of quantitation of the method is 0.004 to 0.200 mg/dL. Samples in which the analyte concentrations exceed the upper limit of quantitation should be diluted with deionized water and reanalyzed.

4. Typical calibrator curves have a correlation coefficient (r^2) >0.99.

5. Typical intra- and inter-assay imprecision is <10%.

6. Quality control: The run is considered acceptable if calculated concentrations of HVA/VMA in the controls are within +/− 20% of target values. Quantifying ion in the sample is considered acceptable if the ratios of qualifier ions to quantifying ion are within +/− 20% of the ion ratios for the calibrators.

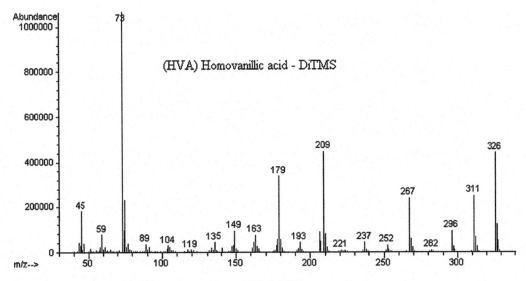

Fig. 24.4. Electron impact ionization mass spectrum of HVA – diTMS.

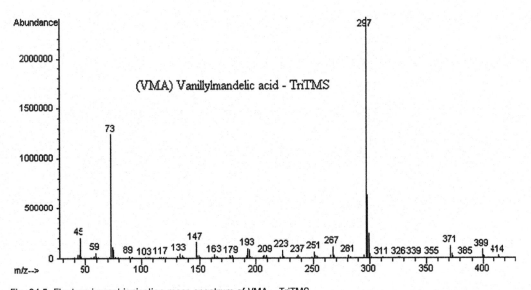

Fig. 24.5. Electron impact ionization mass spectrum of VMA – TriTMS.

7. Patient results: The results are expressed in mg of HVA or VMA per g of creatinine. As the diluted urine samples have concentration of 2 mg/dL and the results are expressed as g of creatinine, the results calculated from the calibration curve are multiplied by 500. Reference intervals for HVA and VMA are given in **Table 24.4** (4, 5).

Table 24.3
Quantitation ions for HVA, VMA, HVA-d3, and VMA-d3

	Quantitation ion	Quantitation ions
HVA	326	209, 296
VMA	297	371, 399
HVA-d3	334	187, 319
VMA-d3	300	374, 402

Table 24.4
Reference Intervals for HVA and VMA

Age	HVA (mg/g creatinine)	VMA (mg/g creatinine)
0–6 months	<39.0	<18.8
6–12 months	<32.6	<18.8
1–4 years	<22.0	<11.0
5–9 years	<15.1	<8.3
10–19 years	<12.8	<8.2
>19 years	<7.6	<6.0

4. Notes

1. Lyophilized controls are stable until expiration on the bottles when stored unopened at 2–8°C. If acidified, reconstituted controls are stable for 30 days. When no acid is used, controls are stable for only 5 days.

2. The amount of creatinine in each sample must be measured prior to sample extraction using a validated method. The amount of creatinine is used to assess the overall concentration of the urine sample and the results are expressed as HVA or VMA to creatinine ratios.

3. The formula for calculating the volume of urine in μL, containing 200 μg creatinine is: $400 \times 50/$(creatinine concentration in mg/dL). The final concentration of the diluted urine is 2 mg/dL.

4. When transferring the ethyl acetate layer, it is important that no aqueous layer is transferred. This will result in failed assay, as HVA/VMA will not derivatize.

5. Do not over-dry samples because this could cause low recovery.

6. Do not over-dry the extract. This will result in poor recovery and failed run.

7. Electron impact ionization spectra are needed in the initial stages of method set-up to establish retention times and later on, if there is a need for change in quantifying or qualifying ions. They are not needed for routine quantitation.

Acknowledgement

We acknowledge the help of David Scott in preparing the figures.

References

1. Strenger, V., Kerbl, R., Dornbusch, H.J., Ladenstein, R., Ambros, P.F., Ambros, I.M. and Urban, C. (2007) Diagnostic and prognostic impact of urinary catecholamines in neuroblastoma patients. *Pediatr Blood Cancer*, **48**, 504–9.

2. Williams, C.M. and Greer, M. (1963) Homovanillic acid and vanilmandelic acid in diagnosis of neuroblastoma. *JAMA*, **183**, 836–40.

3. Nishi, M., Miyake, H., Takeda, T., Takasugi, N., Sato, Y. and Hanai, J. (1986) Vanillylmandelic acid, homovanillic acid, and catecholamines in urine of infants with neuroblastoma 6- to 11-month-old. *Jpn J Clin Oncol*, **16**, 351–5.

4. Soldin, S.J. and Hill, J.G. (1981) Liquid-chromatographic analysis for urinary 4-hydroxy-3-methoxymandelic acid and 4-hydroxy-3-methoxyphenylacetic acid, and its use in investigation of neural crest tumors. *Clin Chem*, **27**, 502–3.

5. Soldin, S.J., Lam, G., Pollard, A., Allen, L.C. and Logan, A.G. (1980) High performance liquid chromatographic analysis of urinary catecholamines employing amperometric detection: references values and use in laboratory diagnosis of neural crest tumors. *Clin Biochem*, **13**, 285–91.

6. Hanai, J., Kawai, T., Sato, Y., Takasugi, N., Nishi, M. and Takeda, T. (1987) Simple liquid-chromatographic measurement of vanillylmandelic acid and homovanillic acid in urine on filter paper for mass screening of neuroblastoma in infants. *Clin Chem*, **33**, 2043–6.

7. Seviour, J.A., McGill, A.C., Dale, G. and Craft, A.W. (1988) Method of measurement of urinary homovanillic acid and vanillylmandelic acid by gas chromatography-mass spectrometry suitable for neuroblastoma screening. *J Chromatogr*, **432**, 273–7.

8. Gleispach, H., Huber, E., Fauler, G., Kerbl, R., Urban, C. and Leis, H.J. (1995) Neuroblastoma screening: labeling of HVA and VMA for stable isotope dilution gas chromatography-mass spectrometry. *Nutrition*, **11**, 604–6.

9. Fauler, G., Leis, H.J., Huber, E., Schellauf, C., Kerbl, R., Urban, C. and Gleispach, H. (1997) Determination of homovanillic acid and vanillylmandelic acid in neuroblastoma screening by stable isotope dilution GC-MS. *J Mass Spectrom*, **32**, 507–14.

Chapter 25

Quantitation of 17-OH-Progesterone (OHPG) for Diagnosis of Congenital Adrenal Hyperplasia (CAH)

Ravinder J. Singh

Abstract

Most (90%) cases of congenital adrenal hyperplasia (CAH) are due to mutations in the steroid 21-hydroxylase gene (*Cyp21*). CAH due to 21-hydroxylase deficiency is diagnosed by confirming elevations of 17-hydroxyprogesterone (OHPG) and androstenedione (ANST) with decreased cortisol. By contrast, in two less common forms of CAH, due to 17-hydroxylase or 11-hydroxylase deficiency, OHPG and ANST levels are not significantly elevated and measurement of progesterone (PGSN) and deoxycorticosterone (DOC), respectively are necessary for diagnosis. Since 21-hydroxyase deficiency is more common and results in remarkable increase in OHPG, this test is most commonly ordered compared to other steroid intermediates. Various methods are used in clinical laboratories for the analysis of OHPG in serum or plasma but mass spectrometric methods are considered gold standard method.

Key words: Tandem mass spectrometry, endocrine, congenital adrenal hyperplasia, steroids, 17-hydroxyprogesterone, 21-hydroxylase.

1. Introduction

The adrenal glands, ovaries, testes, and placenta produce 17-hydroxyprogesterone (OHPG), which is in turn hydroxylated at the 11- and 21-positions to produce cortisol. Deficiency of either 11- or 21-hydroxylase results in decreased cortisol synthesis, and feed-back inhibition of adrenocorticotrophic hormone (ACTH) secretion is lost. Deficiency of the 21-hydroxylase enzyme is the primary cause of CAH. Since the increased pituitary release of ACTH increases production of OHPG, this hormone serves as the best screening test for CAH (1–2). Clinically, OHPG testing provides merely supplementary information and though useful for screening, should never be employed as the sole

U. Garg, C.A. Hammett-Stabler (eds.), *Clinical Applications of Mass Spectrometry*, Methods in Molecular Biology 603, DOI 10.1007/978-1-60761-459-3_25, © Humana Press, a part of Springer Science+Business Media, LLC 2010

diagnostic tool. OHPG analysis can also be useful as part of a battery of tests to evaluate females with hirsutism or infertility which can result from adult-onset CAH.

The majority of OHPG is bound to transcortin and albumin, so total OHPG is measured in the assays performed in clinical laboratories. OHPG is converted to pregnanetriol and conjugated and is finally excreted in the urine. In all instances, OHPG measurement is more sensitive and specific to diagnose disorders of steroid metabolism compared to measurement of pregnanetriol in urine. The OHPG method described here allows the determination of OHPG in 100 μl of serum using liquid chromatography tandem mass spectrometry (LC-MS/MS) methodology (3–5).

2. Materials

2.1. Samples

Collect 1 mL blood in plain redtop tube. Separate serum and store in the refrigerator. Freeze the sample if assay cannot be performed within 7 days. 0.5 mL serum (*see* **Note 1**).

2.2. Reagents and Buffers

1. 50% methanol: Combine 2 L methanol and 2 L deionized H_2O. Mix. Store at room temperature. Reagent is stable for 6 months (*see* **Note 2**).

2. Mobile phase A: To 1 L of HPLC-grade H_2O add 1 mL of formic acid. Mix and degas. Store at ambient temperature. Reagent is stable for 6 months.

3. Mobile phase B: To 1 L of acetonitrile add 1 mL of formic acid. Mix and degas. Store at ambient temperature. Reagent is stable for 6 months.

4. Reconstitution Solvent: Dilute 700 mL of methanol to 1 L with deionized H_2O. Add 1 mL of 1 mg/mL stock estriol for a final concentration of 1 μg/mL. Store at –20°C. Reagent is stable for 6 months.

5. Bovine Serum Albumin Standard Buffer, 0.1 M PBS, 1.0% BSA, 0.9% NaCl, pH 7.4: Dissolve 2.14 g $Na_2HPO_4·7 H_2O$, 0.268 g $NaH_2PO_4·H_2O$, 0.9 g NaCl, 1.0 g bovine serum albumin in 75 ml of deionized H_2O. Adjust pH to 7.4 (with dilute NaOH or HCl) and bring to 100 mL volume with deionized H_2O. Use this buffer to prepare standards described in the calibration section of this procedure. Store at 4°C. Calibrators are stable for 10 years.

6. LC-MS/MS performance Check Sample (15 ng/ml): Dilute the working internal standard 1:2 with reconstitution solvent. Stable for 10 years.

7. 17 α-Hydroxyprogesterone Powder – minimum 98 % purity.

8. Charcoal Stripped Human Serum.

9. Stock Estriol: Dissolve 25 mg of estriol in 25 mL of methanol. Store at −20°C. Stable for 2 years.

2.3. Standards and Calibrators

1. Stock I, 50 µg/mL: Dissolve 10 mg 17 α-Hydroxyprogesterone powder in 200 mL methanol. Store in 5 mL aliquots at −70°C in crimp-cap vials. Reagent is stable for 10 years.

2. Stock II, 500 ng/mL: Dilute 2 mL of 17-hydroxyprogesterone Stock I to 200 mL with reconstitution solvent. Store in 10 mL aliquots at −70°C in crimp-cap vials. Reagent is stable for 10 years.

3. Stock III, 40 ng/mL: Dilute 8 mL of 17-hydroxyprogesterone Stock II to 100 mL with BSA standard buffer. Store at −70°C. Stable for 10 years.

4. Calibrators: Prepare the calibrators according to **Table 25.1**.

Table 25.1
Prepare working standards according to the table below

Standard	BSA standard buffer	Amount of stock III to add	OHPG conc. ng/dL
Std 7	None	50 mL	4,000
Std 6	Fill to 50 mL	31.25 mL	2,500
Std 5	Fill to 50 mL	12.5 mL	1,000
Std 4	Fill to 50 mL	2.5 mL	200
Std 3	Fill to 50 mL	0.5 mL	40
Std 2	Fill to 50 mL	0.2 mL	16
Std 1	50 mL	None	0

2.4. Internal Standard

1. Stock internal standard, 0.1 mg/mL: Weigh 10 mg of OHPG 2,2,4,6,6,21,21,21-d$_8$ (>98% purity) into a 10 ml volumetric flask and bring to volume with methanol. Store at −20°C. Reagent is stable for 10 years.

2. Working internal standard, 30 ng/mL: Add 30 µL of stock internal standard to 1 L of reconstitution solvent. Stable for 2 years when stored at −20°C.

2.5. Quality Controls

1. Three controls low, medium, and high are prepared by spiking Charcoal Stripped Human serum (SeraCare Life Sciences) (*see* **Note 3**).

2. Pour approximately 500 ml of the Charcoal Stripped Human Serum into three, 1 L volumetric flasks. Label the three flasks as low, medium, and high and add 20, 140, and 400 uL of OHPG Stock II (500 ng/mL), respectively.

3. Place a magnetic stir bar in each flask and mix the contents on a stir-plate for 5 min.

4. After testing each pool in an assay, aliquot them into 20 mL plastic scintillation vials. Store at −70°C. Controls are stable for 10 years.

5. Run quality controls with each run (*see* **Note 4**).

2.6. Supplies

1. Autosampler vials with caps.

2. Disposable culture tubes 13 × 100 mm.

3. Disposable culture tubes 12 × 75 mm.

4. Extraction Cartridges: Strata- × 30 mg/mL (Phenomenex).

5. The analytical column is a Synergi Max-RP, 150 × 2 mm, 4 μm (Phenomenex).

6. The pre-column filter, C12 (Max RP) 4 mm L × 2 mm ID (Phenomenex).

2.7. Equipment

1. Tandem Mass Spectrometer with an ESI or APCI.

2. HPLC with autosampler.

3. Evaporator.

3. Methods

3.1. Sample Preparation:

1. Pipet 0.1 mL of each standard, control, and sample to the appropriately labeled 13 × 100 mm glass tubes.

2. Add 50 μL of working internal standard to each tube and vortex. Incubate at least 10 min.

3. Add 1 ml of deionized H_2O to each tube and vortex.

3.2. Solid Phase Extraction

1. Condition extraction cartridges by applying 1.0 mL of methanol to each cartridge at a rate of <2.0 mL/min. Discard eluates.

2. Apply 1.0 mL of deionized H_2O to each extraction cartridge at a rate of <2.0 mL/min and discard eluates.

3. Apply each prepared standard, control, and sample to the appropriate conditioned cartridge at a rate of <2.0 mL/min. Discard eluates.

4. Apply 2.0 mL of 50% methanol to each cartridge at a rate of <2.0 mL/min. Discard eluates. Wipe any hanging drops of fluid from cartridge nipples.

5. Place 12 × 75 mm glass tubes under each cartridge and apply 1.0 mL of methanol to each at a rate of <2.0 mL/min. Collect eluates.

6. Place the tubes containing the eluates into an evaporator and dry under nitrogen.

7. Add 150 µL of reconstitution solvent to each dried eluate tube and gently mix.

8. Transfer the reconstituted samples to the auto sampler vials and inject 30 µL into LC-MS/MS.

3.3. Instrument Operating Conditions

LC-MS/MS is used with an ESI or APCI source to monitor ion pairs in multiple reaction monitoring mode (**Table 25.2**) (*see* **Note 5**). The flow rate of the mobile phase is 0.75 mL/min with no split of the post-column flow to the mass spectrometer. The solvent gradient is as follows: Equilibrate of the HPLC column with 60% mobile phase B for 2 min which is then raised to 72% over 2 min for elution of analytes. Increase to 100% mobile phase B for 1 min to wash out other hydrophobic peaks. Return to 60% mobile phase B for 1 min to recondition before the next injection. The total run time is 6 min per injection. To assure that system is optimized and is working well to run patient samples, the signal-to-noise for LC-MS/MS performance check sample should at least be 1,000 but not less than 500.

Table 25.2
MRM transitions

Analyte	Q1	Q3	Qualifier ion
17-OHP	331.4	109.2	97.3
17-OHP-IS	339.0	113.2	100.3

3.4. Data Analysis

The data are collected and analyzed using mass spectrometer vendor specific software. The quantitation of OHPG is made from a standard curve using the peak area ratio of analyte to that of the internal standard. The calibration curve is constructed using weighting of 1/x and linear regression. The limit of quantification of the method is 15 ng/dL. The linearity is up to 2,000 ng/dL with very high correlation coefficient (>0.999). The inter-day precision should be less than 10% for a control with an OHPG mean value of ~2,000 ng/dL. During method validation it is critical to make sure that relative recovery of OHPG from human serum/plasma matrix is 100 ±10% and carry over is less than 0.01%.

4. Notes

1. Gel barrier phlebotomy tubes are unacceptable as steroids bind to the gel.

2. The organic solvents used in the extraction and mobile phases are flammable. They should be handled with care and used in an exhaust hood. Use eye protection when working with acids or bases.

3. Assay 0.1 ml of the Charcoal Stripped Serum prior to use to make sure that it is negative for OHPG.

4. Mean and acceptable limits are established by assaying each pool level 20 times over multiple days. The means, standard deviations, and %CVs are calculated for each level of control. An established %CV, based on assay historical performance and medical relevance, is used to calculate the ± 2 SD range. All QC values are graphed on paper QC charts and entered into lab information system for each assay, and monitored for being inbounds, trends, and shifts. QC charts are reviewed according to laboratory quality rules.

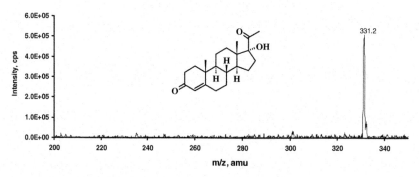

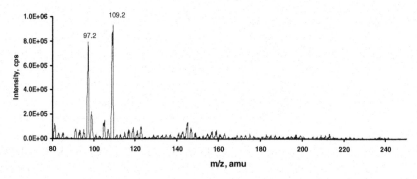

Fig. 25.1. Tandem Mass Spectra of OHPG during infusion of solution into LC-MS/MS. (**A**) Precursor ion spectrum. (**B**) Product ion spectrum.

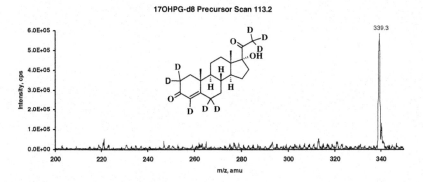

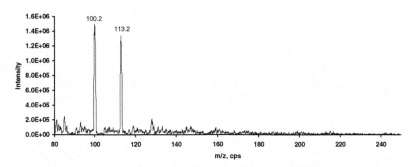

Fig. 25.2. Tandem Mass Spectra of OHPG-IS during infusion of solution into LC-MS/MS. (**A**) Precursor ion spectrum. (**B**) Product ion spectrum.

5. OHPG signal-to-noise is optimized by infusing OHPG solution at a concentration of 10 μg/ml (**Figs. 25.1** and **25.2**). This method is based on off-line SPE clean-up of the serum/plasma sample, to avoid suppression, before injection into the LC-MS/MS.

References

1. Von Schnakenburg K, Bidlingmaier F, Knorr D (1980) 17-hydroxyprogesterone, androstenedione, and testosterone in normal children and in prepubertal patients with congenital adrenal hyperplasia. *Eur J Pediatr* **133**:259–67.

2. Collett-Solberg P (2001) Congenital adrenal hyperplasia: from genetics and biochemistry to clinical practice, part I. *Clin Pediatr* **40**:1–16.

3. Wudy SA, Hartmann M, Svoboda M (2000) Determination of 17-hydroxyprogesterone in plasma by stable isotope dilution/ benchtop liquid chromatography-tandem mass spectrometry. *Horm Res* **53**:68–71.

4. Wudy SA, Hartmann M, Homoki J (2000) Hormonal diagnosis of 21-hydroxylase deficiency in plasma and urine of neonates using benchtop gas chromatography-mass spectrometry. *J Endocrinol* **165**:679–83.

5. Kao P, Machacek DA, Magera MJ, Lacey JM, Rinaldo P (2001) Diagnosis of adrenal cortical dysfunction by liquid chromatography-tandem mass spectrometry. *Ann Clin Lab Sci* **31**:199–204.

Chapter 26

Determination of Hydroxyurea in Serum or Plasma Using Gas Chromatography-Mass Spectrometry (GC-MS)

David K. Scott, Kathleen Neville, and Uttam Garg

Abstract

Hydroxyurea is an antineoplastic drug, which is also widely used in the treatment of sickle cell disease. Various methods including colorimetry, high performance liquid chromatography, and gas chromatography-mass spectrometry (GC-MS) are available for the assay of hydroxyurea. In the gas chromatography method described, the drug is extracted from serum, plasma, or urine using ethyl acetate and phosphate buffer (pH 6). The organic phase containing drug is separated and dried under a stream of nitrogen. After trimethylsilyl derivatization, samples are analyzed using GC-MS. Quantitation of the drug in a sample is achieved by comparing responses of the unknown sample to the responses of the calibrators using selected ion monitoring. Tropic acid is used as an internal standard.

Key words: Hydroxyurea, mass spectrometry, sickle cell, antineoplastic drugs.

1. Introduction

Hydroxyurea is an antineoplastic drug used for the treatment of hematological malignancies such as polycythemia vera and essential thrombocytosis. It is also used in the treatment of sickle cell disease (1–4). The exact mechanisms of action of hydroxyurea remain known, but the drug is known to increase nitric oxide levels, which activates soluble guanylyl cyclase and cyclic GMP and subsequently causes vasodilatation. By activating fetal hemoglobin production and removing rapidly dividing cells that preferentially produce sickle hemoglobin, hydroxyurea reduces the sickling encountered with these cells and thus reduces sickling-related complications (5, 6).

U. Garg, C.A. Hammett-Stabler (eds.), *Clinical Applications of Mass Spectrometry*, Methods in Molecular Biology 603, DOI 10.1007/978-1-60761-459-3_26, © Humana Press, a part of Springer Science+Business Media, LLC 2010

Treatment of hydroxyurea is monitored through various clinical and laboratory parameters. The clinical side effects of hydroxyurea treatment are quite serious and range from neurological and gastrointestinal symptoms to mucositis, anorexia, stomatitis, bone marrow toxicity, and alopecia. The laboratory monitoring includes complete blood counts with differential and liver and renal function tests. For the evaluation of clinical treatment and to study pharmacokinetics, the direct measurement of hydroxyurea is useful (1). The methods of measurement include colorimetry (7), high performance liquid chromatography (HPLC) (8, 9), and gas chromatography-mass spectrometry (GC-MS) (10). Colorimetric assays not only require large sample volumes but also have poor sensitivity and specificity. Similarly HPLC methods are laborious and lack specificity. In contrast, GC-MS methods have proven to be fast, reliable, and sensitive and specific. In this chapter, we describe a GC-MS method for the measurement of hydroxyurea involving acidic extraction of hydroxyurea followed by trimethylsilylation derivatization.

2. Materials

2.1. Sample

Serum or plasma (heparin or EDTA) are acceptable samples for this procedure. Sample is stable for 1 week when refrigerated or 2 month when frozen at −20°C.

2.2. Reagents and Buffers

1. 6 N Hydrochloric acid: Add 50 mL deionized water to a 100 mL volumetric flask and fill to the mark with concentrated HCl.
2. 0.4 M Phosphate solution A: Dissolve 11.0 g monobasic sodium phosphate monohydrate into 200 mL deionized water. Stable for 1 year at room temperature.
3. 0.4 M Phosphate solution B: Dissolve 10.7 g dibasic sodium phosphate heptahydrate into 100 mL deionized water. Stable for 1 year at room temperature.
4. 0.4 M Phosphate buffer: Add phosphate solution A to a 500 mL beaker. Adjust the pH to 6.0 ± 0.1 by slowly adding phosphate solution B (see **Note 1**). Stable for 1 year at room temperature.
5. Derivatizing agent: One part pyridine plus two parts Bis-(Trimethylsilyl)trifluoroacetamide (BSTFA) containing 1% TMCS (trimethylchlorosilane) (UCT Inc., Bristol, PA).
6. Human drug-free plasma or serum (UTAK Laboratories, Inc., Valencia, CA).

2.3. Standards and Calibrators

1. Hydroxyurea primary standard, 1 mg/mL: Add 10.0 mg hydroxyurea to a 10 mL volumetric flask and fill to the mark with deionized water. Stable for 2 months at −20°C.

2. Hydroxyurea secondary standard, 100 μg/mL: Add 1.0 mL hydroxyurea primary standard (1 mg/mL) to a 10 mL volumetric flask and fill to the mark with deionized water. Stable for 2 months at −20°C.

3. Hydroxyurea tertiary standard, 10 μg/mL: Add 100 μl hydroxyurea primary standard (1 mg/mL) to a 10 mL volumetric flask and fill to the mark with deionized water. Stable for 2 months at −20°C.

4. Working calibrators: Prepare according to **Table 26.1**. Calibrators are stable for 2 months at −20°C.

Table 26.1
Preparation of calibrators. Prepare calibrators in 10 mL volumetric flasks according to the table below and fill to the mark with UTAK drug-free human serum

Secondary standard (μL)	Tertiary standard (μL)	Final concentration (μg/mL)
0	100	0.1
0	1000	1.0
200	0	2.0
1000	0	10.0

Calibrators are stable for 2 months at −20°C.

2.4. Internal Standard and Quality Controls

1. Working internal standard, tropic acid, 1 mg/dL: Add 5 mg of d,l-tropic acid (Sigma-Aldrich, St. Louis, MO) to a 500 mL volumetric flask. Add approximately 250 mL deionized water and 100 μL of concentrated HCl. Mix the contents to dissolve tropic acid, and fill to the mark with deionized water. Stable for 2 years at −20°C.

2. Hydroxyurea quality control primary standard, 1 mg/mL: Add 10.0 mg hydroxyurea powder to a 10 mL volumetric flask and fill to the mark with deionized water. Stable for 2 months at −20°C.

3. Hydroxyurea quality control secondary standard, 100 μg/mL: Add 1.0 mL hydroxyurea stock standard (1 mg/mL) to a 10 mL volumetric flask and fill to the mark with deionized water. Stable for 2 months at −20°C.

4. Hydroxyurea quality control tertiary standard, 10 µg/mL: Add 100 µL hydroxyurea stock standard (1 mg/mL) to a 10 mL volumetric flask and fill to the mark with deionized water. Stable 2 months at −20°C.

5. Working quality controls are made according to **Table 26.2** (*see* **Note 2**). Controls are stable for 2 months, stored at −20°C.

Table 26.2
Preparation of in-house controls. Prepare controls in 10 mL volumetric flasks according to the table below and fill to mark with UTAK normal human serum

Primary control (µL)	Secondary control (µL)	Tertiary control (µL)	Final conc. (µg/mL)
0	0	500	0.5
0	500	0	5.0
500	0	0	50.0

Controls are stable for 2 months at −20°C.

2.5. Supplies

1. 13 × 100 mm screw-cap glass tubes (Fisher Scientific, Pittsburgh, PA). These tubes are used for extraction and concentrating drug extracts.

2. Transfer pipettes (Samco Scientific, San Fernando CA).

3. Autosampler vials (12 × 32 mm with crimp caps) with 0.3 mL limited volume inserts (P.J. Cobert Associates, St. Louis, MO).

4. GC column: Zebron ZB-1 with dimensions of 30 m × 0.25 mm × 0.25 µm (Phenomenex, Torrance, California).

2.6. Equipment

1. TurboVap®1 V Evaporator (Zymark Corporation, Hopkinton, MA, USA).

2. A gas chromatograph/mass spectrometer (GC-MS), model 6890 N/5973 operated in electron impact mode (Agilent Technologies, Wilmington, DE).

3. Methods

3.1. Stepwise Procedure

1. Add 0.5 mL calibrator, control, or sample into a labeled 13 × 100 mm screw-cap glass tube.

2. Add 1.0 mL of working phosphate buffer (0.4 M, pH 6.0).

3. Add 0.1 mL of working internal standard (1 mg/dL tropic acid.).

4. Add 4 mL ethyl acetate.

5. Cap and rock the tubes for 5 min.

6. Centrifuge at ~1,600 × g for 5 min.

7. Transfer the upper organic layer to a 13 × 100 mm tube. Take care not to transfer the aqueous layer (*see* **Note 3**).

8. Evaporate the organic layer to dryness under nitrogen stream at 40°C.

9. Reconstitute the residue with 100 µL derivatizing agent.

10. Cap and heat the tubes in dry block at 65°C for 20 min.

11. Remove the tubes from the dry block and cool them to room temperature.

12. Transfer the contents to autosampler vials containing tapered inserts. Crimp the cap the vials and place in the autoinjector of the GC-MS.

13. Inject 1 µL on GC-MS for analysis.

3.2. Instrument Operating Conditions

The instrument's operating conditions are given in **Table 26.3**.

Table 26.3
GC/MS operating conditions

Column pressure	5 psi
Injector temp.	250°C
Mode	Splitless
Purge time on	0.7 min
Detector temp.	280°C
Initial oven temp.	70°C
Initial time	1.0 min
Temperature ramp	25°C/min
Final oven temp.	280°C
Final time	6.0 min
MS source temp.	230°C
MS mode	Electron Impact at 70 eV, selected ion monitoring
MS tune	Autotune

3.3. Data Analysis

1. Representative GC-MS chromatogram of derivatized hydroxyurea and tropic acid (internal standard) is shown in **Fig. 26.1**. GC-MS selected ion chromatograms are shown in **Fig. 26.2**. Electron impact ionization mass spectra of these compounds are shown in **Figs. 26.3 and 26.4** respectively (*see* **Note 4**). Ions used for identification and quantification are listed in **Table 26.4**.

2. Analyze data using Target Software (Thru-Put Systems, Orlando, FL) or similar software. The quantifying ions (**Table 26.4**) are used to construct standard curves of the peak area ratios (calibrator/internal standard pair) vs. concentration. These curves are then used to determine the concentrations of the controls and unknown samples.

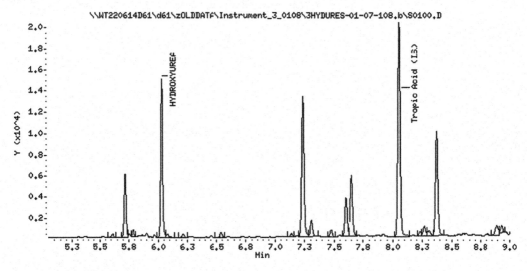

Fig. 26.1 GC-MS chromatogram of a 10 μg/mL calibrator.

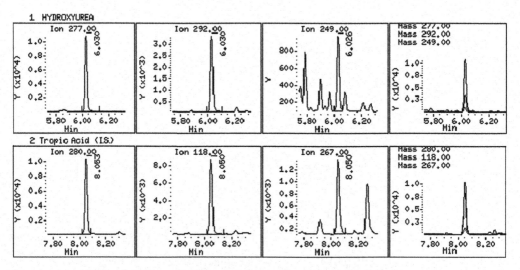

Fig. 26.2. Selected ion chromatograms of hydroxyurea – triTMS and tropic acid – diTMS.

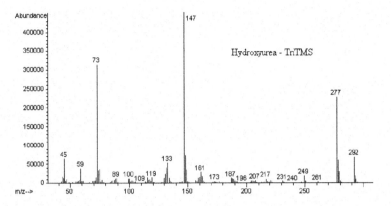

Fig. 26.3. Electron impact ionization mass spectrum of hydroxyurea – triTMS.

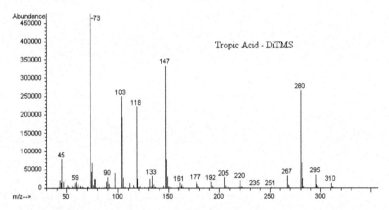

Fig. 26.4. Electron impact ionization mass spectrum of tropic acid – diTMS.

Table 26.4
Quantitation and qualifying ions

	Quantitation ion	Qualifier ions
Hydroxyurea-triTMS	277	292, 249
Tropic Acid-diTMS	280	118, 267

3. The linearity/limit of quantitation of the method is 0.1 to 10 μg/mL (*see* **Note 5**). Samples in which the drug concentrations exceed the upper limit of quantitation should be diluted with negative serum or plasma and retested.

4. Typical calibration curves have a correlation coefficient (r^2) >0.99.

5. Typical intra- and inter-assay imprecision is <10%.

6. Quality control: The run is considered acceptable if calculated concentrations of hydroxyurea in the controls are within +/− 20% of target values. Quantifying ion in the sample is considered acceptable if the ratios of qualifier ions to quantifying ion are within +/− 20% of the ion ratios for the calibrators.

4. Notes

1. It takes approximately 30 mL of solution B to adjust the pH to 6.0.

2. When possible, calibrators and controls should be made from different batches of drugs at another time by different analysts.

3. When transferring the ethyl acetate layer, it is important that no aqueous layer is transferred. This will result in failed assay, as the drugs will not derivatize.

4. Electron impact ionization spectra are needed in the initial stages of method set up to establish retention times and later on, if there is a need for change in quantifying or qualifying ions. They are not needed for routine quantitation.

5. This procedure can be used for urine samples. Since urine generally has higher concentrations of hydroxyurea, standard curve is made using 1, 10, 100 μg/mL calibrators. For urine assay, quantitation ion is 292.

References

1. Bachir, D., Hulin, A., Huet, E., Habibi, A., Nzouakou, R., El Mahrab, M., Astier, A. and Galacteros, F. (2007) Plasma and urine hydroxyurea levels might be useful in the management of adult sickle cell disease. *Hemoglobin*, **31**, 417–25.

2. Ferguson, R.P., Arun, A., Carter, C., Walker, S.D. and Castro, O. (2002) Hydroxyurea treatment of sickle cell anemia in hospital-based practices. *Am J Hematol*, **70**, 326–8.

3. Platt, O.S. (2008) Hydroxyurea for the treatment of sickle cell anemia. *N Engl J Med*, **358**, 1362–9.

4. Svarch, E., Machin, S., Nieves, R.M., Mancia de Reyes, A.G., Navarrete, M. and Rodriguez, H. (2006) Hydroxyurea treatment in children with sickle cell anemia in Central America and the Caribbean countries. *Pediatr Blood Cancer*, **47**, 111–2.

5. Cokic, V.P., Smith, R.D., Beleslin-Cokic, B.B., Njoroge, J.M., Miller, J.L., Gladwin, M.T. and Schechter, A.N. (2003) Hydroxyurea induces fetal hemoglobin by the nitric oxide-dependent activation of soluble guanylyl cyclase. *J Clin Invest*, **111**, 231–9.

6. Ferster, A., Vermylen, C., Cornu, G., Buyse, M., Corazza, F., Devalck, C., Fondu, P., Toppet, M. and Sariban, E. (1996) Hydroxyurea for treatment of severe sickle cell anemia: a pediatric clinical trial. *Blood*, **88**, 1960–4.

7. Bolton, B.H., Woods, L.A., Kaung, D.T. and Lawton, R.L. (1965) A simple method of colorimetric analysis for hydroxyurea (Nsc-32065). *Cancer Chemother Rep*, **46**, 1–5.

8. Manouilov, K.K., McGuire, T.R. and Gwilt, P.R. (1998) Colorimetric determination of hydroxyurea in human serum using high-performance liquid chromatography. *J Chromatogr B Biomed Sci Appl*, **708**, 321–4.

9. Pujari, M.P., Barrientos, A., Muggia, F.M. and Koda, R.T. (1997) Determination of hydroxyurea in plasma and peritoneal fluid by high-performance liquid chromatography using electrochemical detection. *J Chromatogr B Biomed Sci Appl*, **694**, 185–91.

10. James, H., Nahavandi, M., Wyche, M.Q. and Taylor, R.E. (2006) Quantitative analysis of trimethylsilyl derivative of hydroxyurea in plasma by gas chromatography-mass spectrometry. *J Chromatogr B Analyt Technol Biomed Life Sci*, **831**, 42–7.

Chapter 27

Quantitation of Ibuprofen in Blood Using Gas Chromatography-Mass Spectrometry (GC-MS)

Gerry Huber and Uttam Garg

Abstract

Ibuprofen is a non-narcotic, non-steroidal anti-inflammatory drug used for the treatment of pain, fever, and inflammatory diseases such as rheumatoid arthritis, osteoarthritis, and ankylosing spondylitis. It is also used for induction of closure of patent ductus arteriosus (PDA) in neonates. Although the exact mechanism of action of ibuprofen is not known, it is believed to mediate its therapeutic effects through the inhibition of cyclooxygenase and subsequently by the inhibition of prostacyclin production. As the drug has a number of side effects, which correlate to its circulating concentration, monitoring of ibuprofen in plasma or serum is desired for patients receiving high-dose therapy. Chromatographic methods are frequently used for the assay of ibuprofen, as no immunoassays are currently available.

In the method described, the drug is extracted from the serum or plasma using methylene chloride and phosphate buffer (pH 6). Meclofenamic acid is used as an internal standard. The organic phase containing the drug is separated and dried under stream of nitrogen. After trimethylsilyl derivatization, analysis is done using gas-chromatography/ mass spectrometry (GC-MS). Quantification of the drug in a sample is achieved by comparing responses of the unknown sample to the responses of the calibrators using selected ion monitoring.

Key words: Ibuprofen, mass spectrometry, gas chromatography, non-steroidal anti-inflammatory drug.

1. Introduction

Ibuprofen is a non-steroidal, anti-inflammatory drug used in the treatment of inflammatory diseases such as rheumatoid arthritis, gout, cystic fibrosis, osteoarthritis, and ankylosing spondylitis. Ibuprofen lysine is preferred over indomethacin for the closure of patent ductus arteriosus (PDA) in infants due to its lesser side effects (1–3). Ibuprofen can be given intravenously or orally, though oral administration has been shown to be as effective as intravenous administration (4). When taken orally, it is well absorbed in the intestine, reaches maximum plasma concentration

U. Garg, C.A. Hammett-Stabler (eds.), *Clinical Applications of Mass Spectrometry*, Methods in Molecular Biology 603, DOI 10.1007/978-1-60761-459-3_27, © Humana Press, a part of Springer Science+Business Media, LLC 2010

within 1–2 h, and exhibits its maximum effect at 2–4 h. The half-life of ibuprofen is approximately 1–4 h. Ibuprofen is extensively metabolized in the liver through oxidation of the isobutyl group. The two major metabolites 2-hydroxyibuprofen and 2-carboxyibuprofen are excreted by the kidneys.

Ibuprofen is believed to work through the inhibition of cyclooxygenase resulting in the inhibition of prostacyclin production. Therapeutic drug monitoring for patients receiving high-dose therapy is important as high serum or plasma concentrations are associated with many side effects such as nausea, epigastric pain, diarrhea, vomiting, dizziness, blurred vision, tinnitus, and edema. Therapeutic drug monitoring is particularly important for neonates receiving ibuprofen for the treatment of PDA as metabolism in very young children is unpredictable. The therapeutic target for these patients is 10–50 μg/mL.

Various methods using high performance liquid and gas chromatography have been described for the analysis of ibuprofen (5–8), including some that enable the simultaneous determination of the S–(+) ibuprofen and R–(−) ibuprofen enantiomers (8). Another HPLC method describes the simultaneous determination of the major phase I and II metabolites (9). In addition to HPLC methods, several GC-MS methods involving derivatization of ibuprofen have been also described (10, 11). GC-MS methods generally involve acidic extraction of the drug followed by derivatization. In this procedure we describe a GC-MS method for the assay of ibuprofen.

2. Materials

2.1. Sample

Serum or plasma (heparin or EDTA) are acceptable samples for this procedure. Sample is stable for 1 week when refrigerated or 1 month when frozen at −20°C.

2.2. Reagents and Buffers

1. Meclofenamic acid (Sigma Chemical Co., St. Louis, MO).

2. Ibuprofen, 1 mg/mL, methanolic solution (Cerilliant Corporation, Round Rock, TX).

3. Bis-(Trimethylsilyl)trifluoroacetamide (BSTFA) with 1% TMCS (trimethylchlorosilane) (UCT Inc., Bristol, PA).

4. Human drug-free plasma or serum (UTAK Laboratories, Inc., Valencia, CA).

5. 0.4 M Phosphate solution A: Dissolve 11.0 g monobasic sodium phosphate monohydrate into 200 mL deionized water. Stable for 1 year at room temperature.

6. 0.4 M Phosphate solution B: Dissolve 10.7 g dibasic sodium phosphate heptahydrate into 100 mL deionized water. Stable for 1 year at room temperature.

7. Phosphate buffer: Add phosphate solution A to a 500 mL beaker. Adjust the pH to 6.0 ± 0.1 by slowly adding phosphate solution B (*see* **Note 1**). Stable for 1 year at room temperature.

8. Extraction tubes: Add 3.0 mL methylene chloride and 1.0 mL 0.4 M phosphate buffer C into 13×100 mm extraction tubes. Stable for 6 months at room temperature.

2.3. Standards and Calibrators

1. Primary standard: 1 mg/mL methanolic solution.

2. 100 μg/mL ibuprofen secondary standard: Dilute 1 mL of primary standard to 10 mL with methanol. Store in amber bottle at $-20°C$. Stable for 1 year.

3. Working calibrators: prepare as described in **Table 27.1**. Stable for 1 year when stored in amber bottles at $-20°C$.

Table 27.1
Preparation of calibrators. Final volume of each calibrator is 1 mL

1 mg/mL Primary standard (μL)	100 μg/mL Secondary standard (μL)	Drug-free serum (μL)	Final conc. (μg/mL)
0	0	1000	0
0	50	950	5
50	0	950	50
200	0	800	200

2.4. Internal Standard and Quality Controls

1. Primary internal standard (meclofenamic acid 1.0 mg/mL): Add 10.0 mg meclofenamic acid to a 10 mL volumetric flask and bring volume to the mark with methanol. Store in amber bottle. The solution is stable for 1 year at $-20°C$.

2. Working internal standard (Meclofenamic acid 10.0 μg /mL): Add 100 μL of primary internal standard to a 10 mL volumetric flask and bring to volume with methanol. Store in amber bottle. The solution is stable for one year at $-20°C$.

3. Quality control samples: In-house controls are used to validate sample analysis. Prepare controls according to **Table 27.2** (*see* **Note 2**).

2.5. Supplies

1. 13×100 mm screw-cap glass tubes (Fisher Scientific, Pittsburgh, PA). These tubes are used for extraction and concentrating drug extracts.

Table 27.2
Preparation of controls. Final volume of each control is 1 mL

1 mg/mL Primary standard (μL)	100 μg/mL Secondary standard (μL)	Drug-free serum (μL)	Final conc. (μg/mL)
0	250	750	25
100	0	900	100

2. Transfer pipettes (Samco Scientific, San Fernando CA).

3. Autosampler vials (12 × 32 mm with crimp caps) with 0.3 mL limited volume inserts (P.J. Cobert Associates, St. Louis, MO).

4. GC column: Zebron ZB-1, 15 m × 0.25 mm × 0.25 μm. (Phenomenex, Torrance, California).

2.6. Equipment

1. A gas chromatograph/mass spectrometer system (GC-MS; 6890/5975 or 5890/5972) with autosampler and operated in electron impact mode (Agilent Technologies, Wilmington, DE).

2. TurboVap® IV Evaporator (Zymark Corporation, Hopkinton, MA, USA).

3. Methods

3.1. Stepwise Procedure

1. Add 100 μL of calibrator, control, and sample to the appropriately labeled extraction tube.

2. Add 200 μL working meclofenamic acid (internal standard) to each tube.

3. Cap the tubes and rock for 5 min.

4. Centrifuge the tubes for 5 min at ~1,600 × g.

5. Aspirate the top aqueous layer using a vacuum system (*see* **Note 3**).

6. Transfer organic phase to a concentration tube (*see* **Note 4**).

7. Prepare an unextracted standard by combining 100 μL secondary ibuprofen standard and 100 μL of working internal standard into a concentration tube.

8. Evaporate the samples to dryness under nitrogen in the evaporator at 45°C (*see* **Note 5**).

9. Add 100 μL BSTFA with 1% TMCS and 200 μL ethyl acetate to each tube.

10. Cap the tubes and mix by gentle vortexing.

11. Derivatize by incubating samples in a heat block for 15 min at 65°C.

12. Transfer the contents to autosampler vials containing glass inserts, cap, and place onto the GC-MS for analysis.

3.2. Instrument Operating Conditions

Inject 1 μL onto the GC/MS column. The instrument operating conditions are given in **Table 27.3**.

Table 27.3
GC-MS operating conditions

Column pressure	5 psi
Injector temp.	250°C
Purge time on	0.5 min
Detector temp	280°C
Initial oven temp	120°C
Initial time	1.0 min
Temperature ramp	30°C/min
Final oven temp	270°C
Final time	4.0 min
MS source temp.	230 °C
MS mode	Electron Impact at 70 eV, selected ion monitoring
MS tune	Autotune

3.3. Data Analysis

1. Representative GC-MS chromatogram of ibuprofen and meclo-fenamic acid (internal standard) is shown in **Fig. 27.1** GC-MS selected ion chromatograms are shown in **Fig. 27.2** Electron impact ionization mass spectra of these compounds are shown in the **Figs. 27.3 and 27.4** respectively (*see* **Note 6**). Ions used for identification and quantification are listed in **Table 27.4**.

2. Analyze data using Target Software (Thru-Put Systems, Orlando, FL) or similar software. The quantifying ions (**Table 27.4**) are used to construct standard curves of the peak area ratios (calibrator/internal standard pair) vs. concentration. These curves are then used to determine the concentrations of the controls and unknown samples.

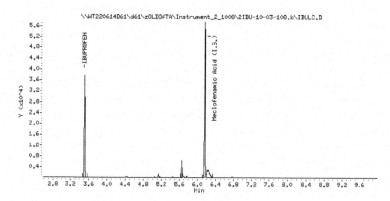

Fig. 27.1. GC-MS chromatogram of TMS-derivatives of ibuprofen and meclofenamic acid (internal standard) for 25 µg/mL calibrator.

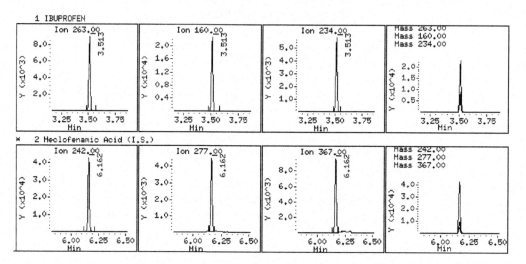

Fig. 27.2. Selected ion chromatograms of TMS-derivatives ibuprofen and meclofenamic acid (internal standard).

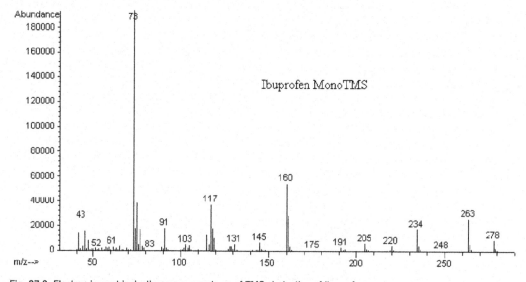

Fig. 27.3. Electron impact ionization mass spectrum of TMS-derivative of ibuprofen.

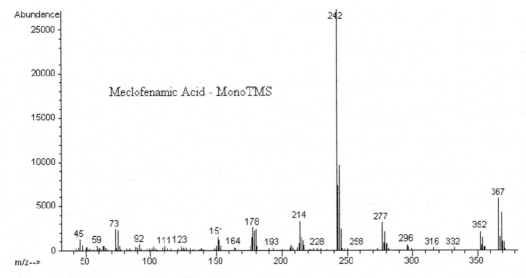

Fig. 27.4. Electron impact ionization mass spectrum of TMS-derivative of meclofenamic acid.

Table 27.4
Quantitation and qualifying ions for ibuprofen and meclofenamic acid (IS)

Analyte	Quantifying ion	Qualifying ions
Ibuprofen	263	160, 234
Meclofenamic acid	242	277, 367

3. The linearity/limit of quantitation of the method is 5–200 µg/mL. Samples in which the drug concentrations exceed the upper limit of quantitation should be diluted with negative serum or plasma and retested.

4. Standard curves should have a correlation coefficient (R^2) >0.99.

5. Intra- and inter-assay imprecision should be <10%.

6. Quality control: The run is considered acceptable if calculated concentrations of drugs in the controls are within +/− 20% of target values. Quantifying ion in the sample is considered acceptable if the ratios of qualifier ions to quantifying ion are within +/− 20% of the ion ratios for the calibrators.

4. Notes

1. It takes approximately 30 mL of buffer B to adjust the pH to 6.0.

2. When possible, calibrators and controls should be made from different batches of drugs and at different times by different analysts.

3. Remove the aqueous phase as much as possible without disturbing the organic phase. It is okay to remove interface between the aqueous and organic layers.

4. Be careful not to transfer any remaining aqueous phase into the concentration tubes. This will result in failed run, as the drugs will not derivatize.

5. Do not over-dry the extract. This will result in poor recovery and failed run.

6. Electron impact ionization spectra are needed in the initial stages of method set up to establish retention times and later on, if there is a need for change in quantifying or qualifying ions. They are not needed for routine quantification.

Acknowledgement

We acknowledge the help of David Scott in preparing the figures.

References

1. Su, P.H., Chen, J.Y., Su, C.M., Huang, T.C. and Lee, H.S. (2003) Comparison of ibuprofen and indomethacin therapy for patent ductus arteriosus in preterm infants. *Pediatr Int*, **45**, 665–70.

2. Keady, S. and Grosso, A. (2005) Ibuprofen in the management of neonatal Patent Ductus Arteriosus. *Intensive Crit Care Nurs*, **21**, 56–8.

3. Aranda, J.V. and Thomas, R. (2006) Systematic review: intravenous Ibuprofen in preterm newborns. *Semin Perinatol*, **30**, 114–20.

4. Supapannachart, S., Limrungsikul, A. and Khowsathit, P. (2002) Oral ibuprofen and indomethacin for treatment of patent ductus arteriosus in premature infants: a randomized trial at Ramathibodi Hospital. *J Med Assoc Thai*, **85(Suppl 4)**, S1252–8.

5. Sochor, J., Klimes, J., Sedlacek, J. and Zahradnicek, M. (1995) Determination of ibuprofen in erythrocytes and plasma by high performance liquid chromatography. *J Pharm Biomed Anal*, **13**, 899–903.

6. Lemko, C.H., Caille, G. and Foster, R.T. (1993) Stereospecific high-performance liquid chromatographic assay of ibuprofen: improved sensitivity and sample processing efficiency. *J Chromatogr*, **619**, 330–5.

7. Castillo, M. and Smith, P.C. (1993) Direct determination of ibuprofen and ibuprofen acyl glucuronide in plasma by high-performance liquid chromatography using solid-phase extraction. *J Chromatogr*, **614**, 109–16.

8. Canaparo, R., Muntoni, E., Zara, G.P., Della Pepa, C., Berno, E., Costa, M. and Eandi, M. (2000) Determination of Ibuprofen in human plasma by high-performance liquid chromatography: validation and application in pharmacokinetic study. *Biomed Chromatogr*, **14**, 219–26.

9. Kepp, D.R., Sidelmann, U.G., Tjornelund, J. and Hansen, S.H. (1997) Simultaneous quantitative determination of the major phase I and II metabolites of ibuprofen in biological fluids by high-performance liquid chromatography on dynamically modified silica. *J Chromatogr B Biomed Sci Appl*, **696**, 235–41.

10. Maurer, H.H., Kraemer, T. and Weber, A. (1994) Toxicological detection of ibuprofen and its metabolites in urine using gas chromatography-mass spectrometry (GC-MS). *Pharmazie*, **49**, 148–55.

11. Heikkinen, L. (1984) Silica capillary gas chromatographic determination of ibuprofen in serum. *J Chromatogr*, **307**, 206–9.

Chapter 28

Quantitation of Indomethacin in Serum and Plasma Using Gas Chromatography-Mass Spectrometry (GC-MS)

Sarah Thomas, Amanda Sutton, and Uttam Garg

Abstract

Indomethacin is a non-narcotic and non-steroidal anti-inflammatory drug used in the treatment of various inflammatory diseases such as rheumatoid arthritis, osteoarthritis, and ankylosing spondylitis. In neonates, it is also used for induction of closure of patent ductus arteriosus (PDA). Its mechanism of action is believed to be through the inhibition of cyclooxygenase. Due to narrow therapeutic window and number of side effects, it's monitoring, particularly in neonates, is recommended. In the gas chromatography method described here, the drug is extracted from serum or plasma using methylene chloride and phosphate buffer (pH 6). The methylene chloride phase containing drug is separated and dried under stream of nitrogen. The drug is derivatized using Bis-(Trimethylsilyl)trifluoroacetamide (BSTFA) with 1% TMCS (trimethylchlorosilane). The derivatized drug is analyzed using gas chromatography- mass spectrometry. Quantitation of the drug in a sample is achieved by comparing responses of the unknown sample to the responses of the calibrators using selected ion monitoring. Meclofenamic acid is used as an internal standard.

Key words: Indomethacin, mass spectrometry, gas chromatography, patent ductus arteriosus, non-steroidal anti-inflammatory drug.

1. Introduction

Indomethacin is a non-narcotic and non-steroidal anti-inflammatory drug used for the treatment of rheumatoid arthritis, gout, cystic fibrosis, ankylosing spondylitis, osteoarthritis, and many other inflammatory diseases. It is also used in neonates for the induction of closure of patent ductus arteriosus (PDA) (1–3). Indomethacin is believed to work through the inhibition of cyclooxygenase resulting in the inhibition of prostacyclin production. The drug is available in various forms and doses. Although it

U. Garg, C.A. Hammett-Stabler (eds.), *Clinical Applications of Mass Spectrometry*, Methods in Molecular Biology 603, DOI 10.1007/978-1-60761-459-3_28, © Humana Press, a part of Springer Science+Business Media, LLC 2010

is completely and readily absorbed after oral administration, indomethacin undergoes appreciable enterohepatic circulation and is mostly eliminated via renal excretion and some by biliary excretion. Peak plasma concentrations are achieved in about 2 h. The mean half-life of indomethacin in adults is 2–6 h, while for infants the half-life is longer, ~20 h (4, 5).

Therapeutic drug monitoring is important for this drug due to its narrow therapeutic window (0.3–3.0 μg/mL) and the number of known adverse effects: gastritis, drowsiness, lethargy, nausea, vomiting, convulsions, paresthesia, headache, dizziness, hypertension, cerebral edema, tinnitus, disorientation, hepatitis, and renal failure. Contraindications for its use in neonates include necrotizing enterocolitis, impaired renal function, active bleeding, and thrombocytopenia. The boxed warnings of indomethacin include its association with an increased risk of adverse cardiovascular events, myocardial infarction, stroke, and new onset or worsening of pre-existing hypertension.

The commonly used methods for quantitation of indomethacin involve high performance liquid chromatography (6–10) and gas chromatography-mass spectrometry (GC-MS) (11, 12). Here we describe a GC-MS method for the determination of indomethacin.

2. Materials

2.1. Sample

Serum or plasma (heparin or EDTA) are acceptable samples for this procedure. Sample is stable for 1 week when refrigerated or for 1 month when frozen at −20°C.

2.2. Reagents and Buffers

1. Bis-(Trimethylsilyl)trifluoroacetamide (BSTFA) with 1% TMCS (trimethylchlorosilane) (United Chemical Technologies, Bristol, PA).

2. Solution A (0.4 M): Dissolve 11.0 g of sodium phosphate monobasic monohydrate in 200 mL deionized water. Solution is stable at room temperature for 3 months.

3. Solution B (0.4 M): Dissolve 10.7 g of sodium phosphate dibasic heptahydrate in 100 mL deionized water. Solution is stable at room temperature for 3 months.

4. Phosphate buffer (0.4 M): Add phosphate solution A to a 500 mL beaker. Adjust the pH to 6.0 ± 0.1 by slowly adding phosphate solution B (*see* **Note 1**). Stable for 1 year at room temperature.

5. Extraction tubes: Add 3.0 mL methylene chloride and 1.0 mL 0.4 M phosphate buffer into 13 × 100 mm extraction tubes. Cap tightly and store at room temperature.

2.3. Standards and Calibrators

1. 1 mg/mL Indomethacin Primary Standard: Prepare in methanol from analytical grade indomethacin powder (Sigma-Aldrich, St. Louis, MO). The indomethacin powder is stable at 4°C for 2 years. The primary standard solution is stable for 1 year when stored at −20°C.

2. 100 μg/mL Indomethacin Secondary Standard: Prepare from primary standard by 1:10 dilution with methanol. The secondary standard is stable for 1 year at −20°C.

3. Indomethacin Working Calibrators: Prepare calibrators according to **Table 28.1**. Calibrators are stable for 6 months at −20°C.

Table 28.1
Preparation of calibrators. Final volume of each calibrator is 10 mL

100 μg/mL secondary standard (mL)	10 μg/mL calibrator (mL)	Drug-free serum (mL)	Final conc. (μg/mL)
0	0	10	0
1	0	9	10
0	2	8	2
0	0.25	9.75	0.25

2.4. Internal Standard and Quality Controls

1. Primary Internal Standard (meclofenamic acid 1.0 mg/mL): Add 10.0 mg meclofenamic acid to a 10 mL volumetric flask and bring volume to the mark with methanol. Store in amber bottle. The solution is stable for 1 year at −20°C.

2. Meclofenamic Acid Working Internal Standard (10.0 μg /mL): Add 100 μL of primary internal standard to a 10 mL volumetric flask and bring volume to the mark with methanol. Store in amber bottle. The solution is stable 1 year at −20°C.

3. Quality control samples: Prepare controls according to **Table 28.2** (*see* **Note 2**).

Table 28.2
Preparation of controls. Final volume of each control is 10 mL

100 μg/mL secondary standard (mL)	5 μg/mL control (mL)	Drug-free serum (mL)	Final conc. (μg/mL)
0.5	0	9.5	5
0	2	8.0	1

2.5. Supplies	1. 13 × 100 mm screw-cap glass tubes (Fisher Scientific, Pittsburgh, PA). These tubes are used for extraction and concentrating drug extracts.
	2. Transfer pipettes (Samco Scientific, San Fernando CA).
	3. Autosampler vials (12 × 32 mm with crimp caps) with 0.3 mL limited volume inserts (P.J. Cobert Associates, St. Louis, MO).
	4. GC column: Zebron ZB-1 with dimensions of 15 m × 0.25 mm × 0.25 μm (Phenomenex, Torrance, California).
2.6. Equipment	1. A gas chromatograph/mass spectrometer system (GC-MS; 6890/5975 or 5890/5972) with autosampler and operated in electron impact mode (Agilent Technologies, Wilmington, DE).
	2. TurboVap® IV Evaporator (Zymark Corporation, Hopkinton, MA, USA).

3. Methods

3.1. Stepwise Procedure	1. Add 200 μL of sample, calibrator, or control to an appropriately labeled acidic extraction tube.
	2. Add 100 μL meclofenamic acid working internal standard to each tube.
	3. Cap the tubes and gently mix by rocking for 5 min.
	4. Centrifuge the tubes for 5 min at ~1,600 × *g*.
	5. Aspirate the top aqueous layer using a vacuum system (*see* **Note 3**).
	6. Transfer organic phase to a concentration tube (*see* **Note 4**).
	7. Prepare an unextracted standard by combining 100 μL secondary indomethacin standard and 200 μL of working internal standard into a concentration tube.
	8. Evaporate the samples to dryness under nitrogen in the evaporator at 40°C (*see* **Note 5**).
	9. Add 75 μL BSTFA with 1% TMCS to each tube.
	10. Cap the tubes and vortex gently.
	11. Incubate the tubes in a heat block for 15 min at 65°C to derivatize.
	12. Transfer the contents of the tubes to autosampler vials containing glass inserts and inject 1 μL on GC-MS for analysis.

3.2. Instrument Operating Conditions

The instrument's operating conditions are given in **Table 28.3**.

Table 28.3
GC-MS operating conditions

Column pressure	5 psi
Injector temp	250°C
Purge time on	0.5 min
Detector temp	280°C
Initial oven temp	120°C
Initial time	1.0 min
Temperature ramp	30°C/min
Final oven temp	270°C
Final time	4.0 min
MS source temp.	230°C
MS mode	Electron impact at 70 eV, selected ion monitoring
MS tune	Autotune

3.3. Data Analysis

1. Representative GC-MS chromatogram of indomethacin and meclofenamic acid (internal standard) is shown in **Fig. 28.1**. GC-MS selected ion chromatograms are shown in **Fig. 28.2**. Electron impact ionization mass spectra of these compounds

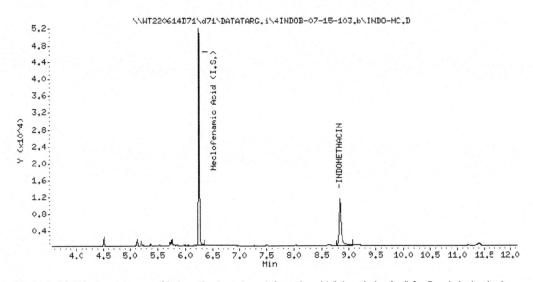

Fig. 28.1. GC-MS chromatogram of indomethacin and meclofenamic acid (internal standard) for 5 μg/mL standard.

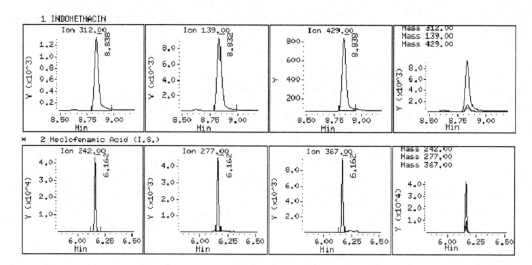

Fig. 28.2. Selected ion chromatograms of indomethacin – monoTMS and meclofenamic acid – momoTMS.

are shown in the **Figs. 28.3 and 28.4**, respectively (*see* **Note 6**). Ions used for identification and quantification are listed in **Table 28.4**.

2. Analyze data using Target Software (Thru-Put Systems, Orlando, FL) or similar software. The quantifying ions (**Table 28.4**) are used to construct standard curves of the peak area ratios (calibrator/internal standard pair) vs. concentration. These curves are then used to determine the concentrations of the controls and unknown samples.

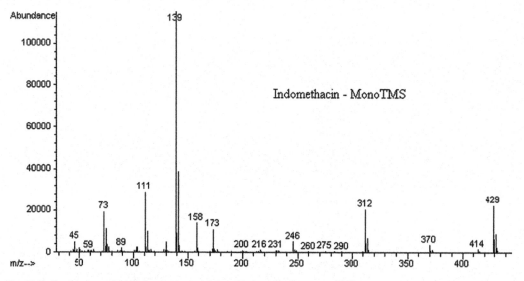

Fig. 28.3. Electron impact ionization mass spectrum of indomethacin – monoTMS.

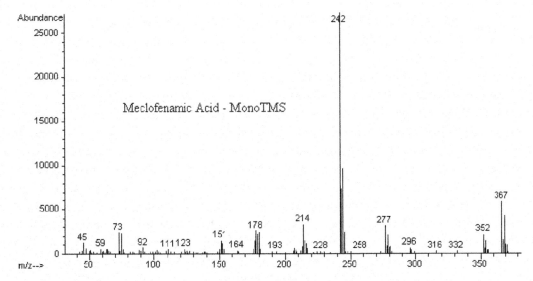

Fig. 28.4. Electron impact ionization mass spectrum of meclofenamic acid – monoTMS.

Table 28.4
Quantitation and qualifying ions for indomethacin and meclofenamic acid

	Quantifying ions	Qualifying ions
Indomethacin	312	139, 429
Meclofenamic acid	242	277, 367

3. The linearity/limit of quantitation of the method is 0.25–10 µg/mL. Samples in which the drug concentrations exceed the upper limit of quantitation should be diluted with negative serum or plasma and retested.

4. Typical calibration curve has correlation coefficient (r^2) of >0.99.

5. Typical intra- and inter-assay imprecision is <10%.

6. Quality control: The run is considered acceptable if calculated concentrations of drugs in the controls are within +/− 20% of target values. Quantifying ion in the sample is considered acceptable if the ratios of qualifier ions to quantifying ion are within +/− 20% of the ion ratios for the calibrators.

4. Notes

1. Approximately 30 mL of buffer B is required to adjust the solution pH to 6.0.

2. When possible calibrators and controls should be made independently using different lots of drugs by different analysts.

3. Remove the aqueous phase as completely as possible without disturbing the organic phase. It is okay to remove interface between the aqueous and organic layers.

4. Be careful not to transfer any remaining aqueous phase into the concentration tubes. This will inhibit the derivitization process and result in analysis failure.

5. Do not over-dry the extract as this will result in poor recovery and result in a run failure.

6. Electron impact ionization spectra are needed in the initial stages of method set up to establish retention times and later on, if there is a need for change in quantifying or qualifying ions. They are not needed for routine quantitaion.

Acknowledgment

We acknowledge the help of David Scott in preparing the figures.

References

1. Lee, K.S. (2008) Indomethacin therapy for patent ductus arteriosus in premature infants. *Pediatr Cardiol*, **29**, 873–5.

2. O'Donovan, D.J., Fernandes, C.J., Nguyen, N.Y., Adams, K. and Adams, J.M. (2004) Indomethacin therapy for patent ductus arteriosus in premature infants: efficacy of a dosing strategy based on a second-dose peak plasma indomethacin level and estimated plasma indomethacin levels. *Am J Perinatol*, **21**, 191–7.

3. Takami, T., Yoda, H., Kawakami, T., Yamamura, H., Nakanishi, T., Nakazawa, M., Takei, Y., Miyajima, T. and Hoshika, A. (2007) Usefulness of indomethacin for patent ductus arteriosus in full-term infants. *Pediatr Cardiol*, **28**, 46–50.

4. Broussard, L.A., McCudden, C.R. and Garg, U. (2007) Therapeutic drug monitoring of analgesics in "Therapeutic Drug Monitoring Data". Eds. Hammett-Stabler C.A. and Dasgupta A., AACC Press, Washington DC, pp. 193–6.

5. Fenton, J. (1998) Indomethacin in "The Laboratory and the Poisoned Patient: A Guide for Interpreting Laboratory Data", AACC Press, Washington DC, pp. 193–5.

6. Al Za'abi, M.A., Dehghanzadeh, G.H., Norris, R.L. and Charles, B.G. (2006) A rapid and sensitive microscale HPLC method for the determination of indomethacin in plasma of premature neonates with patent ductus arteriousus. *J Chromatogr B Analyt Technol Biomed Life Sci*, **830**, 364–7.

7. Sato, J., Amizuka, T., Niida, Y., Umetsu, M. and Ito, K. (1997) Simple, rapid and sensitive method for the determination of indomethacin in plasma by high-performance liquid chromatography with ultraviolet detection. *J Chromatogr B Biomed Sci Appl*, **692**, 241–4.

8. Singh, A.K., Jang, Y., Mishra, U. and Granley, K. (1991) Simultaneous analysis of flunixin, naproxen, ethacrynic acid, indomethacin, phenylbutazone, mefenamic acid and thiosalicylic acid in plasma and urine by high-performance liquid chromatography and gas chromatography-mass spectrometry. *J Chromatogr*, **568**, 351-61.

9. Taylor, P.J., Jones, C.E., Dodds, H.M., Hogan, N.S. and Johnson, A.G. (1998) Plasma indomethacin assay using high-performance liquid chromatography-electrospray-tandem mass spectrometry: application to therapeutic drug monitoring and pharmacokinetic studies. *Ther Drug Monit*, **20**, 691-6.

10. Vree, T.B., van den Biggelaar-Martea, M. and Verwey-van Wissen, C.P. (1993) Determination of indomethacin, its metabolites and their glucuronides in human plasma and urine by means of direct gradient high-performance liquid chromatographic analysis. Preliminary pharmacokinetics and effect of probenecid. *J Chromatogr*, **616**, 271-82.

11. Dawson, M., Smith, M.D. and McGee, C.M. (1990) Gas chromatography/negative ion chemical ionization/tandem mass spectrometric quantification of indomethacin in plasma and synovial fluid. *Biomed Environ Mass Spectrom*, **19**, 453-8.

12. Nishioka, R., Harimoto, T., Umeda, I., Yamamoto, S. and Oi, N. (1990) Improved procedure for determination of indomethacin in plasma by capillary gas chromatography after solid-phase extraction. *J Chromatogr*, **526**, 210-4.

Chapter 29

Quantitation of Iothalamate in Urine and Plasma Using Liquid Chromatography Electrospray Tandem Mass Spectrometry (HPLC-ESI-MS/MS)

Ross J. Molinaro and James C. Ritchie

Abstract

The following chapter describes a method to measure iothalamate in plasma and urine samples using high performance liquid chromatography combined with electrospray positive ionization tandem mass spectrometry (HPLC-ESI-MS/MS). Methanol and water are spiked with the internal standard (IS) iohexol. Iothalamate is isolated from plasma after IS spiked methanol extraction and from urine by IS spiked water addition and quick-spin filtration. The plasma extractions are dried under a stream of nitrogen. The residue is reconstituted in ammonium acetate–formic acid–water. The reconstituted plasma and filtered urine are injected into the HPLC-ESI-MS/MS. Iothalamate and iohexol show similar retention times in plasma and urine. Quantification of iothalamate in the samples is made by multiple reaction monitoring using the hydrogen adduct mass transitions, from a five-point calibration curve.

Key words: Iothalamate, iohexol, mass spectrometry, liquid chromatography, glomerular filtration rate, renal function.

1. Introduction

Iothalamate, a triiodinated derivative of benzoic acid, is a diagnostic agent commonly used in radiology to enhance or create the necessary visual contrast in a radiographic image. First introduced for clinical purposes in 1954, it is available as the sodium salt, the meglumine salt, or a combination of the two. Other triiodinated benzoic acid derivatives that are also used as contrast agents include iohexol, iodamide, ioxithalamate, and ioglicate (1).

The pharmacodynamic properties of iothalamate create an optimal compound for assessing renal function (2). Plasma and urine iothalamate values are used clinically to calculate glomerular filtration rate (GFR) and assess kidney disease in infants and adults because of

U. Garg, C.A. Hammett-Stabler (eds.), *Clinical Applications of Mass Spectrometry*, Methods in Molecular Biology 603, DOI 10.1007/978-1-60761-459-3_29, © Humana Press, a part of Springer Science+Business Media, LLC 2010

these characteristics (3–7). Iothalamate clearance correlates well with other compounds used for measuring renal clearance (8–11); however, the early methods used to measure the compound in body fluids relied upon the use of isotope-labeled iothalamate and are less suited for use in the modern clinical laboratory (8, 9, 11–15). Alternatively, methods using HPLC-UV, fluorescence excitation, and capillary electrophoresis are described (6, 16–18). We describe a rapid, accurate, and sensitive HPLC-ESI-MS/MS method for quantifying iothalamate in plasma and urine samples that correlate well with the aforementioned methods and promises to permit the administration of smaller doses to patients for renal assessment.

2. Materials

2.1. Reagents and Buffers

1. Formic acid and methanol (Fisher Scientific, Fair Lawn, NJ or Sigma-Aldrich, St. Louis, MO), and were of analytical or chromatography grade.

2. 1 M Ammonium Acetate (77 g anhydrous ammonium acetate in 1 L water). Store at room temperature 18–24°C. Stable for 1 year.

3. Mobile Phase A: (2 mM ammonium acetate-1 mL/L formic acid in water). Store at room temperature 18–24°C. Stable for 1 year.

4. Mobile Phase B: (2 mM ammonium acetate–1 mL/L formic acid in methanol). Store at room temperature 18–24°C. Stable for 1 year.

5. Human drug-free pooled plasma.

6. Human drug-free pooled urine.

2.2. Standards and Calibrators

1. Primary iothalamate standard: Conray® (Iothalamate Meglumine Injection U.S.P., 282 mg/mL) (see **Note 1**).

2. Secondary iothalamate standard (50 mg/mL): Prepare by transferring 886.5 μLs of iothalamate primary standard into a 5 mL volumetric flask and diluting to volume with analytical grade water. Secondary standard is stable for 1 year at −80°C.

3. Tertiary urine standard (5 mg/mL): Prepare by transferring 1000 μLs of secondary standard to a 10 mL volumetric flask and diluting with analytical grade water. Tertiary standard is stable for 1 year at −80°C.

4. Tertiary plasma standard (1 mg/mL): Prepare by transferring 200 μLs of secondary combo standard to a 10 mL volumetric flask and diluting with analytical grade water. Tertiary standard is stable for 1 year at −80°C.

Table 29.1
Preparation of standards

A. Plasma

Standard	Drug-free plasma (mL)	Tertiary plasma standard (mL)	Concentration (μg/mL)
1	10.0	–	0
2	9.970	0.030	3
3	9.940	0.060	6
4	9.875	0.125	12.5
5	9.750	0.250	25

B. Urine

Standard	Drug-free urine (mL)	Tertiary urine standard (mL)	Concentration (μg/mL)
1	10.0	–	0
2	9.940	0.060	30
3	9.875	0.125	60
4	9.750	0.250	125
5	9.500	0.500	250

5. Working plasma and urine standards are made according to **Table 29.1** using 10 mL volumetric flasks. The standards are stable for 6 months when stored at −20°C.

2.3. Internal Standard and Quality Controls

1. Iohexol primary internal standard (IS): Omnipaque™ (iohexol) Injection (240 mgI/mL) (*see* **Note 1**).

2. Secondary IS (50 mg/mL): Prepare by transferring 1.04 mL of iohexol IS to a 5 mL volumetric flask and diluting with analytical-grade water. Secondary standard is stable for 1 year at −80°C.

3. Tertiary IS (1 mg/mL): Prepare by transferring 200 µLs of secondary IS to a 10 mL volumetric flask and diluting with analytical-grade water. Tertiary IS is stable for 1 year at −80°C.

4. Working plasma and urine ISs are made according to **Table 29.2** using 100 mL volumetric flasks. Stable for 1 month when stored at 4°C.

Table 29.2
Preparation of working internal standards

A. Working Plasma IS

Tertiary Plasma IS (1 mg/mL)	Methanol
800 µL	QS to 100 mL

B. Working Urine IS

Tertiary urine IS (1 mg/mL)	Methanol
8.00 mL	QS to 100 mL

Table 29.3
Preparation of in-house controls

A. Plasma

Control	Negative plasma (mL)	Tertiary plasma standard (mL)	Concentration (µg/mL)
Low	9.980	0.020	2
Medium	9.900	0.100	10
High	9.800	0.200	20

B. Urine

Control	Negative urine (mL)	Tertiary urine standard (mL)	Concentration (µg/mL)
Low	9.060	0.040	20
Medium	9.800	0.200	100
High	9.600	0.400	200

5. Quality control samples: Prepare according to **Table 29.3** using 10 mL volumetric flasks. Use tertiary iothalamate standards of a different lot from that used to prepare standards. The controls are stable for 6 months when stored at −20°C.

2.4. Supplies

1. Target DP^{TM} Vials C4000-1 W autosampler vials, 0.300 mL limited volume Target Poly inserts, DP^{TM} Blue Cap (T/RR Septa) (National Scientific, Rockwood, TN, USA)

2. Seal-Rite® 1.5 mL tubes (USA Scientific, Inc., Ocala, FL, USA)

3. 15 × 75 mm glass tubes (Fisher Scientific, Fair Lawn, NJ, USA)

4. Microcon® YM-10 Centrifugal Filter Unit (Sigma-Aldrich, St. Louis, MO, USA)

2.5. Equipment	1. Microcentrifuge
	2. Waters Alliance HT 2795 Separation Module with Micromass Quatro Micro API and Masslink software
	3. Zymark TurboVap® IV Evaporator
	4. Multi-tube vortexer
	5. Agilent dC18 column, 5 µm, 2.1 × 20 mm

3. Methods

3.1. Plasma – Stepwise Procedure

1. Label 1.5 mL microcentrifuge tubes for each standard, control, and sample to be analyzed.

2. Pipette 400 µL of working plasma IS to each microcentrifuge tube.

3. Add 100 µL of plasma standard, plasma control, or plasma sample into appropriate tubes.

4. Vortex for 2 min (*see* **Note 2**).

5. Centrifuge samples at 11,356 × g for 6 min.

6. Label a 15 × 75 mm glass tube for each standard, control, and sample.

7. Accurately transfer 300 µL supernatant from step 5 to labeled 15 × 75 mm glass tubes (*see* **Note 3**).

8. Concentrate eluate to dryness under N_2 in a water bath at 40°C.

9. Add 300 µL mobile phase A to residue in glass tubes.

10. Vortex for 15 sec to ensure adequate reconstitution.

11. Label autosampler vials and add glass inserts into labeled vials.

12. Transfer entire amount of reconstituted sample into appropriately labeled vial inserts.

3.2. Urine – Stepwise Procedure

13. Label a 1.5 mL microcentrifuge tube for each standard, control, and sample to be analyzed. Pipette 400 µL of working urine IS to each.

14. Add 100 µL of urine standard, control, or sample into appropriate tubes.

15. Adequately mix by aspirating and dispensing by pipette 5X.

16. Transfer entire volume to appropriately labeled YM-10 filter units.

17. Centrifuge samples at 11,356 × g for 30 min (*see* **Note 4**).

18. Label autosampler vials. Add glass inserts into labeled vials.

19. Accurately transfer 300 µL filtrate from spun samples into appropriately labeled vial inserts.

3.3. Instrument
Operating Conditions

1. Inject 10 μLs plasma/serum filtrate onto HPLC-ESI-MS/MS.

2. Inject 5 μL urine filtrate onto HPLC-ESI-MS/MS (*see* **Note 5**).

3. The instrument's operating conditions are given in **Table 29.4**.

Table 29.4
HPLC-ESI-MS/MS operating conditions

A. HPLC[a]

Column temp. (°C)	45	
Flow (mL/min)	0.6	
Gradient	Time (min)	Percent of mobile phase A
	0.0	100 (Start)
	0.7	100
	0.9	0
	2.0	0
	4.0	100 (End)

B. MS/MS Tune Settings[b]

Capillary (kV)	1.5
Cone (V)	35
Source temp. (°C)	140
Desolvation temp (°C)	350
Cone gas flow (L/hr)	10
Desolvation gas flow (L/hr)	650
LM 1 resolution	14.0
LM 2 resolution	14.0
HM 1 Resolution	10.0
HM 1 Resolution	10.0

[a]Mobile phase A, 2 mM ammonium acetate–1 mL/L formic acid in water; mobile phase B, 2 mM ammonium acetate–1 mL/L formic acid in methanol.
[b]Tune settings may vary slightly between instruments.

3.4. Data Analysis

1. Iothalamate and iohexol precursor and product ions used for quantification are described in **Table 29.5**. These ions were optimized based on tuning parameters, also listed in **Table 29.5**, and can change based on tuning parameters.

2. The data are analyzed using QuanLynx Software (Waters Corp., Milford, MA, USA) or similar software. Multiple reaction monitoring and peak area ratios of analyte vs IS of

Table 29.5
Precursor and primary and secondary product ions for iothalamate and iohexol

	Precursor ion (H+ transitions)	Product ions[a] (primary, secondary)	Dwell (sec)	Cone (V)	Collision (eV)	Delay (sec)
Iothalamate	614.5	360.5[b], 486.5	0.2	35	20	0.030
Iohexol	821.6	652.3[b], 730	0.2	34	30	0.030

[a]Optimized *m/z* may change based on tuning parameters.
[b]Quantitation trace for primary ion.

quantifying ions is used to quantify the concentration of iothalamate in samples. Ratio limits of primary and secondary product ions were set at 10% to confirm the presence of the iothalamate and iohexol. The run is considered acceptable if calculated concentrations of iothalamate in the controls are within two standard deviations of the target values. Liquid chromatography retention time window limits for iothalamate and iohexol were set at 0.4 and 0.2 min, respectively (all compounds with transitions specific for iothalamate and iohexol within the stated window limit will be labeled iothalamate and iohexol). HPLC-ESI-MS/MS selected ion chromatogram is shown in **Fig. 29.1**.

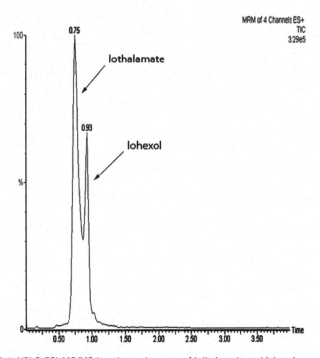

Fig. 29.1. HPLC-ESI-MS/MS ion chromatograms of iothalamate and iohexol.

3. The linearity/limits of quantitation of the plasma method is 0.0188–25 µg/mL, and 0.0188–250 µg/mL for the urine method. Samples in which the iothalamate concentrations exceed the upper limit of quantitation can be diluted with appropriate pooled urine or serum.

4. Typical coefficient of correlation of the standard curve is >0.99.

5. Typical intra- and inter-assay variations are <10%.

4. Notes

1. Contrasting agents are light sensitive and should be protected by wrapping all standard, control, and IS solutions with aluminum foil or another material and secure from light exposure.

2. Ensure the microcentrifuge tube caps are secured properly.

3. Be careful not to transfer any of the pellets from the bottom of the tubes after microcentrifugation.

4. Entire volume of sample should be filtered. Length of spin may exceed 30 min in some cases.

5. The exchange of urine and serum standards and controls for testing has not been validated.

References

1. Forte, J.S. (2006) Contrast Media: Chemistry, Pharmacology and Pharmaceutical Aspects. *Course of Study for the Certificate of Competence in Administering Intravenous Injection.*

2. Sigman, E.M., Elwood, C., Reagan, M.E., Morris, A.M. and Catanzaro, A. (1965) The renal clearance of I-131 labeled sodium iothalamate in man. *Invest Urol*, **2**, 432–8.

3. Agarwal, R., Bills, J.E., Yigazu, P.M., Abraham, T., Gizaw, A.B., Light, R.P., Bekele, D.M. and Tegegne, G.G. (2009) Assessment of iothalamate plasma clearance: Duration of study affects quality of GFR. *Clin J Am Soc Nephrol*, **4**, 77–85.

4. Lewis, J., Greene, T., Appel, L., Contreras, G., Douglas, J., Lash, J., Toto, R., Van Lente, F., Wang, X. and Wright, J.T., Jr. (2004) A comparison of iothalamate-GFR and serum creatinine-based outcomes: acceleration in the rate of GFR decline in the African American Study of Kidney Disease and Hypertension. *J Am Soc Nephrol*, **15**, 3175–83.

5. Agarwal, R. (2003) Ambulatory GFR measurement with cold iothalamate in adults with chronic kidney disease. *Am J Kidney Dis*, **41**, 752–9.

6. Wilson, D.M., Bergert, J.H., Larson, T.S. and Liedtke, R.R. (1997) GFR determined by nonradiolabeled iothalamate using capillary electrophoresis. *Am J Kidney Dis*, **30**, 646–52.

7. Holliday, M.A., Heilbron, D., al-Uzri, A., Hidayat, J., Uauy, R., Conley, S., Reisch, J. and Hogg, R.J. (1993) Serial measurements of GFR in infants using the continuous iothalamate infusion technique. Southwest Pediatric Nephrology Study Group (SPNSG). *Kidney Int*, **43**, 893–8.

8. de Vries, P.A., Navis, G., de Jong, P.E. and de Zeeuw, D. (1998) Can continuous intraperitoneal infusion of 125I-iothalamate and 131I-

hippuran be used for measurement of GFR in conscious rats? *Ren Fail*, **20**, 249–55.

9. Wiener, S.N., Shah, Y.P., Mares, R.M. and Flynn, M.J. (1982) Correlation of I-125 iothalamate and tc-99m DPTA measurements of GFR using the single injection method. *Clin Nucl Med*, **7**, 359–63.

10. Rootwelt, K., Falch, D. and Sjokvist, R. (1980) Determination of glomerular filtration rate (GFR) by analysis of capillary blood after single shot injection of 99mTc-DTPA. A comparison with simultaneous 125I-iothalamate GFR estimation showing equal GFR but difference in distribution volume. *Eur J Nucl Med*, **5**, 97–102.

11. Anderson, C.F., Sawyer, T.K. and Cutler, R.E. (1968) Iothalamate sodium I 125 vs cyanocobalamin Co 57 as a measure of glomerular filtration rate in man. *JAMA*, **204**, 653–6.

12. Adefuin, P.Y., Gur, A., Siegel, N.J., Spencer, R.P. and Hayslett, J.P. (1976) Single subcutaneous injection of iothalamate sodium I 125 to measure glomerular filtration rate. *JAMA*, **235**, 1467–9.

13. Malamos, B., Dontas, A.S., Koutras, D.A., Marketos, S., Sfontouris, J. and Papanicolaou, N. (1967) 125I-sodium iothalamate in the determination of the glomerular filtration rate. *Nucl Med (Stuttg)*, **6**, 304–10.

14. Elwood, C.M., Sigman, E.M. and Treger, C. (1967) The measurement of glomerular filtration rate with 125I-sodium iothalamate (Conray). *Br J Radiol*, **40**, 581-3.

15. Sigman, E.M., Elwood, C.M. and Knox, F. (1966) The measurement of glomerular filtration rate in man with sodium iothalamate 131-I (Conray). *J Nucl Med*, **7**, 60–8.

16. Farthing, D., Sica, D.A., Fakhry, I., Larus, T., Ghosh, S., Farthing, C., Vranian, M. and Gehr, T. (2005) Simple HPLC-UV method for determination of iohexol, iothalamate, p-aminohippuric acid and n-acetyl-p-aminohippuric acid in human plasma and urine with ERPF, GFR and ERPF/GFR ratio determination using colorimetric analysis. *J Chromatogr B Analyt Technol Biomed Life Sci*, **826**, 267–72.

17. Isaka, Y., Fujiwara, Y., Yamamoto, S., Ochi, S., Shin, S., Inoue, T., Tagawa, K., Kamada, T. and Ueda, N. (1992) Modified plasma clearance technique using nonradioactive iothalamate for measuring GFR. *Kidney Int*, **42**, 1006–11.

18. Guesry, P., Kaufman, L., Orloff, S., Nelson, J.A., Swann, S. and Holliday, M. (1975) Measurement of glomerular filtration rate by fluorescent excitation of non-radioactive meglumine iothalamate. *Clin Nephrol*, **3**, 134–8.

Chapter 30

Identification and Quantitation of Ketamine in Biological Matrices Using Gas Chromatography-Mass Spectrometry (GC-MS)

Timothy P. Rohrig, Megan Gamble, and Kim Cox

Abstract

Ketamine hydrochloride, a Schedule III drug under the Controlled Substances Act, is a dissociative anesthetic that has a combination of stimulant, depressant, hallucinogenic, and analgesic properties. This procedure utilizes solid phase extraction of blood, tissue, or urine samples to isolate ketamine (special K) and its metabolite norketamine. The extract is then assayed using selected ion monitoring gas chromatography/mass spectrometry for absolute structural confirmation of the compound and the compounds quantified by comparing responses of the unknown samples to the responses of standards.

Key words: Ketamine, mass spectrometry, gas chromatography, Solid Phase Extraction (SPE).

1. Introduction

Ketamine is a weakly basic amino compound, available as the hydrochloride salt. The drug has been used primarily as a preoperative veterinary anesthetic in the United States since 1972 (1, 2). It is structurally and pharmacologically related to phencyclidine, a failed clinical anesthetic turned popular street drug, and is capable of producing some of the same hallucinogenic side effects (1). However, unlike phencyclidine, liquid ketamine is most often administered by ingesting with drink or through intravenous or intramuscular injection (1, 2). Powdered ketamine is either snorted or smoked using marijuana or tobacco for delivery (2). The production of ketamine is a complex and time-consuming process, therefore, making clandestine production impractical and street sales rare (2). It is typically distributed in social settings

U. Garg, C.A. Hammett-Stabler (eds.), *Clinical Applications of Mass Spectrometry*, Methods in Molecular Biology 603,
DOI 10.1007/978-1-60761-459-3_30, © Humana Press, a part of Springer Science+Business Media, LLC 2010

among friends and acquaintances, and because of its sedative and dissociative properties may be used in drug-facilitated sexual assaults (2).

Once ingested, ketamine rapidly exerts its anesthetic effects on the central nervous system (CNS). It is redistributed into peripheral tissues and metabolized in the liver to two compounds of pharmacological interest, norketamine and dehydronorketamine (1, 3), as seen in **Fig. 30.1**. Ketamine has a very short plasma half-life of about 2–3 h and ~90% of a dose is excreted in the urine over a 72-h period (1, 3).

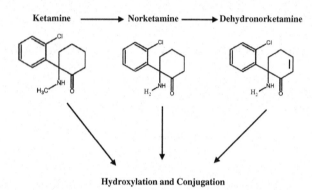

Fig. 30.1. Metabolism of ketamine, norketamine, and dehydroketamine.

Ketamine and its metabolites are measured in biological specimens for toxicological and forensic purposes. Samples are generally screened by immunoassay for the presence of ketamine followed by confirmation of positive samples by gas chromatography mass-spectrometry. The drug and metabolites are extracted from samples using liquid or solid phase extraction (4, 5).

2. Materials

2.1. Sample

Preferred samples include postmortem tissues, urine, and fluoridated blood from the cardiac and femoral area collected during autopsies, and antemortem urine and fluoridated blood collected from outside agencies intended for human performance testing. Aliquots should be stored in glass tubes or plastic jars at a temperature range of 2–8°C until the time of analysis.

2.2. Reagents and Buffers

1. Sodium phosphate buffer: 100 mM, pH 6.0. Prepare using monobasic and dibasic sodium phosphate. Dissolve 13.84 g of sodium phosphate (monobasic) in 500 mL of DI water; dilute to 1,000 mL. Buffer is stable at room temperature for 6 months.

2. 1 N Acetic Acid. Stable at room temperature for 6 months.

3. Elution Solvent: Methylene chloride:isopropanol:ammonium hydroxide (78:20:2) is stored in an amber bottle at room temperature and is prepared fresh daily.

4. 0.1% Methanolic HCl is prepared by adding 1 mL of concentrated hydrochloric acid to 100 mL of methanol and then diluting 1:10 with methanol. Stable in an amber glass bottle at room temperature for 3 months.

5. 50/50 Hexane/Ethyl Acetate is stable at room temperature for 1 year.

6. Bovine drug-free blood (Lampire Biological Laboratories, Piperville PA).

7. Human drug-free urine.

2.3. Calibrators/ Calibration

1. Primary standards: Ketamine and norketamine, 1 mg/mL (Cerilliant Corporation,Round Rock, TX).

2. Secondary standard (100 μg/mL): Prepare by transferring 1.0 mL of each primary standard (ketamine and norketamine) into a 10 mL volumetric flask and diluting to volume with methanol. Secondary standards are stable for 1 year at 4°C.

3. Tertiary standard (10 μg /mL): Prepare by transferring 1.0 mL of each secondary standard to a 10 mL volumetric flask and diluting with methanol. Tertiary standard is stable for 1 year at 4°C.

4. Working standard (1 μg/mL): Prepare by transferring 1.0 mL of each tertiary standard to a 10 mL volumetric flask and diluting with methanol. Working standards are stable for 1 year at 4°C.

5. Working calibrators. Prepare according to **Table 30.1**. The calibrators are prepared fresh for each analytical run.

Table 30.1
Preparation of calibrators

Calibrator	Negative blood or urine (mL)	Tertiary standard (mL)	Concentration (ng/mL)
1	1.98	0.020	10
2	1.95	0.050	25
3	1.90	0.100	50
4	1.80	0.200	100

2.4. Internal Standard and Quality Controls

1. Primary internal standard: Phencyclidine-D5, 1 mg/mL (Cerilliant Corporation, Round Rock, TX).

2. Secondary internal standard (10 µg /mL): Prepare by transferring 1.0 mL of primary internal standard to a 100 mL volumetric flask and diluting with methanol. Secondary internal standard is stable for 1 year at 4°C.

3. Working internal standard (1 µg /mL): Prepare by transferring 1.0 mL of secondary internal standard to a 10 mL volumetric flask and diluting with methanol. Working internal standard is stable for 1 year at 4°C.

4. Positive control (25 ng/mL): Prepare by adding 2.5 mL of the working standard to 98.5 mL of the appropriate drug-free matrix. Positive control is stable for 1 year at 4°C (**Table 30.2**).

Table 30.2
Preparation of in-house control

Control	Negative blood or urine (mL)	Working standard (mL)	Concentration (ng/mL)
Positive	98.5	2.5	25

5. Negative control: In-house negative urine or negative blood or equivalent.

6. Unextracted retention time standard (optional): Prepare by adding 50 µL of the working standard to 50 µL of the working internal standard to a tube and drying down under a gentle stream of nitrogen.

2.5. Supplies

1. Clean Screen solid phase extraction (SPE) columns (United Chemical Technologies, Bristol, PA).

2. Autosampler vials (12 × 32 mm; clear crimp) 0.300 mL conical with spring inserts, and 11 mm crimp seals with FEP/NatRubber (MicroLiter Analytical Supplies, INC, Suwanee, GA).

3. 16 × 100 mm glass tubes (Fisher Scientific, Fair Lawn, NJ).

2.6. Equipment

1. Cerex System 48 Positive Pressure Manifold for solid phase extraction (SPEware Corporation, Baldwin Park, CA).

2. A gas chromatograph/mass spectrometer system (GC/MS; 6890/5973 Agilent Technologies, Wilmington, DE).

3. RTX-5MS GC column, 15 m × 0.25 mm × 0.25 µm (Restek International, Bellefonte, PA).

4. Pierce 18835 Reacti-Therm III™ Heating Module with Reacti-Vap Evaporator, (Rockford, IL).

3. Methods

3.1. Preparation of Tissue Homogenates

1. Portions of tissue samples are weighed (approximately 20 g if available) to generate one part tissue to three parts DI water creating a 1:4 tissue homogenate.

3.2. Stepwise Procedure

1. Prepare an unextracted standard as described above (optional).

2. Add 2 mL (4 g 1:4 tissue homogenate) of samples calibrator and controls to appropriately labeled 16 × 100 mm glass tubes.

3. Add 0.100 mL of internal standard to each tube.

4. Add 2 mL of 100 mM phosphate buffer and 2 mL DI water and vortex.

5. Centrifuge samples at ~1,200 × g for 10 min.

6. Label and insert a SPE cartridge into the positive pressure manifold for every calibrator, control, and sample.

7. Start a vacuum and draw 3 mL of methanol through each column at ~ 1–2 mL/min.

8. Load and draw 3 mL of DI water through each column at ~ 1–2 mL/min.

9. Load and draw 1 mL of 100 mM phosphate buffer through each column being careful not to let the cartridge go dry.

10. Load sample and draw through at ~ 1–2 mL/min.

11. Add 3 mL of deionizer water and draw through at ~ 1–2 mL/min.

12. Add 1 mL of 1.0 M acetic acid and draw through at ~ 1–2 mL/min.

13. Dry the cartridges for 5 min under full vacuum.

14. Load and draw 2 mL of hexane through each column at ~ 1–2 mL/min.

15. Load and draw 3 mL of 50/50 hexane/ethyl acetate through each column at ~ 1–2 mL/min.

16. Load and draw 3 mL of methanol through each column at ~ 1–2 mL/min.

17. Dry the cartridges for 5 min under full vacuum.

18. Place appropriately labeled concentration tubes in appropriate position for eluate collection.

19. Add 3 mL elution solvent and collect eluate at ~ 1–2 mL/min.

20. Concentrate eluate under a gentle stream of N_2 at 40°C to approximately one-half its original volume.

21. Add one drop (~ 20 µL) 0.1% methanol hydrochloric acid.

22. Concentrate eluate to dryness under a gentle stream of N_2 at 40°C.

23. Reconstitute with 0.100 mL ethyl acetate and inject 0.002 mL on GC-MS for analysis.

3.3. Instrument Operating Conditions

The instrument's operating conditions are given in **Table 30.3**.

Table 30.3
GC operating conditions

Initial oven temp	100°C
Initial time	0.0 min
Ramp 1	25°C/min
Final temp	300°C
Final time	0.0 min
Injector temp	200°C
Detector temp	300°C
Purge time on	1.0 min

3.4. Data Analysis

1. The data are analyzed using ChemStation Software (Agilent Technologies). The concentrations of ketamine and its metabolite are determined by constructing standard curves using selected ion monitoring and peak ratios of the analyte to internal standard using the quantifying ions given in **Table 30.4**. GC-MS selected ion chromatograms are shown in the **Fig. 30.2A** and **B**. The run is considered acceptable if calculated concentrations of drugs in the controls are within 20% of target values. Quantifying ion in the sample is considered acceptable if the ratios of the qualifier ions are within 20% of the ion ratios for the calibrators. Mass spectra of ketamine and norketamine is shown in **Fig. 30.3A** and **B**.

Table 30.4
Quantitation and qualifying ions for cocaine and its metabolites

	Quantitation ion	Qualifier ions
Ketamine	180	209.1, 152
Norketamine	166	168, 195
Phencyclidine-D5	205.1	248.2

```
QUANTITATION REPORT FOR Ketamine ON : Milhous

Data File              : C:\MSDChem\1\DATA\1229\29Dec08A.03p\bldket02.D
Tune File Name         : C:\MSDChem\1\5973N\atune.u
Tune Date              : 29 Dec 2008    2:52 pm               Mult : 0
Acq Method Name        : KET.M Calib date : 30 Dec 2008 1:10 pm
Sample Name            : Std 2
Acquisition date       : 29 Dec 2008    3:30 pm

Retention Time    4.30 Ketamine          +/- 2.00% =   4.22 -   4.39 min
Retention Time    4.44 PCP-D5            +/- 2.00% =   4.35 -   4.53 min
R.R.T. =   0.969  Unknown target ion / ISTD target ion =   0.22

Ketamine        => 180.0 =   297167  209.0 =    80435  152.0 =    38966
PCP-D5          => 205.0 =  1338825  248.0 =   366683

Ketamine        => 209.0/180.0 =    27.1 +/-  20.0% rel =   21.5 -   32.3
Ketamine        => 152.0/180.0 =    13.1 +/-  20.0% rel =   11.8 -   17.6
PCP-D5          => 248.0/205.0 =    27.4 +/-  20.0% rel =   21.0 -   31.6

Concentration =    42.81 ** CONCENTRATION > 0.00 ng LIMIT **<==
```

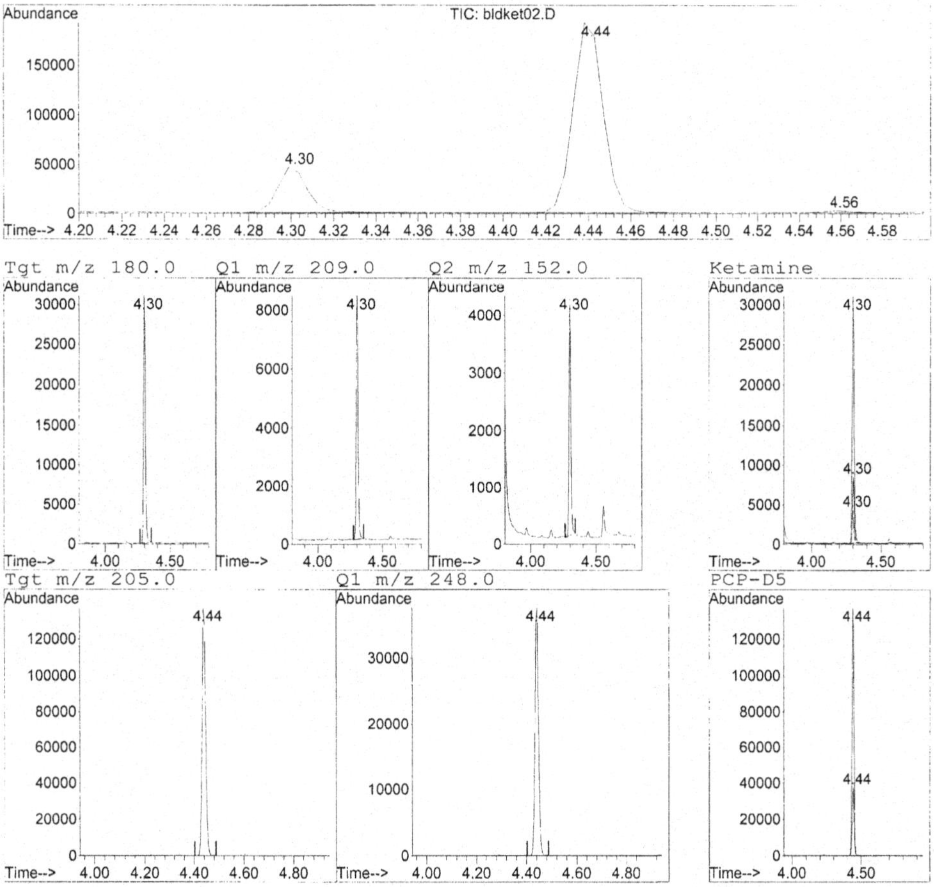


```
Ketamine          : RT extraction window from    3.80 to    4.80 min
PCP-D5            : RT extraction window from    3.94 to    4.94 min

Page 143                       Printed on : 12 Mar 2009 2:24 pm
```

Fig. 30.2. (**A and B**) GC-MS selected ion chromatograms for ketamine and norketamine.

QUANTITATION REPORT FOR Norketamine ON : Milhous

Data File	:	C:\MSDChem\1\DATA\1229\29Dec08A.03p\bldket02.D
Tune File Name	:	C:\MSDChem\1\5973N\atune.u
Tune Date	:	29 Dec 2008 2:52 pm Mult : 0
Acq Method Name	:	KET.M Calib date : 30 Dec 2008 1:10 pm
Sample Name	:	Std 2
Acquisition date	:	29 Dec 2008 3:30 pm

Retention Time 4.16 Norketamine +/- 2.00% = 4.07 - 4.24 min
Retention Time 4.44 PCP-D5 +/- 2.00% = 4.35 - 4.53 min
R.R.T. = 0.936 Unknown target ion / ISTD target ion = 0.28

Norketamine => 166.0 = 375860 168.0 = 122552 195.0 = 100781
PCP-D5 => 205.0 = 1338825 248.0 = 366683

Norketamine => 168.0/166.0 = 32.6 +/- 20.0% rel = 26.2 - 39.2
Norketamine => 195.0/166.0 = 26.8 +/- 20.0% rel = 19.8 - 29.8
PCP-D5 => 248.0/205.0 = 27.4 +/- 20.0% rel = 21.0 - 31.6

Concentration = 44.68 ** CONCENTRATION > 0.00 ng LIMIT **<==

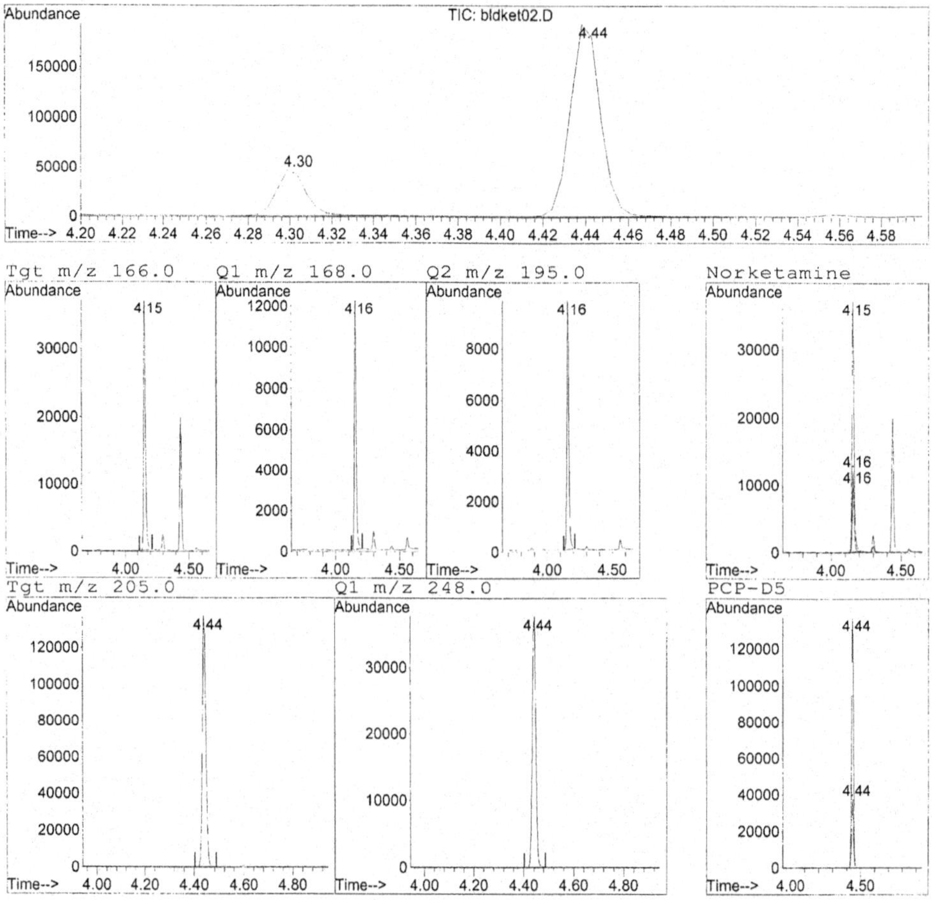

Norketamine : RT extraction window from 3.65 to 4.66 min
PCP-D5 : RT extraction window from 3.94 to 4.94 min

Page 142 Printed on : 12 Mar 2009 2:24 pm

Fig. 30.2. (continued)

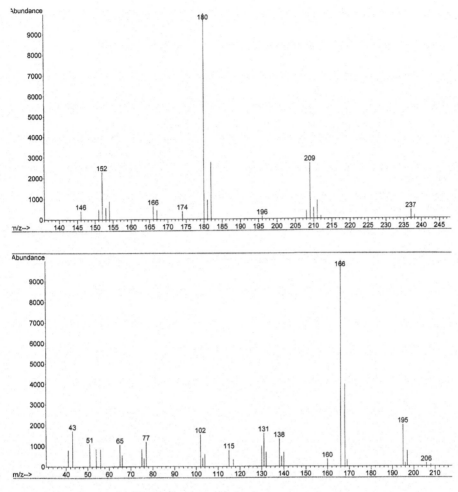

Fig. 30.3. **(A and B) Mass spectra of ketamine and norketamine.**

2. The presence of ketamine or norketamine is confirmed when the concentration of the analyte is greater than or equal to the calibrator (10 ng/mL) (*see* **Note 1**).

3. The linearity/limit of quantitation of the method is 10–100 ng/mL (*see* **Note 2**).

4. Typical coefficient of correlation of the standard curve is >0.99.

5. Typical intra- and inter-assay variations are <10%

4. Notes

1. Non-urine samples may be reported as positive, less than the lowest calibrator as long as identification criteria are met and the concentration is greater than the limit of quantitation of the assay.

2. Samples in which the drug concentrations exceed the upper limit of quantitation should be diluted with the appropriate matrix and retested.

References

1. Baselt, R. C. (2000) Disposition of Toxic Drugs and Chemicals in Man, 5th ed. Chem. Tox. Inst., pp. 456–458.

2. Moffatt, A. C., Osselton, M. D. and Widdop, B. (2004) Clarke's Analysis of Drugs and Poisons, 3rd ed. Pharm Press, pp. 1152–1153.

3. Drug-Facilitated Sexual Assault Resource Guide. 2003-R0502-001. NDIC and George Mason University, pp. 9–11, May 2003.

4. Applications Manual (1991) CLEAN SCREEN Extraction Column. Worldwide Monitoring Corp. Horsham, PA.

5. Trace Application Manual; SPEware Corporation; San Pedro, CA (2004).

Chapter 31

Measurement of Filter Paper Bloodspot Lead by Inductively Coupled Plasma-Mass Spectrometry (ICP-MS)

Denise M. Timko and Douglas F. Stickle

Abstract

The potential for adverse effects of lead (Pb) exposure on development in children remains a health concern in the U.S., and programmatic screening of children for elevated blood lead levels ([Pb] >10 µg/dL) is widespread. With sufficiently sensitive technology for the measurement of lead such as ICP-MS, it is possible to utilize filter paper bloodspots as a specimen suitable for lead screening. Filter paper bloodspot specimens are relatively inexpensive, easy to collect, and stable during transport. For these reasons they are preferred by many program clinics for child subjects. We describe measurement of Pb from filter paper bloodspots using ICP-MS and bloodspot standards.

Key words: Lead, ICP-MS, filter paper, bloodspot.

1. Introduction

Exposure to lead in infancy and childhood is known to have adverse effects on mental development (1–3). Exposure occurs primarily from contact of contaminated surfaces of man-made products and transfer of the residuals to the mouth. In particular, the former use of lead in paint means that many poorly maintained properties might yet be sources of lead exposure for children (4). For these reasons, screening for elevated blood lead level (EBLL, [Pb] >10 µg/dL) in at-risk children is a recommendation of the Centers for Disease Control (5–8), and programmatic screening is common in the U.S. More than 200 laboratories participated in the national Wisconsin Blood Lead Proficiency Testing Program for whole blood samples in 2008 (9).

U. Garg, C.A. Hammett-Stabler (eds.), *Clinical Applications of Mass Spectrometry*, Methods in Molecular Biology 603, DOI 10.1007/978-1-60761-459-3_31, © Humana Press, a part of Springer Science+Business Media, LLC 2010

ICP-MS is a gold standard for measurement of trace metals such as Pb in fluids (10). With an analytically sensitive method such as ICP-MS, bloodspot specimens are an alternative to capillary or venipuncture specimens for purposes of screening for EBLL (11). Despite this, only 25 lead testing laboratories participated in 2008 proficiency testing report, using bloodspots as a sample type (9). It is not known, however, what overall number or what fraction of samples are processed for Pb testing by these laboratories per year for each sample type. In our institution, with availability of bloodspot testing, 55% of samples submitted are bloodspot samples. The relatively low cost coupled with the ease of collection, storage, and shipment of specimens are advantages of bloodspots compared to whole blood specimens (12). Another advantage to the use of this specimen is a low sample rejection rate in comparison to that observed for capillary or venous specimens, where coagulation due to poor anticoagulant mixing upon collection is a common cause of rejection for up to 5% of capillary specimens.

We describe here a method for measurement of blood [Pb] from bloodspot specimens using inductively coupled plasma-mass spectrometry (ICP-MS). In this method, bloodspot standards are used, where the bloodspots are prepared from Pb-spiked whole blood samples. The method therefore also describes use of ICP-MS for measurement of Pb in whole blood. The methods follow in outline those of previously published methods (13–15).

2. Materials

2.1. Reagents and Buffers

1. Multi-element daily performance check solution (SmartTune Solution, Elan DRC Plus II, PerkinElmer, Waltham, MA)

2. Spex Certiprep Lead (Pb) Standard: ICP-AES grade, (Fisher Scientific, Pittsburgh, PA), 1000 mg/L in 2% Nitric Acid.

3. Lead-free ETDA anticoagulated – whole blood.

4. 4% Nitric acid (HNO_3): 70% double-distilled nitric acid, (GFS Chemicals, Powell, OH). Add 20 mL nitric acid into 480 mL type I water. Stable at room temperature for 6 months.

5. 10% Triton Solution: Add 5 mL of Triton X-100 Wetting Agent (Fisher Scientific, Pittsburgh, PA) to 45 mL of type I water. Triton is a very viscous solution, pipette slowly. This solution will take a long time to dissolve. Stable at room temperature for 6 months.

6. Diluent (0.05% Triton/5 mM EDTA): Add 10 mL of 10% Triton solution and 3.8 g EDTA (Sigma ED4SS) into 1990 mL of type I water. Stable at room temperature for 6 months (see **Note 1**).

7. 1% Nitric acid (HNO_3): 70% double-distilled nitric acid, (GFS Chemicals, INC). Add 10 mL of nitric acid into 990 mL of type I water. Stable at room temperature for 6 months.

2.2. Internal Standards

1. Internal Standard, (10 mg/L Terbium): Add 1 mL of Spex Certiprep Terbium (Tb) Standard (1000 mg/L) to 100 mL of 1% nitric acid. Stable for 6 months at room temperature. This is an intermediate solution used to separately prepare whole blood and bloodspot working internal standard solutions.

2. Whole Blood Working Internal Standard, (0.5 µg/dL Tb): Add 1 mL of the 10 mg/L Tb internal standard into 1980 mL of type I water, and add 20 mL of nitric acid. Mix well. Stable for 6 months at room temperature.

3. Blood Spot Working Internal Standard (0.25 µg/dL Tb): Add 250 µL of 10 mg/L Tb internal standard into 990 mL of type I water, and add 10 mL of nitric acid. Stable for 6 months at room temperature.

4. Whole Blood Diluent/IS Reagent: Add 1200 mL of the diluent (0.05% Triton/5 mM EDTA, #4 above) to 600 mL of whole blood working internal standard. Stable at room temperature for 6 months.

2.3. Working Aqueous (AQ) Standards: 0.0, 6.25, 12.5, 25.0, and 50.0 µg/dL Standards

Prepare aqueous standards by diluting the certified Spex Certiprep Lead (Pb) Standard into 4% nitric acid using the mixtures shown in **Table 31.1**. Stable at 6±5°C for 6 months. Verify linearity of the materials using the whole blood ICP-MS procedure. A correlation coefficient of $r^2 \geq 0.99$ should be obtained for the aqueous standard curve. Prepare new aqueous standards if $r^2 \leq 0.98$. Quality control materials should also be run during the linearity assessment in order to rule out any systematic preparation error that could produce a perfectly linear but inaccurate standard curve.

Table 31.1
Preparation of aqueous standards

Final concentration (µg/dL)	Volume of CertiPrep (Pb) std (µL)	Volume of 4% nitric acid (mL)	Volume of AQ (50 µg/dL) std (mL)
50.0	50	100	–
25.0	–	5.0	5.0
12.5	–	7.5	2.5
6.25	–	8.75	1.25
0.0	–	10.0	–

2.4. 10.0, 20.0, 30.0, and 40.0 μg/dL Whole Blood Standards

1. Intermediate 4000 μg/dL Pb solution: Prepare adding 40 μL of Spex CertiPrep Lead Standard to 960 μL of type I water.

2. Prepare whole blood standards by mixing the 4000 μg/dL Pb solution and ETDA anticoagulated – whole blood as shown in **Table 31.2**. Allow the spiked whole blood samples to sit refrigerated for 24 h before measurement by ICP-MS. Verify the whole blood lead concentrations in triplicate using the ICP-MS whole blood procedure. The final lead concentration assigned for each standard is the average of the triplicate measurements. Assure the standards are reasonably evenly distributed according to the intended range of Pb concentrations (*see* **Note 2**).

Table 31.2
Preparation of whole blood standards

Final added Pb concentration (μg/dL)	Volume of AQ (4,000 μg/dL) std (μL)	Volume of whole blood (mL)
40.0	40	4.0
30.0	30	4.0
20.0	20	4.0
10.0	10	4.0

2.5. Bloodspot Standards: 10.0, 20.0, 30.0, and 40.0 μg lead/dL Whole Blood on Filter Paper

1. Prepare and validate whole blood standards as described in **Section 2.4**.

2. Prepare bloodspot standards by successive spotting of 50 μL of the whole blood standards onto filter paper. The bloodspots should be prepared at the same time that the [Pb] in the whole blood standard is measured. After spotting, allow bloodspots to dry overnight. The standard curve using bloodspot standards with their assigned Pb concentrations should then be checked for linearity using the bloodspot ICP-MS procedure. A correlation coefficient of $r^2 \geq 0.95$ should be obtained for a bloodspot standard curve (*see* **Note 3**). As described for the aqueous Pb standards, measurement of bloodspot QC samples should also be assessed at the time of assessment of linearity of the bloodspot standard curve. Bloodspot standards are stable for 2 months when stored in "ziplock" bags at $6 \pm 5°C$.

2.6. Controls

1. Bio-Rad Lyphocheck Whole Blood Control, Levels 1, 2, and 3. Store at $6 \pm 5°C$ until expiration date on label. Reconstitute each vial with 2.0 mL of Type I water, allow to stand for 20 min, swirling occasionally. Before sampling, gently invert the vial several times to ensure homogeneity. Store at $6 \pm 5°C$ tightly capped for 14 days.

2. Blood spot QC samples: Prepare by applying 50 μL of the reconstituted Bio-Rad Lyphocheck Whole Blood Control onto filter paper. Allow to dry overnight, then store in "ziplock" bags at 6 ± 5°C. Bloodspot QC samples are stable for 6 months (*see* **Note 4**).

2.7. Supplies and Equipment

1. ICP-MS: Perkin-Elmer, ELAN 6100 DRC fitted with a Meinhard nebulizer and cyclonic spray chamber. Nickel cones are used.

2. Cetac ASX-520 Autosampler

3. 1/4" (6 mm) manual disk punch

4. Isolab incubator/shaker, Isolab NC-1034

5. Rocker/mixer, Isolab NC-1030

6. 17 × 200 mm 15 mL polystyrene conical tubes

7. Pipettors capable of dispensing 2.5 μL, 5 μL, 7.5 μL, 10 μL, 20 μL, 50 μL, 1 mL, and 2 mL

8. Schleicher & Schuell 903 filter paper or equivalent

3. Methods

3.1. Preparation of Whole Blood Specimens for Measurement Using Aqueous Standards

1. Allow controls and samples to come to room temperature before pipetting.

2. Label 17 × 200 mm polystyrene 15 mL conical tubes for each sample to be measured: blank (1 sample: H20), calibrators (5 samples: 0, 6.25, 12.5, 25.0, and 50.0 μg/dL), controls (3 samples: Q1, Q2, Q3), diluent (1 sample), and *n* whole blood specimens (1 through *n* samples).

3. Pipette 3 mL of the combined diluent/IS reagent into each labeled tube.

4. Pipet 50 μL of blank, aqueous calibrator, control, or sample into the appropriate labeled tube. Do not pre-wet pipet tip with whole blood. Rinse pipet tip with tube mixture back into tube before withdrawal and ejection of tip.

5. Vortex tubes for 10 sec.

6. Place uncapped tubes into the auto sampler.

7. Measure ^{208}Pb and ^{159}Tb using the ICP-MS procedure (see below) (*see* **Notes 5** and **6**).

3.2. Preparation of Bloodspot Specimens for Measurement Using Bloodspot Standards

1. Allow controls and samples to come to room temperature before punching.

2. Label 17 × 200 mm polystyrene 15 mL conical tubes for each sample to be measured: blank (1 sample: filter paper blank), calibrators (4 samples: nominally 10, 20, 30, and 40 μg/dL), controls (3 samples: Q1, Q2, Q3), diluent (1 sample), and *n* bloodspot specimens (1 through *n* samples).

3. Pipette 2 mL of diluent into each tube.

4. Punch one 1/4" (6 mm) diameter disk for each blank, blood-spot calibrator, control, or patient sample into the appropriate labeled tube (*see* **Note 7**).

5. Check to ensure that all bloodspots have been placed into and are wetted by the diluent solution.

6. Shake bloodspot tubes for 60 min using the Lab-Line Plate-shaker set on 6.

7. Pipette 1 mL of the working bloodspot Tb internal standard reagent to each tube.

8. Vortex tubes for 10 sec. Ensure that each bloodspot is settled to the bottom of the tube. If not, using a wooden applicator stick to push the bloodspot to the bottom.

9. Place uncapped tubes into the autosampler.

10. Measure ^{208}Pb and ^{159}Tb using the ICP-MS procedure (see below).

3.3. ICP Measurement of Pb and Tb

Analysis is performed using the ELAN 6100 DRCII ICP-MS or equivalent using the instrument operating conditions given in **Table 31.3**. The instrument is taken through a 30 min warmup period while pumping rinse solution before conducting the daily

Table 31.3
ICP-MS instrument conditions

Isotopes: ^{208}Pb, ^{159}Tb
Dwell: 50 ms
Peak hopping
Pulse detector
Average spectral peak
Smoothing factor: 5
Auto lens: On
Sample flush: 40 sec, speed −20
Read delay: 15 sec, speed −20
Wash: 60 sec, speed −24
Sweeps/read: 30
Readings/replicate: 1
Replicates: 3

performance check. Analyze the multi-element daily performance check solution and assure instrument specifications are met. Verify and document any additional performance checks.

There are numerous choices for obtaining a report of the primary data from the ICP-MS instrument. In our laboratory, primary data are recorded in a results report text file on the instrument computer using the report options within the ELAN software. The results report text file is programmed to include average counts-per-second (cps) for ^{208}Pb and ^{159}Tb for each sample, as well as the time variation (RSD, or root square deviation) in cps readings for each element. Sample results are identified in the text file by a sample identifier as well as by the autosampler position.

3.4. Data Analysis

Transfer data from the primary data text file to a preprogrammed Excel spreadsheet for analysis. Primary ^{208}Pb and ^{159}Tb data for each sample are transformed into the cps ratio ^{208}Pb/^{159}Tb (R). The Pb concentration for QC and unknown samples is computed by interpolation of (R-R (blank)) onto the linear (R-R (blank)) vs. [Pb] curve computed for standards. An example of standard curve for bloodspot standards is shown in **Fig. 31.1**. The standard curve is acceptable if the linear correlation coefficient (r^2) is greater than 0.950 and QC criteria are met (*see* **Note 8**). Each sample should be reviewed to assure the internal standard (Tb) cps are within 25% of the mean Tb cps observed for all samples and the RSD for Pb for each sample should be <10% of the average cps (*see* **Note 9**). Results for sample [Pb] measurement are reported in µg/dL using one decimal place. Results <2 µg/dL are reported as "<2 µg/dL" (*see* **Note 10**). A fixed reference range is used for reporting of [Pb] in the U.S.: results are flagged as elevated for [Pb] >10 µg/dL.

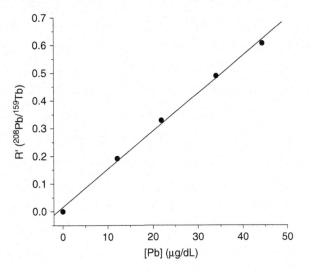

Fig. 31.1. Example bloodspot standard curve. R' (*y* axis) is R − R$_{blank}$, where R = cps ratio (^{208}Pb/^{159}Tb). In this example, r^2 = 0.997.

3.5. Repeat Sample Analysis for Bloodspot [Pb] >25 μg/dL

Because of the opportunity for sample contamination by Pb as a pre-analytical variable due to handling of filter paper and filter paper specimens, it is recommended that high values ([Pb] >10 μg/dL) for measured bloodspot [Pb] be repeated before reporting. Sample analysis should follow that of the original sample analysis, with two important exceptions. First, in order to rule out the possibility of Pb contamination of sample tubes rather than of sample, sample tubes used in the repeat analysis should be pre-rinsed with 4% nitric acid. Second, additional samples from the non-blood portions of the filter paper sample should also be measured to evaluate for possible general Pb contamination of the filter paper collection card.

3.6. Classification of [Pb] Results

[Pb] results reports may be appended with CDC classifications for [Pb] in children as follows (5) (*see* **Note 11**):

Class I: <10 μg/dL = Not considered to be lead-poisoned.

Class IIA: 10–14 μg/dL = May need to be rescreened more frequently.

Class IIB: 15–19 μg/dL = Nutritional and educational intervention and more frequent screening. If persists, environmental investigation and intervention.

Class III: 20–44 μg/dL = Environmental evaluation, remediation, and medical evaluation. May need pharmacologic (Succimer) treatment.

Class IV: 45–69 μg/dL = Medical and environmental intervention, including chelation therapy.

Class V: > = 70 μg/dL = Medical emergency (critical result).

4. Notes

1. For reuse of any reagent bottles, wash using type I water only, then rinse with 4% nitric acid, then rinse three times with type I water. Only 4% nitric acid prepared from double-distilled nitric acid should be used throughout these procedures.

2. The intended Pb concentration range of whole blood standards is 10–40 μg/dL, with an intended relatively even distribution of concentrations separated by 10 μg/dL. However, note that the whole blood samples used to produce bloodspot standards will generally have measurable but low (e.g., <4 μg/dL) Pb concentration. Assigned values to the spiked whole blood standards account for this by measurement of the spiked whole blood. Thus, measured values for Pb concentration will generally be in the vicinity of (slightly

greater than) the spiked Pb concentration. However, if an individual-spiked sample does not have this property, making for an uneven distribution of values for standards, then a second Pb-spiked whole blood sample may be produced and substituted for the original.

3. Unusual hematocrit of one of the whole-blood specimens is one reason why results for an individual bloodspot standard may not align perfectly with remainder of the bloodspot standard curve. If repetition of the standard curve consistently shows the same pattern of departure from the line for an individual bloodspot standard, then the concentration of the bloodspot standard may be readjusted according to the evidence of such data. Alternatively, a second Pb-spiked bloodspot standard may be produced and substituted for the original.

4. Commercial quality control materials (e.g., Bio-Rad Lyphochek) are appropriate materials for use in the aqueous standards whole blood ICP-MS procedure. This is particularly true because, in our experience, ICP-MS measurements almost always correspond exactly to the published lot mean for Pb concentration (generally established by atomic absorption, which is in greater use than ICP-MS). Use of the same materials for production of bloodspots for use as QC is also appropriate. However, it should be noted that there is a clear expectation, as verified by experience, that QC values obtained by measurement of bloodspots thus produced should be less than that obtained for the whole blood measurements. This is because the spread on filter paper of a given volume of reconstituted lyophilized whole blood is significantly greater than that of whole blood, due in part to the fact that the lyophilized material no longer contains red blood cells. Commercial quality control materials are therefore used to produce QC bloodspots as a matter of convenience of having large quantities of ready-made materials of uniform composition. As for all QC procedures, the use of the QC bloodspots is only to assess correspondence of the results of a given run with expected results as established by prior runs.

5. Pb is one of few elements with more than two stable isotopes of significant relative abundance. We use measurement only of the most abundant isotope, ^{208}Pb.

6. Complete matrix matching of aqueous and whole-blood sample may be achieved by adding 50 μL of a whole-blood sample to each aqueous sample tube (viz., to the blank and calibrator sample tubes) and 50 μL of H_2O to each whole-blood tube (viz., to each QC and other sample tubes). In practice, we have found no difference in comparison of final Pb measurements obtained with and without the use of exact matrix matching of samples and standards. However, a difference

may be observed in absolute recovery of internal standard, which is often marginally greater in aqueous samples than in whole blood samples. The difference may be apparent in a metric plot (counts per second of ^{159}Tb vs. sample sequence number) and is presumably due to consistent protein-associated line losses for whole-blood samples, affecting both Pb and Tb.

7. Samples should be punched from the most central area of the bloodspot and away from the fringes and printed marks. Samples that have insufficient area for the punch, that appear caked, discolored or incompletely saturated, or that show evidence of contamination should be rejected. Refer to CLSI bloodspot collection guidelines (16) for elaboration of rejection criteria. Whereas bloodspot specimens are desirable from the standpoint of programs and clinic collectors, in part precisely because of the perceived ease of collection, it should be emphasized that it is also easy to collect such specimens improperly. Numerous variables affect bloodspot area per volume of blood. These variables include hematocrit (an uncontrollable patient-dependent variable) but also whether the volume is applied continuously or discontinuously (a controllable variable). Bloodspot collection instructions are for finger-stick single continuous application of blood to fill a pre-printed filter paper collection card circle (approximately 50 μL of sample). These instructions parallel those of the CLSI guidelines for collection of heel-stick bloodspots for newborn screening (16). In our experience, trained collectors can achieve the instructed collection objectives readily and easily. Nonetheless, it was found in a recent study conducted at our institution that many submitted specimens are non-ideal, both with respect to the total volume of blood applied and with respect to the continuity of application of that volume (17). It is therefore important that proper specimen collection guidelines be periodically reviewed with collecting facilities. It is a detail worth noting that the composition of bloodspots is not uniform; there is significant accretion of red cells at the bloodspot perimeter (18, 19), likely due to a chromatographic effect of fluid separation from serum by the filter paper (20). For this reason, it is important that bloodspots be large enough for sampling to avoid the region of the perimeter of the bloodspot. Patient sample punches should be obtained from an area that is judged as near as possible to the approximate center of the best available bloodspot.

8. CAP criterion for acceptance of results of proficiency testing of samples is ±4 μg/dL, or ±10%, whichever is greater. Note, however, that the CDC recommends that laboratories meet criterion of ±2 μg/dL in the range of [Pb] <10 μg/dL (8).

QC criteria of $\pm$ 2 µg/dL is practical throughout the range of [Pb] used in standards. Note additionally that, whereas the majority of patient samples are likely to have results below 10 µg/dL, it may not be necessary to utilize the entire range of the standard curve, or the higher [Pb] QC results, in order to accept a given run.

9. Deviation of internal standard (Tb) cps from the mean is evaluated by a "metric plot" of Tb cps vs. sample number in order of the run. The use of a ±25% acceptance criterion is somewhat arbitrary and is intended to be able to indicate a gross failure either in sample preparation, sample introduction, or instrument response. It should be noted that Tb cps over the course of very long (4–5 h) runs may show a progressive change due either to progressive change in instrument response or to progressive change in input line losses over time. In such cases, it is still unlikely that variation in Tb will for any one sample exceed 25% of the mean value; however, where progressive change in Tb cps is observed, criterion for acceptance of Tb cps may practically be based on the running mean for Tb cps rather than on the all-samples mean, especially if controls with acceptable computed [Pb] measurements have similar Tb cps.

10. Functional sensitivity of the bloodspot assay is less than that used as the lower limit of reporting; however, there is at present no practical or clinical value in reporting of [Pb] <2 µg/dL. For the aqueous standards whole-blood Pb assay, a lower limit for quantitation in the range of 0.2 µg/dL can be achieved for research purposes by using lower Pb concentration range of Pb standards.

11. At our institution, currently approximately 3% of first sample submissions meet criterion for designation as an elevated blood lead level (EBLL; [Pb] >10 µg/dL). This is rare enough that any EBLL designation may be of immediate concern to a clinician. It is important therefore that the laboratory be immediately ready to answer inquiries as to recommendations for followup of EBLL. We post and utilize followup recommendations per relative extent of elevation of measured [Pb] as given by the CDC (6).

References

1. Surkan, P. J., Zhang, A., Trachtenberg, F., Daniel, D. B., McKinlay, S., Bellinger, D. C. (2007) Neuropsychological function in children with blood lead levels <10 microg/dL. *Neurotoxicology* 28, 1170–1177.

2. Meyer, P. A., Brown, M. J., Falk, H. (2008) Global approach to reducing lead exposure and poisoning. *Mutat Res* 659, 166–175.

3. Bellinger, D. C. (2008) Very low lead exposures and children's neurodevelopment. *Curr Opin Pediatr* 20, 172–177.

4. American Academy of Pediatrics Committee on Environmental Health. (2005) Lead exposure in children: prevention, detection, and management. *Pediatrics* 116, 1036–1046.

5. Centers for Disease Control and Prevention (CDC). (1997) Screening Young Children for Lead Poisoning: Guidance for State and Local Public Health Officials. (http://www.cdc.gov/nceh/lead/guide/guide97.htm).

6. Centers for Disease Control and Prevention (CDC). (2000) Recommendations for blood lead screening of young children enrolled in medicaid: targeting a group at high risk. *MMWR Recomm Rep* **49**, 1–13.

7. Meyer, P. A., Pivetz, T., Dignam, T. A., Homa, D. M., Schoonover, J., Brody, D. (2003) Surveillance for elevated blood lead levels among children – United States, 1997–2001. *MMWR Surveill Summ* **52**, 1–21.

8. Centers for Disease Control and Prevention (CDC). (2007) Interpreting and managing blood lead levels < 10 microg/dL in children and reducing childhood exposures to lead: recommendations of CDC's Advisory Committee on Childhood Lead Poisoning Prevention. *MMWR Recomm Rep* **56**, 1–16.

9. Wisconsin State Laboratory of Hygiene, Environmental Health Division, Toxicology Section. (2008) Blood Lead and Erythrocyte Protoporphyrin Proficiency Testing Programs. (http://www.slh.wisc.edu/wps/wcm/connect/extranet/ehd/toxicology/blept.php).

10. Nuttall, K. L., Gordon, W. H., Ash, K. O. (1995) Inductively coupled plasma mass spectrometry for trace element analysis in the clinical laboratory. *Ann Clin Lab Sci* **25**, 264–271.

11. Verebey, K. (2000) Filter paper-collected blood lead testing in children. *Clin Chem* **46**, 1024–1026.

12. McDade, T. W., Williams, S., Snodgrass, J. J. (2007) What a drop can do: dried blood spots as a minimally invasive method for integrating biomarkers into population-based research. *Demography* **44**, 899–925.

13. Paschal, D. C., Caldwell, K. L., Ting, B. G. (1995) Determination of lead in whole blood using inductively coupled argon plasma mass spectrometry with isotope dilution. *J Anal At Spectrom* **10**, 367–370.

14. Schutz, A., Bergdahl, I. A., Ekholm, A., Skerfving, S. (1996) Measurement by ICP-MS of lead in plasma and whole blood of lead workers and controls. *Occup Environ Med* **53**, 736–740.

15. Di Martino, M. T., Michniewicz, A., Martucci, M., Parlato, G. (2004) EDTA is essential to recover lead from dried blood spots on filter paper. *Clin Chim Acta* **350**, 143–150.

16. Hannon, W., Baily, C., Bartoshesky, L., Davin, B., Hoffman, G., King, P., et al. Blood collection of filter paper for newborn screening programs, approved standards. National Committee for Clinical Laboratory Standards Document LA4-249. Wayne, PA, 2003.

17. Peck, H. R., Timko, D. M., Landmark, J. D., Stickle, D. F. (2009) A survey of apparent blood volumes and sample geometries among filter paper bloodspot samples submitted for lead screening. *Clin Chim Acta* **400**, 103–106.

18. Cernik, A. A. (1974) Determination of blood lead using a 4-0 mm paper punched disc carbon sampling cup technique. *Br J Ind Med* **31**, 239–244.

19. El-Hajjar, D. F., Swanson, K. H., Landmark, J. D., Stickle, D. F. (2007) Validation of use of annular once-punched filter paper bloodspot samples for repeat lead testing. *Clin Chim Acta* **377**, 179–184.

20. Stickle, D. F., Rawlinson, N. J., Landmark, J. D. (2009) Increased perimeter red cell concentration in filter paper bloodspot samples is consistent with constant-load size exclusion chromatography occurring during application. *Clin Chim Acta* **401**, 42–45.

Chapter 32

Multiplex Lysosomal Enzyme Activity Assay on Dried Blood Spots Using Tandem Mass Spectrometry

X. Kate Zhang, Carole S. Elbin, Frantisek Turecek, Ronald Scott, Wei-Lien Chuang, Joan M. Keutzer, and Michael Gelb

Abstract

Deficiencies in any of the 50 degradative enzymes found in lysosomes results in the accumulation of undegraded material and subsequently cellular dysfunction. Early identification of deficiencies before irreversible organ and tissue damages occur leads to better clinical outcomes. In the method which follows, lysosomal α-glucosidase, α-galactosidase, β-glucocerebrosidase, acid sphingomyelinase, and galactocerebrosidase are extracted from dried blood spots and incubated individually with an enzyme-specific cocktail containing the corresponding substrate and internal standard. Each enzyme cocktail is prepared using commercially available mixture of substrate and internal standard at the predetermined optimized molar ratio. After incubation, the enzymatic reactions are quenched using an ethyl acetate/methanol solution and all five enzyme solutions are combined. The mixtures of the reaction products are prepared using liquid–liquid and solid-phase extractions and quantified simultaneously using selected ion monitoring on LC-MS-MS system.

Key words: Dried blood spot, β-glucocerebrosidase, acid sphingomyelinase, galactocerebrosidase, lysosomal α-glucosidase, α-galactosidase, lysosomal storage disorders, mass spectrometry, solid phase extraction, enzyme activity assay.

1. Introduction

Lysosomal storage disorders (LSDs) are genetic diseases caused by a deficiency in one of the 50 degradative enzymes that hydrolyze proteins, DNA, RNA, polysaccharides, and lipids in lysosomes. In LSDs, undegraded material accumulates in the lysosomes of affected individuals, causing an increase in the size and number of lysosomes within cells, cellular dysfunction, and progressive clinical manifestations (1). Most LSDs result from deficiencies in a single lysosomal enzyme. The prevalence of lysosomal storage

U. Garg, C.A. Hammett-Stabler (eds.), *Clinical Applications of Mass Spectrometry*, Methods in Molecular Biology 603, DOI 10.1007/978-1-60761-459-3_32, © Humana Press, a part of Springer Science+Business Media, LLC 2010

disorders as a group is ~1/5,000 (2). Enzyme replacement therapy and umbilical stem cell transplantation are available for some of LSDs (3, 4). Better clinical outcomes are achieved if patients are treated early, before irreversible organ and tissue damages occur (3–5). Given these realizations, there is now widespread interest in newborn screening or high-risk population screening for treatable LSDs.

This chapter focuses on the use of tandem mass spectrometry (MS) to measure the activity of five lysosomal enzymes, lysosomal α-glucosidase (GAA), α-galactosidase (GLA), β-glucocerebrosidase (GBA), acid sphingomyelinase (ASM), and galactocerebrosidase (GALC). These enzymes are deficient in Pompe disease, Fabry disease, Gaucher disease, Niemann-Pick type A/B, and Krabbe disease, respectively (6–10). The enzymes are extracted from dried blood spots on filter paper and incubated individually and concurrently with the enzyme-specific substrate and internal standard. The substrates are the structural analogues to the natural

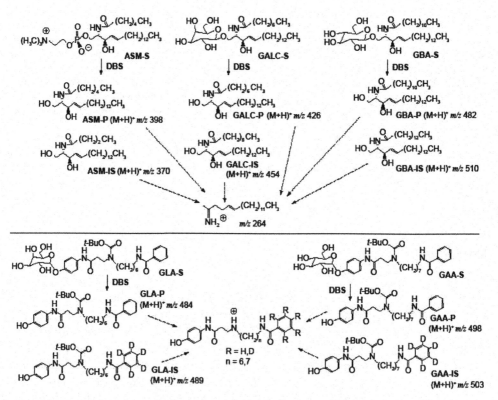

Fig. 32.1. Substrates, products, and internal standards for the five lysosomal enzyme activity assays. The masses of the products and internal standards are indicated. The ceramide products and internal standards ASM-P, ASM-IS, GALC-P, GALC-IS, GBA-P, and GBA-IS undergo CID to give the common ammonium ion shown (m/z = 264). GLA-P, GLA-IS, GAA-P, and GAA-IS undergo neutral–loss CID to give four different secondary ammonium ions. Enzymatic reactions with DBS are shown by solid arrows, and CID is shown by dashed arrows. Reproduced from Ref. (7) with permission from the American Association for Clinical Chemistry (AACC).

substrates. This avoids the problem that some mutations may render the enzyme less active on an artificial substrate while activity on the natural substrate is maintained or vice versa.

The enzymatic reactions are shown in **Fig. 32.1**. Each enzyme cocktail is prepared directly from the commercially available mixture of substrate and internal standard at the predetermined and optimized molar ratio. The enzymatic reactions are quenched using an ethyl acetate/methanol solution and subsequently the five reaction solutions are combined. The mixtures of the reaction products are prepared using liquid–liquid and solid-phase extraction and quantified simultaneously using selected ion monitoring on a LC-MS-MS system. Recently the enzyme activity assay for mucopolysaccharidosis I (11) is being incorporated into the multiplex assay; and efforts are underway to develop tandem MS assays for other lysosomal enzymes and other diseases caused by deficiency in specific enzymes.

2. Materials

2.1. Samples: Dried Blood Spot (DBS) Preparation from Venous Whole Blood

1. Collect venous blood in EDTA tube(s). If blood is collected the blood samples, ship at a remote location on cold packs for overnight delivery and hold at 4°C upon arrival.

2. Mix the blood by inverting the tube several times.

3. Pipet 75 µL of well-mixed whole blood onto the filter paper. The five LSD enzymes tested in this assay are stable in whole blood held at 4°C for up to 72 h.

4. Dry the spotted cards horizontally without touching each other for 4–8 h at ambient room temperature.

5. Store the dried cards in sealed plastic bags containing desiccant and a humidity indicator card at 4°C for up to 1 week or at −20°C for extended periods.

2.2. Reagents and Buffers (see Note 1)

1. 0.1 M zinc chloride solution

2. N-Acetylgalactosamine (GALNAc), and acarbose (Toronto Research Chemicals, Toronto, Ontario, Canada).

3. Analyte specific reagent (ASR) of each of the following enzyme substrates (S) and internal standards (IS) mixtures, (CDC Foundation, Atlanta, GA, USA) (*see* **Note 2**):

   i. Acid sphingomyelinase (ASM) ASR: Each vial typically contains a mixture of ASM-S and ASM-IS at a molar ratio of 50:1 with the total weight of 3.41 mg.

   ii. β-glucocerebrosidase (GBA) ASR: Each vial typically contains a mixture of GBA-S and GBA-IS at a molar ratio of 50:1 with the total weight of 7.84 mg.

iii. Lysosomal α-glucosidase (GAA) ASR: Each vial typically contains a mixture of GAA-S and GAA-IS at a molar ratio of 100:1 with the total weight of 7.97 mg.

iv. α-galactosidase (GLA) ASR: Each vial typically contains a mixture of GLA-S and GLA-IS at a molar ratio of 500:1 with the total weight of 38.78 mg.

4. Galactocerebrosidase (GALC) substrate (S) and internal standard (IS) (Genzyme Pharmaceutical, Liestal, Switzerland). Each vial contains a mixture of GALC-S and GALC-IS at a molar ratio of 150:1 with the total weight of 10.6 mg.

5. Quench solution: Mix equal volumes of ethyl acetate and methanol (both analytical grade). Stable when stored at room temperature for 1 month.

6. Wash solution: Mix 190 mL of ethyl acetate and 10 mL of methanol (analytical grade). Stable when stored at room temperature for 1 month.

7. Mobile phase: 80% acetonitrile and 20% water with 0.2% formic acid.

8. Extraction buffer solution: 20 mM sodium phosphate, pH 7.1. Add 2.40 g of sodium phosphate monobasic (NaH_2PO_4) to a beaker containing about 900 mL of water and stir to dissolve the powder. Adjust the pH to 7.1 with sodium hydroxide. Bring the solution to 1000 mL with water. Sterile filter and store at room temperature for up to 6 months.

9. Acid sphingomyelinase (ASM) buffer solution: Add 28.45 g of sodium acetate trihydrate to a beaker containing about 200 mL of water and stir to dissolve the powder. Add 1.34 mL of glacial acetic acid, and 1.51 mL of 0.1 M zinc chloride solution to the beaker. Adjust pH to 5.7 with sodium hydroxide or acetic acid (*do not use hydrochloric acid*). Bring the solution to 250 mL with water. The final acetate concentration is 0.92 M (pH 5.7). Sterile filter and store at 2–8°C for up to 6 months.

10. GAA, GBA, and GALC buffer solutions: Add 10.20, 21.45, or 5.99 g of sodium phosphate monobasic, respectively, to beakers containing about 200 mL of water. Add 12.5, 26.29, or 7.35 g of sodium citrate tribasic dehydrate, respectively, to the above solutions and stir until dissolved. Adjust the pH of the buffers to 4.0, 5.1, or 4.4, respectively, with hydrochloric acid or sodium hydroxide. Bring each solution to 250 mL with water. The final citrate–phosphate concentration is 0.3 M (pH 4.0) for GAA, 0.62 M (pH 5.1) for GBA, and 0.18 M (pH 4.4) for GALC buffer solution. Sterile filter and store at 2–8°C for up to 6 months.

11. GLA buffer solution: Add 2.45 g of sodium acetate trihydrate to a beaker containing about 200 mL of water and stir to dissolve the powder. Add 1.46 mL of glacial acetic

acid. Adjust pH to 4.6 with sodium hydroxide or acetic acid (*do not use hydrochloric acid*). Bring the solution to 250 mL with water. The final acetate concentration is 0.174 M (pH 4.6). Sterile filter and store at 2–8°C for up to 6 months.

2.3. Detergent and Inhibitor Solutions

1. 120 g/L sodium taurocholate in water. Store at -20°C for up to 6 months.

2. 96 g/L sodium taurocholate plus 12 g/L oleic acid in water. Store at -20°C for up to 3 months.

3. 100 g/L 3-[(3-cholamidopropyl)dimethylammonio]-1-propanesulfonate (CHAPS) in water. Store at -20°C for up to 3 months.

4. 0.8 mM acarbose in water. Make fresh on the day that assay cocktails are prepared.

5. 1.0 M N-Acetylgalactosamine (GALNAc) in water. Make fresh on the day that assay cocktails are prepared.

2.4. Assay Cocktails (see Notes 3, 4, and 5)

1. ASM assay cocktail: Add 0.15 mL of 120 g/L sodium taurocholate in water to one vial of ASM ASR and vortex briefly. Add 17.85 mL of ASM buffer solution to the vial and vortex again. The final ASM assay solution contains 0.33 mM of ASM-S, 6.67 μM of ASM-IS, 1.0 g/L sodium taurocholate, 0.6 mM of zinc chloride, and 0.92 M sodium acetate, pH 5.7.

2. GBA assay cocktail: Add 2.40 mL of 120 g/L sodium taurocholate in water to one vial of GBA ASR and vortex briefly. Add 15.6 mL of GBA buffer to the vial and vortex again. The final GBA assay solution contains 0.67 mM of GBA-S, 13.33 μM of GBA-IS, 16.0 g/L sodium taurocholate, and 0.62 M of phosphate plus 0.31 M of citrate, pH 5.1.

3. GALC assay cocktail: Add 1.8 mL of 96 g/L sodium taurocholate plus 12 g/L oleic acid solution in water to one vial of GALC ASR and vortex briefly. Add 16.2 mL of GALC buffer to the vial and vortex again. The final GALC assay solution contains 1 mM of GALC-S, 6.67 μM of GALC-IS, 9.6 g/L sodium taurocholate, 1.2 mM of oleic acid, and 0.18 M phosphate plus 0.09 M citrate, pH 4.4.

4. GAA assay cocktail: Add 1.8 mL of 100 g/L CHAPS in water to one vial of GAA ASR and vortex briefly. Add 15.9 mL of GAA buffer followed by 0.3 mL of 0.8 mM acarbose in water to the vial and vortex again. The final GAA assay solution contains 0.67 mM of GAA-S, 6.67 μM of GAA-IS, 10 g/L CHAPS, 13.3 μM acarbose, and 0.3 M phosphate plus 0.15 M citrate, pH 4.0.

5. GLA assay cocktail: Add 0.45 mL of 120 g/L sodium taurocholate in water to one vial of GLA ASR and vortex briefly. Add 14.67 mL of GLA buffer followed by 2.88 mL of 1.0 M GALNAc in water to the vial and vortex again. The final GLA assay solution contains 3.33 mM of GLA-S, 6.67 μM of GLA-IS, 3 g/L sodium taurocholate, 160 mM GALNAc, and 0.142 M sodium acetate, pH 4.6.

6. Assay cocktails are stable for up to 24 weeks at 4°C or up to 3 month at –20°C. The stability may be extended after further testing.

2.5. Calibration Solutions

1. P and IS calibration solutions for each enzyme (CDC Foundation, Atlanta, GA) (*see* **Note 2**).

2. Prepared solutions with P/IS ratios of 0, 0.05, 0.1, 0.5, 1.0, and 2.0. Verify ratios using mass spectrometry. The amount of IS is fixed at 0.1 μM for ASM, GAA, GLA, and GALC assays and 0.2 μM for GBA assay.

3. Example of regression analysis of the calibration curves is illustrated in **Table 32.1**.

Table 32.1
Regression analysis of the calibration curves

	N	Slope	Intercept	r^2	Sdy	Sdslope	Sdintercept
ASM	150	1.019	−0.0079	0.9988	0.0255	0.0029	0.0028
GBA	150	1.058	−0.0033	0.9994	0.0178	0.0021	0.0019
GALC	150	0.959	0.0008	0.9994	0.0164	0.0019	0.0018
GLA	150	0.941	−0.0001	0.998	0.0301	0.0035	0.0033
GAA	84	0.745	0.0044	0.9965	0.0653	0.005	0.0091

r^2: correlation coefficient
Sdy: standard deviation of measured product/internal standard ratio
Sdslope: standard deviation of slope
Sdintercept: standard deviation of intercept

2.6. Quality Control Samples

Quality control samples with low, medium, and high enzyme activities (CDC foundation, Atlanta, GA) (10) (*see* **Note 2**).

2.7. Equipment and Supplies

1. Protein Saver 903 specimen collection filter paper card (Whatman, Kent, UK or equivalent)

2. Deep 96-well polypropylene plates

3. Polypropylene 96-well plates

4. Aluminum plate sealer

5. Shaking microplate incubator capable of ~875 rpm

6. Temperature controlled orbital shaker capable of 225 ± 25 rpm

7. 96-well plate evaporator

8. 0.45 μm 96-well filter plates (Seahorse Labware, Chicopee, MA)

9. 96-well PVC plates (Falcon 353912,BD Biosciences, Bedford, MA)

10. Silica gel (230–400 mesh, 60 Å, Merck, Grade 9385) (Sigma Aldrich, St Louis, MO)

11. Solid-phase extraction MultiScreen$_{HTS}$ vacuum manifold and plate adaptors (Millipore, Billerica, MA)

12. Leap HTC PAL autosampler (Leap Technology, Carrboro, NC)

13. Agilent 1100 binary HPLC system (Agilent, Santa Clara, CA)

14. API 4000 triple quadrupole mass spectrometer (Applied Biosystems, Ontario, Canada)

---

3. Methods

3.1. Extraction of Dried Blood Spots

1. Punch two spots from each DBS sample using a 1/8th inch (3.2 mm) hole puncher and place one spot in one polypropylene 96-well plate and the other spot into the corresponding well of the second plate. When all samples have been punched, cover but do not seal the second plate and set it aside at room temperature to be used for the GALC assay. Do not use a plate sealer to cover the second plate as static electricity may cause the spots to stick to the sealer.

2. To the first plate, add 70 μL of extraction buffer to each well containing spots and seal the plate with an aluminum plate sealer.

3. Centrifuge the plate at $2,000 \times g$ for 1 min (optional). Incubate the plate at 37°C for 60 min while shaking at about 875 RPM.

4. Remove the plate from the incubator after 1 h and centrifuge at $2,000 \times g$ for 1 min to collect the extract to the bottom of the wells. Proceed immediately to the enzymatic reactions step.

3.2. Enzymatic Reactions

1. Thaw the assay cocktails at room temperature. The step needs to be done prior to completion of the DBS extraction. Sonicate all five thawed assay cocktails.

2. Label four 96-well plates as ASM, GBA, GAA, or GLA. To each plate add 15 μL of ASM, GBA, GAA, or GLA assay cocktail into each well. To limit evaporation, set up assay cocktail plates no more than 10 min prior to the completion of the DBS extraction step.

3. Add 10 μL of the DBS extract to the corresponding well of each of the four plates containing assay cocktail. Cover each plate with an aluminum plate sealer.

4. To the plate previously set aside containing one whole punch per sample, add 30 μL of GALC assay cocktail. Cover the plate with an aluminum plate sealer.

5. The compositions of each enzymatic reaction mixtures are summarized in **Table 32.2**.

6. Centrifuge all five plates at 2,000 × g for 1 min to ensure sample and assay cocktail mix together at the bottom of each well. Incubate at 37°C for 20–24 h with orbital shaking at 225 ± 25 RPM.

7. When incubation is complete add 100 μL of quenching solution into each well of all five plates. Work with one assay plate at a time.

8. Transfer the contents of the wells associated with the same sample on all five plates into one corresponding well of a deep-well polypropylene plate.

Table 32.2

Summary of the assay cocktail compositions used in the multiplex assay reaction mixtures

Assay	DBS punch	Assay volume μL	[S] mM	[IS] μM	Buffer	Detergent	Additional components
GAA	1/7[a]	25	0.4	4	0.18 M citrate-phosphate, pH 4.0	6 g/L CHAPS	8 μM acarbose
GLA	1/7[a]	25	2	4	0.09 M sodium acetate, pH 4.6	1.8 g/L sodium taurocholate	96 mM GALNAc
GBA	1/7[a]	25	0.4	8	0.37 M citrate-phosphate, pH 5.1	9.6 g/L sodium taurocholate	none
ASM	1/7[a]	25	0.2	4	0.55 M sodium acetate, pH 5.7	0.6 g/L sodium taurocholate	0.36 mM zinc chloride
GALC	1[b]	30	1	6.7	0.18 M citrate-phosphate, pH 4.4	9.6 g/L sodium taurocholate	1.2 g/L oleic acid

[a]10 out of 70 μL of DBS extractant was used in the assay, equivalent to 1/7 of DBS.
[b]One whole DBS was used in the assay.
Reproduced from Ref. (7) with permission from AACC.

3.3. Sample Clean-Up

1. To limit dripping, pre-wet pipette tips with EA. Add 400 μL of EA followed by 400 μL of water into each well of the deep-well plate. Aspirate and dispense the solutions three times. Seal the plate with an aluminum plate sealer.

2. Centrifuge the plate for 5 min at $2,000 \times g$ to form the two-phase separation.

3. Transfer 300 μL of the top organic layer of each sample to a new deep-well plate. Bring to complete dryness under a stream of nitrogen on a 96-well plate evaporator. To speed the drying process, the plate may be gently heated to 25°C.

4. Reconstitute the plate by adding 100 μL of wash solution into each well.

5. Cover the plate with an aluminum plate sealer and shake it at about 200 RPM for 5 min on a plate shaker.

6. Prepare a silica gel filter plate. Load a 96-well PVC plate with silica to the top of each well (contains ~100 mg of silica gel per well). Scrape off excess silica with a glass slide. Invert an empty filter plate onto the silica-filled plate, taking care to align all the wells. With gloved hands hold the two plates tightly together and flip them over to transfer the silica into the wells of the filler plate.

7. Place a deep-well plate in the lower chamber of a vacuum manifold.

8. Condition the silica gel in the filter plate by adding 250 μL of wash solution. Apply gentle vacuum to draw the solvent through the silica gel (*see* **Note 6**). Discard the wash solution collected in the lower chamber.

9. After conditioning the silica gel, place a clean deep-well plate in the lower chamber of the vacuum manifold.

10. Transfer the reconstituted and mixed samples from step 5 to the corresponding well of the filter plate. Apply gentle vacuum to draw the samples through the silica gel and collect each sample in the deep-well plate in the lower chamber of the vacuum manifold.

11. Wash each well of the filter plate four times with 400 μL of wash solution, collecting each wash in the deep-well plate for a total wash volume of 1,600 μL.

12. Dry the deep well plate under a stream of nitrogen.

13. When dry, cover the plate with an aluminum plate sealer and store it at −20°C for future mass spectrometry analysis.

14. Prior to the mass spectrometry analysis, reconstitute the plate with 200 μL of mobile phase into each well.

15. Cover the plate with an aluminum plate sealer and shake it at about 200 RPM for 5 min on a plate shaker.

16. Remove the plate sealer and cover the plate with aluminum foil for mass spectrometry analysis. Do not use heavy weight aluminum foil.

3.4. MS-MS Analysis

1. MS-MS analysis is performed on an API 4000 triple quadrupole mass spectrometer (*see* **Note 7**). The instrument is operated in positive ion mode. All analytes are monitored by selected reaction monitoring (SRM). The typical experimental set-up on the API 4000 is listed in **Table 32.3**.

Table 32.3
Typical parameters used on the API 4000 triple quadrupole mass spectrometer

LC Conditions	
Mobile Phase:	80/20 acetonitrile/water with 0.2% formic acid
Flow Rate:	0.2 ml/min
Gradient:	Isocratic
Washes:	Methanol with 0.1% formic acid and 50/50 isopropanol/methanol
Filter:	HAI*FILTER* from Higgins Analytical Part #: HF-FKIT
Injection Volume:	20 µl

MS/MS Conditions	
Instrument:	MDS SCIEX API 4000 triple quadrupole
ES Source:	Turbo spray ES+
Dwell Time:	100 msec
Source Temp.:	200°C
CAD Gas:	4
Curtain Gas:	23 psi
Gas 1:	23 psi
Gas 2:	32 psi
IonSpray Voltage:	4500 V
Declustering Potential:	20 V
Collision Cell Exit Potential:	20 V

Name	MRM	CE (eV)	EP (V)
ASM-IS	370.32>264.27	23	2
ASM-P	398.36>264.27	23	2
GALC-IS	454.42>264.27	29	2
GALC-P	426.39>264.27	29	2
GBA-IS	510.48>264.27	33	2
GBA-P	482.45>264.27	31	2
GLA-IS	489.30>389.30	24	7
GLA-P	484.27>384.27	18	7
GAA-IS	503.32>403.32	21	7
GAA-P	498.29>398.29	27	7

2. Thirty μL of sample solution is injected by an autosampler and delivered at a flow rate of 200 μL/min with the HPLC system. The total run time on the LC-MS system is 3 min per sample.

3.5. Data Analysis

1. The enzyme activity is calculated using the formula below:

Activity $(\mu mol/hr/L$ of whole blood$) = 1000*(P/IS)*[IS]/(RF*V*T)$

Where,

P/IS = measured product to internal standard ratio

[IS] = amount of internal standard injected (nmol);
0.1 nmol for GAA-IS, GLA-IS, and ASM-IS;
0.2 nmol for GALC-IS and GBA-IS.

RF   = response factor ratio of product to internal standard

V   = volume of whole blood used in the assay (μL)

T   = incubation time (hour)

2. RF is the slope of the linearity curve of P and IS using P/IS calibration solutions.

3. The volume of blood used in the assay is 0.457 (one-seventh of 3.2) μL whole blood in GAA, GLA, GBA, and ASM assays and 3.2 μL in GALC assay.

4. The blank activity is subtracted from the measured activity to obtain the final activity of each sample. The blank activity is the average activity of blank filter paper samples ($n>=4$) on the same plate with samples.

4. Notes

1. All solutions and buffers should be prepared with HPLC-grade water. Sterile filter buffers to remove particulates and debris. Assay buffers do not need to be handled as sterile solutions. Discard any solutions showing evidence of contamination.

2. Obtain ASRs and QC samples from the CDC Foundation:
Dr. Hui Zhou/Dr. Victor De Jesus

CDC Foundation

Newborn Screening and Molecular Biology Branch

Centers for Disease Control and Prevention

4770 Buford Hwy NE, Mail stop F19

Atlanta, GA 30341

3. All assay cocktail solutions prepared to 18 mL are sufficient for approximately 600 tests (wells) for GALC or 1200 tests for the other four enzymes. It is important to add the detergent first when making assay cocktails. Aliquots sufficient for a single day's use may be prepared and stored frozen to avoid repeated freeze thaw cycles. If more than one vial of an assay cocktail is prepared, pool reconstituted vials of the same assay cocktail together before making aliquots.

4. If an assay cocktail looks hazy or particulates are visible, heat it briefly in warm tap water and/or sonicate it at room temperature for 5 min.

5. Assay cocktails are stable for up to three freeze-thaw cycles.

6. Keep the vacuum gentle enough to avoid forming channels through the silica beds of the filter plate.

7. The assay has been performed on other mass spectrometers, including, Sciex 3200, Quattro Micro, and Quattro Premier as a part of an investigation for assay comparability in different laboratories. Comparable results have been obtained on all instruments.

References

1. Scriver C. R., Beaudet A. L., Sly W. S., Valle D. eds. (2001) *The Metabolic and Molecular Bases of Inherited Disease*, 8th ed. New York, NY: McGraw-Hill.

2. Meikle P. J., Hopwood J. J. (2003) Lysosomal storage disorders: emerging therapeutic options require early diagnosis. *Eur J Pediatr*, **162**: S34–S7.

3. Clarke L. A. (2008) The mucopolysaccharidoses: a success of molecular medicine. *Expert Rev Mol Med* **10(1)**: e1.

4. Rohrbach M, Clarke J. T. (2007) Treatment of lysosomal storage disorders: progress with enzyme replacement therapy. *Drugs* **67**: 2697–716.

5. Chien Y.-H., Chiang S.-C., Zhang X. K., Keutzer J. M., Lee N.-C., Huang A.-C., Chen C.-A., Wu M.-H., Huang P.-H., Tsai F.-J., Chen Y.-T., Hwu W.-L. (2008) Early detection of Pompe disease by newborn screening is feasible: results from the Taiwan screening program. *Pediatrics* **122**: e39–e45.

6. Li Y, Brockman K, Turecek F, Scott C. R., and Gelb M. H. (2004) Tandem mass spectrometry for the direct assay of enzymes in dried blood spots: application to newborn screening for Krabbe disease. *Clin Chem* **50**, 638–40.

7. Li Y, Scott C. R., Chamoles N. A., Ghavami A, Pinto B. M., Turecek F, Gelb M. H.

(2004) Direct multiplex assay of lysosomal enzymes in dried blood spots for newborn screening. *Clin Chem* **50**: 1785–96.

8. Zhang X. K., Elbin C. S., Chuang W-L, Cooper S. K., Marashio C. A., Beauregard C, and Keutzer J. M. (2008) Multiplex enzyme assay screening of dried blood spots for lysosomal storage disorders by using tandem mass spectrometry. *Clin Chem* **54**: 1725–28.

9. Dajnoki A, Mühl A, Fekete G, Keutzer J, Orsini J, DeJesus V, Zhang X. K., Bodamer O. A. (2008) Newborn screening for Pompe disease by measuring acid -glucosidase activity using tandem mass spectrometry. *Clin Chem* **54**: 1624–29.

10. De Jesus R., Zhang X. K., Keutzer J. M., Bodamer O. A., Mühl A., Orsini J. J., Caggana M., Vogt R. F., Hannon W. H. (2009) Development and evaluation of quality control dried blood spot materials in newborn screening for lysosomal storage disorders. *Clin Chem* **55**: 158–64.

11. Blanchard S., Sadilek M., Scott C. R., Turecek T., Gelb M. H. (2008) Tandem mass spectrometry for the direct assay of lysosomal enzymes in dried blood spots: Application to screening newborns for Mucopolysaccharidosis I *Clin Chem* **54** 2067–70.

Chapter 33

Gas Chromatography-Mass Spectrometry Method for the Determination of Methadone and 2-Ethylidene-1,5-Dimethyl-3, 3-Diphenylpyrrolidine (EDDP)

Christine L.H. Snozek, Matthew W. Bjergum, and Loralie J. Langman

Abstract

Methadone is a synthetic opioid used to relieve pain, treat opioid withdrawal, and wean heroin addicts. Measurement of methadone and its major metabolite EDDP in urine is useful for assessing compliance with addiction rehabilitation and pain management programs. This method quantitatively measures methadone and its metabolite EDDP in urine. Methadone and EDDP are recovered from urine by solid phase extraction at pH 6.0. Analysis is performed by gas chromatography with mass spectrometry detection, using selective ion monitoring.

Key words: Methadone, EDDP, opiate addiction, pain management, gas chromatography, GC-MS.

1. Introduction

Methadone is a synthetic opioid, that is structurally dissimilar from natural opiates such as morphine, yet possesses the ability to activate opioid receptors to produce analgesia and sedation (1). The drug does not produce feelings of euphoria and has substantially fewer withdrawal symptoms compared to other opiates, such as heroin (1). Methadone is used clinically to relieve pain, to treat opioid abstinence syndrome (withdrawal), and to wean heroin addicts from illicit drug use (1, 2).

Oral delivery of methadone makes it subject to first-pass metabolism by the liver, creating inter-individual differences in bioavailability ranging from 80 to 95%. Metabolism of methadone to inactive compounds is the main form of elimination; the most prevalent inactive metabolite is 2-ethylidene-1,5-dimethyl-3,

U. Garg, C.A. Hammett-Stabler (eds.), *Clinical Applications of Mass Spectrometry*, Methods in Molecular Biology 603, DOI 10.1007/978-1-60761-459-3_33, © Humana Press, a part of Springer Science+Business Media, LLC 2010

3-diphenylpyrrolidine, or EDDP (1, 3–5). The efficiency of methadone metabolism is prone to wide inter- and intra-individual variability, due both to inherent differences in enzymatic activity as well as enzyme induction or inhibition by numerous drugs. Excretion of methadone and its metabolites (including EDDP) occur primarily through the kidneys (1, 5).

Measurement of methadone and EDDP in urine is useful for assessing compliance with addiction rehabilitation and pain management programs. Patients who are taking methadone for therapeutic purposes excrete both parent methadone and EDDP in their urine. It is important to measure levels of both methadone and EDDP, as urine methadone levels vary widely between patients, depending on factors such as dose, metabolism, and urine pH (6). Acidification of urine will result in increased excretion of unchanged methadone. In contrast pH has less of an influence on EDDP excretion, making it a more useful measure for assessing compliance with therapy (6, 7).

In addition, the presence of EDDP provides confidence that the parent drug was administered appropriately rather than spiked into the urine sample to falsely pass compliance testing. Although absolute concentrations of methadone and EDDP in urine can vary greatly between patients, individuals known to be compliant with therapy typically have EDDP:methadone ratios >0.60 (8). Spiked compliance samples, in contrast, contain large amounts of methadone but little or no EDDP; in such samples the EDDP: methadone ratio is often <0.09 (9, 10).

This method is designed to quantify methadone and its metabolite EDDP in urine. Methadone and EDDP exist in urine as free (unconjugated) drugs, which are extracted at pH 6.0 using a solid phase extraction (SPE). The SPE eluent is evaporated to dryness and the resulting residue reconstituted in n-butyl chloride; analysis is performed using gas chromatography with mass spectrometry detection (GC-MS) and selective ion monitoring. Deuterated compounds (methadone-D_9 and EDDP-D_3) are used as internal standards.

2. Materials

2.1. Solvents and Chemicals

1. Isopropyl alcohol, methyl alcohol, dichloromethane, 1-Chlorobutane (N-butyl chloride), acetonitrile, ammonium hydroxide, sodium phosphate monobasic, sodium phosphate dibasic, and hydrochloric acid are of HPLC or analytical grade.

2. Methadone, Methadone-D_9, EDDP perchlorate, and EDDP-D_3 (Cerilliant Corp, Round Rock, TX).

2.2. Prepared Reagents

1. Extraction solvent: Mix dichloromethane, isopropyl alcohol, and ammonium hydroxide in a volume ratio of 78:20:2. Prepare fresh daily.

2. Sodium phosphate buffer, 0.1 M, pH 6.0, prepare in type I water.

3. Hydrochloric acid, 0.1 M, prepare with type I water.

4. 20.0 ug/mL spiking standard: Quantitatively transfer two ampules of 1.0 mg/ml methadone and two ampules of 1.0 mg/mL EDDP perchlorate to a 100.0 mL volumetric flask, bring to volume with methanol and mix well. Stable up to 2 years, stored at –20°C in screw-cap amber vials with rubber/Teflon septa.

5. 2.0 ug/mL spiking standard: Aliquot 200 uL of 1.0 mg/mL methadone and 1.0 mg/mL EDDP perchlorate into a 100.0 mL volumetric flask, bring to volume with methanol and mix well. Stable up to 2 years, stored at –20°C in screw-cap amber vials with rubber/Teflon septa.

6. Working internal standard: Quantitatively transfer two ampules of methadone-D9 and 20 ampules of EDDP-D3 into a 200.0 mL flask, bring to volume with methanol and mix well. Stable up to 2 years, stored at –20°C in screw-cap amber vials with rubber/Teflon septa.

2.3. Supplies and Analytical Equipment

1. Clean Screen® Extraction Columns (United Chemical Technologies, Inc.).

2. Gas chromatograph, e.g., Agilent Model 6890 or similar (Agilent Technologies).

3. Interfaced mass spectrometer, e.g., Agilent Model 5973 or similar (Agilent Technologies).

4. Analytical column: DB-5MS capillary column (J & W Scientific), 15 m length, 0.25 mm inner diameter, 1.0 um film thickness.

---

3. Methods

3.1. Preparation of Working Standards, Controls, and Unknown Samples (Each Run)

1. Prepare the unextracted cutoff control
   1.1 Add 50 uL of the 2.0 μg/mL spiking standard to a labeled 16 × 100 mm test tube.
   1.2 Add 50 uL of working internal standard.
   1.3 Add 5 mL extraction solvent.
   1.4 Evaporate to dryness under a gentle nitrogen flow with heat ≤35°C.
   1.5 Reconstitute in 75 uL of *n*-butyl chloride.

Table 33.1
Preparation of standards

Standard (ng/mL)	Volume 2.0 ug/mL spiking standard (mL)	Volume 20.0 ug/mL spiking standard (mL)	Final volume (mL)
100	0.05	0	1.0
500	0.25	0	1.0
1,000	0	0.05	1.0
2,000	0	0.1	1.0
4,000	0	0.2	1.0

2. Prepare working calibration standards for the run

  2.1 Aliquot the appropriate amount of 2.0 ug/mL OR 20.0 ug/mL spiking standard, as shown in **Table 33.1**, into a labeled 16 × 125 mm test tube.

  2.2 Dilute to 1.0 mL with drug-free urine.

  2.3 Aliquot 1.0 mL of drug-free urine for the negative (carry-over) control.

  2.4 Add 50 uL of working internal standard to each tube.

3. Prepare working quality controls for the run

  3.1 Dilute spiking standards to desired concentrations, using procedure similar to preparation of calibration standards.

  3.2 Bring to 1.0 mL volume with drug-free urine.

  3.3 Add 50 uL of working internal standard.

4. Prepare unknown samples

  4.1 Add 1.0 mL of each sample to appropriately labeled 16 × 125 mm test tubes. (*See* **Note 1** for samples expected to have high values.)

  4.2 Add 50 uL of working internal standard to each tube.

5. Add 4 mL of 0.1 M phosphate, pH 6.0 to ALL tubes: standards, controls, and unknowns.

6. Vortex to mix.

3.2. Extraction and Derivatization

Note – preconditioning and extraction steps may vary between SPE column manufacturers. Follow recommendations for the column used.

1. Column Extraction – do not allow columns to dry out until stated.

  1.1 Place one labeled column per sample (standard, control, or unknown) in vacuum manifold.

1.2 Under low vacuum (1-2 mL/min), precondition the columns with 3 mL of methanol.

1.3 Precondition the columns with 3 mL of distilled water.

1.4 Precondition the columns with 2 mL of 0.1 M phosphate, pH 6.0 buffer.

1.5 Apply each sample to the appropriate column under low vacuum (1–2 mL/min).

1.6 Wash the columns with 3 mL of distilled water.

1.7 Wash the columns with 2 mL of 0.1 M HCl.

1.8 Wash the columns with 1 mL of acetonitrile.

1.9 Allow the columns to dry under full vacuum for 5 min.

2. Return to low vacuum (1–2 mL/min) and add 5 mL of extraction solvent.

3. Elute under low vacuum into clean, labeled 16 × 100 mm glass tubes.

4. Evaporate to dryness under a gentle nitrogen flow with heat ≤35°C.

5. Reconstitute each sample in 75 uL of *n*-butyl chloride.

6. Transfer to glass autosampler vials.

3.3. Analysis

1. Place the sample extracts on the GC-MS autosampler in the following order:

Unextracted cutoff control

Calibration standards, in order of lowest-to-highest concentration

Negative urine (carryover) control

Quality controls, lowest-to-highest

Unknown samples and additional quality controls (*see* **Note 2**)

2. Set GC-MS method to the following parameters:

2.1 Oven temperature: Pre-run 150°C, ramp 20°C per min to 280°C. Hold 280°C for 1 min, then ramp down to 150°C post-run.

2.2 Inlet temperature: 190°C (*see* **Note 3**)

2.3 Pressure: 3.25 psi

2.4 Total flow: 16 mL/min

3. Inject 1 uL sample at a 15:1 split ratio

4. Detect each compound (**Fig. 33.1**) according to GC-MS parameters detailed in **Table 33.2**. Fragmentation spectra for methadone and EDDP are shown in **Fig. 33.2**.

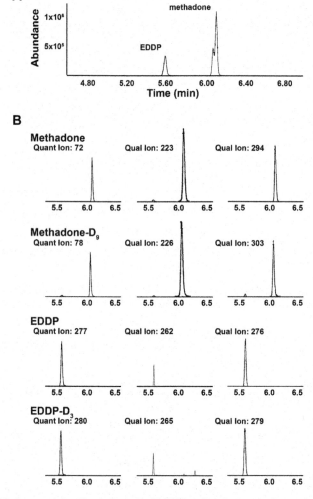

Fig. 33.1. Representative chromatography. **(A)** EDDP and methadone peaks are shown; the respective deuterated internal standards co-elute with each compound. **(B)** the quantitating and two qualifying ions are shown for methadone, EDDP, and the deuterated internal standards. Relative heights are scaled to the quantitating ion for each compound.

Table 33.2
GC-MS parameters

Compound	Quant ion	Qualifier 1 (rel. response)	Qualifier 2 (rel. response)	Retention time (min)	Integration window (min)
Methadone	72	223 (1.70)	294 (1.30)	6.101	5.601–6.601
Methadone-D9	78	226 (1.60)	303 (1.20)	6.068	5.568–6.568
EDDP	277	262 (49.1)	276 (108.5)	5.594	5.094–6.094
EDDP-D3	280	265 (46.7)	279 (121.3)	5.583	5.083–6.083

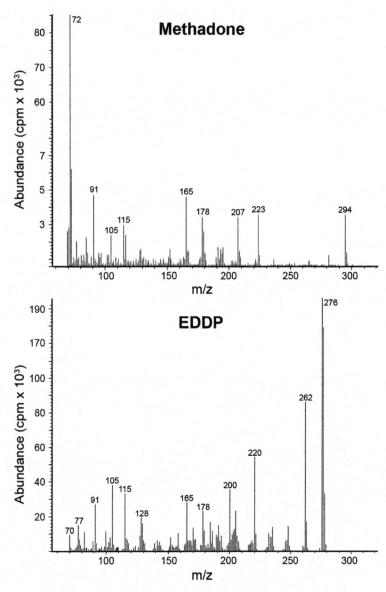

Fig. 33.2. Mass spectra. Fragmentation spectra are shown for methadone (**top**) and EDDP (**bottom**). Note, scale is adjusted to enhance detail of lower-intensity peaks.

4. Notes

1. It may be useful to run samples expected to have high results (e.g., through use of a screening assay) at a dilution in addition to the "neat" analysis described. For example, a 1:20 dilution can be prepared by adding an aliquot of 50 uL of a sample to 1.0 mL drug-free urine.

2. Sample order is at the discretion of the user; the rationale for this order is as follows: The unextracted cutoff control confirms proper functioning of the GC-MS, independently of the success of sample preparation. The standards are run in order of increasing concentration, followed by a blank sample to assess any carryover. Quality control samples are interspersed with unknown samples to monitor the success of analysis throughout the run. To ensure at least 10% of each clinical run is comprised of quality controls and calibrators, we run one control after every nine patient samples.

3. Injector port temperature is intentionally set low to decrease artifactual conversion of methadone to EDDP (9).

References

1. Gutstein, H.B. and Akil, H. (2001) Opioid analgesics. In: Hardman, J., Limbird, L. and Gilman, A. (eds.). *Goodman & Gilman's: The Pharmacological Basis of Therapeutics*, Vol. 10th ed. New York, NY: McGraw-Hill, 569–619

2. DOLOPHINE® HYDROCHLORIDE (Methadone hydrochloride tablets, USP), (2006) Package Insert. Columbus, OH: Roxane Laboratories, Inc.

3. Eap, C.B., Buclin, T. and Baumann, P. (2002) Interindividual variability of the clinical pharmacokinetics of methadone: implications for the treatment of opioid dependence. *Clin Pharmacokinet*, **41**, 1153–93.

4. Ferrari, A., Coccia, C.P., Bertolini, A. and Sternieri, E. (2004) Methadone – metabolism, pharmacokinetics and interactions. *Pharmacol Res*, **50**, 551–9.

5. Baselt, R.C. (2005) *Disposition of Toxic Drugs and Chemicals in Man*, 7th ed. Chemical Toxicology Institute, Foster City, CA.

6. Kerrigan, S. and Goldberger, B.A. (2003) Opioids. In: Levine, B., ed. *Principles of Forensic Toxicology*, Vol. 2nd ed. Washington DC: AACC Press, 187–205.

7. Baselt, R.C. (2004) *Disposition of Toxic Drugs and Chemicals in Man*, 7th ed. Biomedical Publications, Foster City, CA.

8. George, S. and Braithwaite, R.A. (1999) A pilot study to determine the usefulness of the urinary excretion of methadone and its primary metabolite (EDDP) as potential markers of compliance in methadone detoxification programs. *J Anal Toxicol*, **23**, 81–5.

9. Galloway, F.R. and Bellet, N.F. (1999) Methadone conversion to EDDP during GC-MS analysis of urine samples. *J Anal Toxicol*, **23**, 615–9.

10. George, S., Gill, L. and Braithwaite, R.A. (2000) Simple high-performance liquid chromatographic method to monitor vigabatrin, and preliminary review of concentrations determined in epileptic patients. *Ann Clin Biochem*, **37(Pt 3)**, 338–42.

Chapter 34

Liquid Chromatography-Tandem Mass Spectrometry (LC-MS-MS) Method for Monitoring Methotrexate in the Setting of Carboxypeptidase-G2 Therapy

Vikram S. Kumar, Terence Law, and Mark Kellogg

Abstract

Medications that have a narrow therapeutic window must be regularly monitored to assist in clinical dosing. Methotrexate (MTX) is one such chemotherapeutic agent that is closely monitored using various standard assays. Patients who have renal failure and who are treated using high-dose protocols are often given carboxypeptidase-G2 ($CPDG_2$) as a chemoprotective agent. In this setting an inactive metabolite of MTX, 2, 4-diamino-N^{10}-methylpteroic acid (DAMPA) is produced. DAMPA cross-reacts with MTX in most immunoassays, thus making them unsuitable for the monitoring of MTX in the setting of $CPDG_2$ therapy. We describe a rapid LC-MS-MS method that can be used to determine MTX levels in these cases.

Key words: Methotrexate, carboxypeptidase-G2 ($CPDG_2$), 2, 4-diamino-N^{10}-methylpteroic acid (DAMPA), liquid chromatography tandem mass spectrometry (LC-MS-MS).

1. Introduction

Methotrexate (MTX) is an antimetabolite and folic acid antagonist with widespread use in chemotherapy and autoimmune disease. High-dose MTX (HDMTX) therapy is often used to treat osteosarcoma, childhood leukemia, and head and neck cancers (1, 2), because at such doses, MTX can enter cells through passive diffusion and overcome the resistance that develops in standard therapy via saturation of active MTX transport across the cellular membrane (3). Higher doses of MTX also promote the formation of polyglutamated MTX derivatives that show reduced cellular efflux and increased inhibition of the target dihydrofolate reductase (DHFR).

U. Garg, C.A. Hammett-Stabler (eds.), *Clinical Applications of Mass Spectrometry*, Methods in Molecular Biology 603,
DOI 10.1007/978-1-60761-459-3_34, © Humana Press, a part of Springer Science+Business Media, LLC 2010

Leucovorin, a naturally occurring folate, is given following HDMTX infusion to regenerate intracellular folate concentrations to levels required for DNA synthesis in healthy cells. By competing with MTX for active transport into cells, leucovorin also provides a small level of protection from MTX toxicity. MTX is excreted mainly through the kidney, with some metabolism in the liver to 7-hydroxymethotrexate (7OH-MTX). Delayed MTX elimination in patients with poor renal function can lead to toxicities that include renal failure, myelosuppression, dermatitis, and even death (3). If leucovorin rescue is not sufficient to protect from HDMTX-related renal failure, carboxypeptidase-G2 ($CPDG_2$), a chemoprotective agent, can be used in conjunction with leucovorin. $CPDG_2$ is a recombinant bacterial enzyme that converts MTX into the inactive metabolites 2,4-diamino-N^{10}-methylpteroic acid (DAMPA) and glutamate (4–7). DAMPA further undergoes non-renal elimination (8).

$CPDG_2$ enzymatic action is rapid and converts more than 98% of circulating MTX to DAMPA within 15 min. Due to its size, $CPDG_2$ cannot enter the cells and is unable to initially affect intracellular MTX concentrations; however, as extracellular MTX concentrations decline, intracellular MTX gradually effluxes across the gradient resulting in increased serum MTX concentrations after a few hours. Thus leucovorin rescue must be continued after $CPDG_2$ and MTX concentrations are followed to determine when leucovorin can be stopped. Most immunoassays for MTX measurement show a significant cross-reactivity between DAMPA and MTX, making such assays unsuitable following treatment with $CPDG_2$ (2, 9). While reported values will trend downward as DAMPA and MTX are metabolized the contribution of each to signal is unknown. Described is a rapid and sensitive LC-MS-MS method to measure MTX without interference from DAMPA.

2. Materials

2.1. Reagents and Solutions

1. Mobile phase A (0.1% formic acid): Add 1 mL of formic acid to 1,000 mL of deionized water.
2. Mobile phase B: 100% methanol

2.2. Calibrators and Controls

1. Methotrexate calibrators of 0, 0.05, 0.15, 0.30, 0.60, and 1.00 μmol/L: Abbott TDx/TDxFlx Methotrexate II assay kit (Abbott Diagnostics, Abbott Park, IL). (see **Notes 1 and 2**)
2. Controls: Three concentrations spanning measuring range (Abbott Diagnostics, Abbott Park, IL) or equivalent source.

2.3. Supplies	1. 1.5 mL microcentrifuge tubes (Fisher Scientific, Waltham, MA)
	2. Autosampler vials: 12 × 32 mm, clear glass Qsert vials with 300μL fused insert, snap cap, and preslit septa (Waters Corporation, Milford, MA)
2.4. Equipment	1. Liquid Chromatograph/Tandem Mass Spectrometer (Shimadzu Corporation, Columbia, MD) LC-20 AD pumps (2), CMB-20A communication module, SIL-20AC autosampler, and an API 5000 with Turbo V ion source (Applied Biosystems, Foster City, CA).
	2. Hypersil BDS 1 C8, 2.1 mm x 50 mm, 3 um particle size column (Thermo-Fisher Scientific, Waltham, MA).
	3. Microcentrifuge capable of 13,400 × g
	4. Fisher vortex Genie (Thermo-Fisher Scientific, Waltham, MA).

3. Methods

3.1. Stepwise Procedure	1. Label microcentrifuge tubes for each standard, control, and patient sample to be run.
	2. Pipette 180 μL of methanol into all tubes.
	3. Slowly pipette 20 μL of each standard, control, and patient sample into the appropriately labeled tube containing methanol. Cap the tube and vigorously vortex mix for 30 sec.
	4. Centrifuge the standards, controls, and samples for 2 min at 13,400 × g.
	5. After centrifugation, transfer 150 μl of each supernatant into a labeled autosampler vial. Place vials into autoinjector.
	6. Inject 10 μl of sample.
3.2. Liquid Chromatography Conditions	1. Chromatography conditions are given in **Table 34.1**.
	2. Equilibrate the column for 1.0 min with 100% Buffer A between injections.
	3. The total time between sample injections is 5 min and all compounds elute in <3 min.
3.3. Mass Spectrometry	1. Ion source: Turbo V electrospray in positive ion mode.
	2. The molecular ions transition for MTX is m/z^+ 455.1/308.2, that for DAMPA is m/z^+ was 326.1/175.0, and that for 7OH-MTX is m/z^+ was 471.0/323.9.
	3. The instrument setting should be as follows: Ionspray 5,500 V, declustering potential 129 V, entrance potential 10 V, collision energy 28.8 V, collision exit potential 23.0 V, temperature 450°C, nebulizer gas 30, exhaust gas 15, curtain gas 10, and collision gas 5.

Table 34.1
HPLC-ESI-MS/MS operating conditions

Column temp. Room temp			
Flow (mL/min)	0.6		
Gradient	Time (min)	% Mobile phase A	% Mobil phase B
	0.0	100 (Start)	0
	0.9	80	20
	2.7	0	100
	3.6	100	0
	4.0	100 (End)	

3.4. Data Analysis

1. Representative chromatograms are shown in **Fig. 34.1**.

2. Analyze data using Analyst software (Applied Biosystems, Foster City, CA) or similar software. The quantitation of MTX is made from standard curves using multiple reaction monitoring (MRM) and peak area against the calibration curve, from the six standards that show a linear regression through zero with $1/x$ weighting.

3. The linearity of the method is 0.01–1 μmol/L. Expected intra- and inter-assay coefficients of variation are <11%.

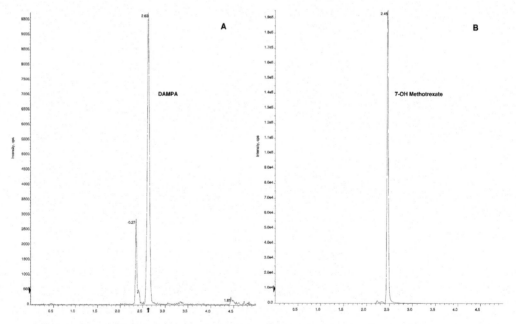

Fig. 34.1. Chromatograms showing DAMPA daughter ion (A)(m/z 175.0) and 7-OH Methotrexate (B)(m/z 323.9) in sample from a patient who has received Carboxypeptidase G2.

4. Samples with MTX levels above the highest standard should be rerun in serial dilutions as needed, 1:10, 1:100, 1:1000 from the extract supernatant with 20% methanol in water.

4. Notes

1. The objective of the method is to be able to rapidly and sensitively measure MTX levels in the setting of DAMPA cross-reactivity. To make the assay useful in a busy clinical environment, the sample preparation is simple and direct. This is also the rationale for the use of commercially available standards and controls.

2. This method does not incorporate an internal standard, again to simplify the procedure and utilize readily available materials. Because sample preparation is a simple protein precipitation and does not incorporate any concentration or reconstitution steps, the lack of an internal standard does not compromise assay quality.

References

1. Ackland SP, Schilsky RL (1987) High-dose methotrexate: a critical reappraisal. J Clin Oncol 5: 2017–2031.
2. Widemann BC, Balis FM, Murphy RF, Sorensen JM, Montello MJ, et al. (1997) Carboxypeptidase-G2, thymidine, and leucovorin rescue in cancer patients with methotrexate-induced renal dysfunction. J Clin Oncol 15: 2125–2134.
3. Ahmad S, Shen FH, Bleyer WA (1978) Methotrexate-induced renal failure and ineffectiveness of peritoneal dialysis. Arch Intern Med 138: 1146–1147.
4. DeAngelis LM, Tong WP, Lin S, Fleisher M, Bertino JR (1996) Carboxypeptidase G2 rescue after high-dose methotrexate. J Clin Oncol 14: 2145–2149.
5. Widemann BC HM, Murphy RF, Balis FM, Adamson PC (1995) Carboxypeptidase-G2 rescue in a patient with high dose methotrexate-induced nephrotoxicity. Cancer 76: 521–526.
6. Chabner BA, Johns DG, Bertino JR (1972) Enzymatic Cleavage of Methotrexate provides a Method for Prevention of Drug Toxicity. Nature 239: 395–397.
7. Buchen S, Ngampolo D, Melton RG, Hasan C, Zoubek A, et al. (2005) Carboxypeptidase G2 rescue in patients with methotrexate intoxication and renal failure. Br J Cancer 92: 480–487.
8. Widemann BC, Sung E, Anderson L, Salzer WL, Balis FM, et al. (2000) Pharmacokinetics and metabolism of the methotrexate metabolite 2, 4-diamino-N(10)-methylpteroic acid. J Pharmacol Exp Ther 294: 894–901.
9. Albertioni F, Rask C, Eksborg S, Poulsen JH, Pettersson B, et al. (1996) Evaluation of clinical assays for measuring high-dose methotrexate in plasma. Clin Chem 42: 39–44.

Chapter 35

Quantitation of Methylmalonic Acid in Serum or Plasma Using Isotope Dilution-Selected Ion Gas Chromatography-Mass Spectrometry

Jie Chen and Michael J. Bennett

Abstract

The measurement of plasma or serum methylmalonic acid (MMA) is useful for monitoring therapy in patients with methylmalonic acidemia due to methylmalonyl-CoA mutase deficiency, defects of vitamin B12 metabolism, and also for the determination of functional vitamin B12 deficiency. This method utilizes a stable-isotope labeled internal standard (trideuterated MMA), which is added to the standards, controls, and patient samples prior to extraction. MMA is derivatized by butylation and the amount quantified against a seven-point standard curve using specific selected ions from both the labeled and unlabeled MMA.

Key words: Methylmalonic academia, vitamin B12 deficiency, isotope dilution mass spectrometry.

1. Introduction

Methylmalonyl coenzyme A (MMA-CoA) is a metabolic intermediate in the pathways by which the amino acids valine, isoleucine, methionine, and threonine, odd-chain-length fatty acids, the cholesterol side-chain, bacterially derived propionic acid, and the nucleotide bases thymine and uracil are eventually converted into succinic acid. Succinic acid is eventually utilized as a fuel within the Krebs TCA cycle (1, 2). All of these diverse metabolic pathways eventually converge toward the production of propionyl-CoA, which is then converted by a biotin-requiring carboxylation to MMA-CoA. MMA-CoA is converted to succinyl-CoA A by the action of MMA-CoA mutase, a mitochondrial matrix enzyme.

This enzyme is one of only two enzymes known to utilize metabolites of vitamin B12 (cobalamin) as a cofactor. Genetic defects of MMA-CoA mutase, defects of cobalamin transport, and intracellular

U. Garg, C.A. Hammett-Stabler (eds.), *Clinical Applications of Mass Spectrometry*, Methods in Molecular Biology 603, DOI 10.1007/978-1-60761-459-3_35, © Humana Press, a part of Springer Science+Business Media, LLC 2010

metabolism of cobalamin, and vitamin B12 deficiency, all result in impaired conversion of MMA-CoA to succinyl-CoA. In these conditions there is intramitochondrial accumulation of MMA-CoA. Cleavage of the CoA moiety results in release of the more water soluble free MMA, which can be measured in a variety of biological fluids. For clinical diagnostic purposes, MMA is measured primarily in plasma or serum for the evaluation of vitamin B12 deficiency and for monitoring patients with cobalamin and mutase defects. A variety of methods involving both GC-MS and LC-MS-MS have been described and all are acceptable as long as stable-isotope labeled internal standards are used and there is adequate separation of MMA from succinate, which are isomeric, and consequently isobaric compounds that would co-interfere in a non-separation system (3–7).

In this method MMA is measured in plasma or serum following addition of a fixed amount of a trideuterated MMA as internal standard. Proteins are precipitated from the samples and the MMA and trideuterated MMA are analyzed as their butyl-esters using GC-MS and selected ion monitoring. The fragmentation pattern for MMA shows much more significant target ions for quantitation than those of traditional trimethylchlorosilane (TMS) derivatives. The sensitivity of butyl-esters of MMA is considerably higher than TMS derivatives.

2. Materials

2.1. Reagents

1. Methylmalonic acid (Sigma-Aldrich, St Louis, MO)
2. Methylmalonic acid d3, 98% pure (Cambridge Isotope Laboratories, Andover, MA)
3. Butanol: hydrochloric acid (Regis Technologies Inc, Morton Grove, IL)

2.2. Working Reagents

1. Stock methylmalonic acid standard. 1 μmol/L stored at –40°C in 200 μL aliquots (stable for 5 years when frozen).
2. Stock deuterated methylmalonic acid standard. 0.1 μmol/L stored at –040°C in 100 μL aliquots (stable for 5 years when frozen).
3. 5 M potassium carbonate. Stable for 2 years at room temperature.
4. Dialyzed pooled plasma for the production of MMA-free plasma for generation of a standard curve is made in-house using a Slide-A-Lyzer Dialysis Cassette according to the manufacturers conditions (Pierce, Rockford, IL).
5. Control materials are in-house generated, high and normal value controls from pooled, previously analyzed samples that are stored in aliquots at –40°C.

3. Methods

3.1. Stepwise Procedure

1. Remove an aliquot of the frozen standard solution of MMA (#1 above) and allow to thaw. Use this standard to generate a 6-point standard curve plus blank with final concentrations of MMA from 0.01 to 5.00 μM (10–5,000 nM).

2. Aliquot 100 μL of control and patient samples to labeled glass tubes. If the patient is a known patient with mutase deficiency and very high baseline MMA levels, aliquot 10 μL of sample.

3. Remove an aliquot of the deuterated internal standard and allow to thaw.

4. Aliquot 10 μL of the 0.1 μmol/L internal standard to each standard, control, and patient tube and vortex well.

5. Precipitate proteins in the tubes by the addition of 900 μL of ethanol. Vortex well and centrifuge at $18,000 \times g$ for 5 min. Transfer the supernatant to a second set of tubes and dry down the supernatant under a steady stream of nitrogen at room temperature.

6. Add 100 μL of butanol: 3 M HCl. Cap and vortex mix the tubes. Incubate at 65°C for 15 min.

7. Add 30 μL of 5 M potassium carbonate to each tube (to neutralize and enhance phase separation) and vortex-mix.

8. Remove the upper butanol layer and transfer to a third set of tubes. Dry down under a steady stream of nitrogen at room temperature.

9. Add 75 μL ethyl acetate to each tube and vortex until the residue has dissolved.

10. Transfer to GC sample vials and inject 1.0 μL into the GC.

3.2. Gas Chromatography Conditions

1. Instrument: An Agilent 6890/5973 GC-MSD instrument was used with a HP-5MS column (30 m × 250 μm × 0.25 μm) cross-linked with 5% phenylmethylsilane (Agilent Technologies, Wilmington, DE).

2. The gas chromatographic running conditions are provided in **Table 35.1**.

Table 35.1
Gas chromatography running conditions for MMA assay

Ramp	°C/min	Next °C	Hold (min)	Run time (min)
Initial		70	2.00	2.00
Ramp 1	5.00	165	0.00	21.00
Ramp 2	60.00	300	10.00	33.25

3.3. Data Acquisition and Analysis

1. Two pairs of ions are acquired for both the internal standard and unknown; MMA at m/z 119 and 157 for the unknown and 122 and 160 for the internal standard. The ratios of each pair are monitored to ensure that there is no contamination. The m/z 157 and 160 ions are used for quantification (**Fig. 35.1**). Fragmentation spectra of MMA and internal standard are shown in **Fig. 35.2**.

2. Data are collected in the time-range of 17.6–18.0 min for quantification and the areas under the respective peaks are integrated for the calculation.

3. The standards are used to generate a standard curve for which the slope k and intercept b are utilized in the calculation of control and unknown values.

4. The MMA concentration is calculated by the following equation and adjusted for sample volume when 10 or 100 μL of sample is used:

$$MMA\ (nmol) = \frac{mz\,157/mz\,160 - b}{K}$$

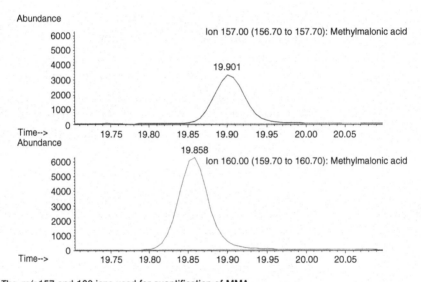

Fig. 35.1. The m/z 157 and 160 ions used for quantification of MMA.

3.4. Quality Control, Proficiency Testing

1. Pooled plasma and serum samples with low and high MMA values are run with every batch. The high values are taken from known patients with methylmalonic academia.

2. There is no accredited proficiency program for plama MMA measurement. Proficiency testing is accomplished by biannually sending samples to a second laboratory that also measures MMA as a clinical service and comparison of the matched data.

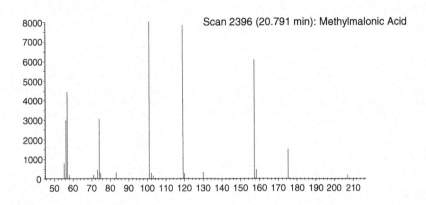

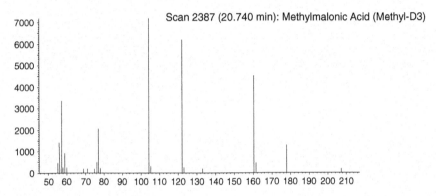

Fig. 35.2. Mass spectra of MMA and d3-MMA.

3.5. Reference Values, Analytical Measurement Range, and Clinical Reportable Range

1. *Reference range*: The reference range determined from 38 normal individuals was determined to be 245+/−100 nmol/L of plasma or serum. Reference values in pregnancy are higher (8).

2. *Analytical measurement range*: This was determined to be 100–500,000 nmol/L using both aqueous standards and serial dilution of an extremely elevated MMA sample (patient with MMA-CoA mutase deficiency in catabolic crisis).

3. *Clinical reportable range*: This was determined to be 100–500,000 nmol/L. Values below 100 should be reported as less than 100 and values greater than 500,000 should be re-evaluated by serial dilution.

References

1. Fenton WA, Gravel RA, Rosenblatt DS. (2001) Disorders of propionate and methyl-malonate metabolism. In: Scriver CR, Beaudet AL, Sly WS, Valle D. (eds.) The metabolic and molecular bases of inherited disease. New York: McGraw-Hill, pp. 2165–2193.

2. Rosenblatt DS, Fenton WA. (2001) Inherited disorders of folate and cobalamin transport and metabolism. In: Scriver CR, Beaudet AL, Sly WS, Valle D. (eds.) The metabolic and molecular bases of inherited disease. New York: McGraw-Hill, pp. 3897–3934.

3. Stabler SP, Marcell PD, Podell ER, Allen RH, Lindenbaum J. (1986) Assay of methylmalonic acid in the serum of patients with cobalamin deficiency using capillary gas chromatography-mass spectrometry. *J Clin Invest* **77**, 1606–1612.

4. Allen RH, Stabler SP, Savage DG, Lindenbaum J. (1990) Diagnosis of cobalamin deficiency I: usefulness of serum methylmalonic acid and total homocysteine concentrations. *Am J Hematol* **34**, 90–98.

5. Windelberg A, Arseth O, Kvalheim G, Ueland PM. (2005) Automated assay for the determination of methylmalonic acid, total homocysteine, and related amino acids in human serum or plasma by means of methychloroformate derivatization and gas chromatography- mass spectrometry. *Clin Chem* **51**, 2103–2109.

6. Magera MJ, Helgeson JK, Matern D, Rinaldo P. (2000) Methylmalonic acid measured in plasma and urine by stable isotope dilution and electrospray tandem mass spectrometry. *Clin Chem* **45**, 1804–1810.

7. Lakso H-A, Appelblad P, Schneede J. (2008) Quantification of methylmalonic acid in human plasma with hydrophilic interaction liquid chromatography separation and mass spectrometric detection. *Clin Chem* **54**, 2028–2035.

8. Metz J, McGrath K, Bennett MJ, Hyland K, Bottiglieri T. (1995) Biochemical indices of vitamin B12 nutrition in pregnant patients with sub-normal serum vitamin B12 levels. *Am J Hematol* **48**, 251–255.

Chapter 36

Quantitation of Methylmalonic Acid in Plasma Using Liquid Chromatography – Tandem Mass Spectrometry

Claudine Fasching and Jasbir Singh

Abstract

Methylmalonic acid (MMA), a biochemical marker for vitamin B12 deficiency, is commonly measured in serum and urine using liquid chromatography-tandem mass spectrometry (LC-MS-MS) with electrospray ionization. Improvements in sample preparation and the ability to analyze MMA without derivatization have further simplified the analysis. In the method we describe, 200 μl of plasma spiked with deuterated d3-MMA internal standard is deproteinized using ultrafiltration. After acidification, the ultrafiltrate is injected into the LC-MS-MS system and MMA eluted under isocratic conditions. MMA and d3-MMA are monitored in multiple reaction monitoring (MRM) in the negative ion mode.

Key words: Methylmalonic acid, ultrafiltration, liquid chromatography, tandem mass spectrometry.

1. Introduction

The measurement of methylamlonic acid (MMA) in plasma is used to assess the status of vitamin B_{12} (cobalamin) (1, 2). MMA is a metabolic intermediate in the conversion of propionyl-CoA to succinyl-CoA. Vitamin B12 is an essential cofactor that plays an important role in the conversion of L-methylmalonyl-CoA to succinyl-CoA. If insufficient vitamin B12 is available, then there is decreased conversion of L-methylmalonyl-CoA to succinyl-CoA. Elevated L-methylmalonyl CoA raises the concentration of D-methylmalonyl CoA, which in the presence of a hydrolase is converted into MMA. Thus, in vitamin B12 deficiency, MMA levels are elevated.

Several methods for MMA determination in serum and urine have been published. Until recently, most of these utilized gas chromatography-mass chromatography (GC-MS) (3–5). With the

U. Garg, C.A. Hammett-Stabler (eds.), *Clinical Applications of Mass Spectrometry*, Methods in Molecular Biology 603,
DOI 10.1007/978-1-60761-459-3_36, © Humana Press, a part of Springer Science+Business Media, LLC 2010

increased availability of liquid chromatography–tandem mass spectrometry (LC-MS-MS), this technique is now the preferred method for the measurement of MMA.

Methodological features common to many LC-MS-MS methods include the use of a deuterated internal standard, i.e., d3-MMA, and sample preparation using either solid phase extraction with strong anion exchange columns (6, 7) or liquid–liquid extraction with organic solvents (8). Sample extraction is followed usually by a derivatization step using butanolic HCl. The MMA derivatives are then injected into the liquid chromatographic system, and the MMA is quantified by electrospray ionization and multiple reaction monitoring mode of tandem mass spectrometry.

A recent publication (9) describing the determination of MMA by LC-MS-MS without the need of prior derivatization has further simplified the analysis of MMA. This method involves a simple deproteination by ultrafiltration of the plasma sample spiked with the internal standard d3-MMA. This ultrafiltrate is then injected into the HPLC system and the MMA is detected and quantified using tandem mass spectrometry. This method has been adopted in our laboratory and it offers a simple and reliable alternative to the more laborious methods.

2. Materials

2.1. Reagents and Buffers

1. Dialyzed Plasma (endogenous methylmalonic acid removal): In this step, endogenous methylmalonic acid is removed. Obtain outdated plasma from a local blood bank or alternate source. Freeze in 10 mL aliquots. Thaw approximately 30 mL and filter, using filter columns (Fisher Scientific, Pittsburgh, PA). Soak dialysis apparatus (Pierce, Slide-a-lyser) for 2 min in 0.9% saline. Place 30 mL of plasma in slide-a-lyser and place unit in a 3 L polypropylene beaker containing 2.2 L of 0.9% saline dialysis solution and a stir bar. Stir at 4°C for 2 h. Replace the old saline solution for fresh saline and continue stirring at 4°C. Repeat the exchange of solutions every 2 h at least five times. The dialysis may be carried out overnight. Removal of methylmalonic acid equals an 85^6 dilution factor ($2{,}200 \text{ mL}/30 \text{ mL} = 85$, six dialysis rounds $= 85^6$ or 3.77×10^{11} dilution factor) (see **Note 1**).

2. 0.9% Saline dialysis solution: Dissolve 19.8 g NaCl in 2.2 L of type 1 water.

3. 0.1% Formic acid in type 1 water: Add 400 mL of type 1 water to a 500 mL volumetric flask. To this add 0.5 mL of 90% formic acid. Mix well and adjust final volume to 500 mL with

type 1 water. Filter by suction through a 0.45-micron filter before using on the analyzer. Store at room temperature for 6 months.

4. 0.1% Formic acid in methanol: Add 400 mL of methanol to a 500 mL volumetric flask. To this add 0.5 mL of 90% formic acid. Mix well. Adjust to final volume of 500 mL with methanol. Filter by suction through a 0.45-micron filter before using on the analyzer. Store at room temperature for 6 months.

5. Formic acid (4%): Combine 960 µl of type 1 water and 40 µl of 90% formic acid. Mix well.

2.2. Calibrators/ Calibration

1. Stock standards (10 mM): Dissolve 11.8 mg methylmalonic acid (Sigma-Aldrich, St. Louis, MO) in methanol, quantitatively transfer to a 10 mL volumetric flask and adjust to a final volume of 10 mL with methanol. The stock solution is stable for 2 years when stored at –10 to –20°C (**Table 36.1**).

Table 36.1
Preparation of calibrators

Calibrator (nM)	Intermediate std (µL)	Dialyzed plasma (µL)
50	25 III	475
100	50 III	450
250	25 II	475
500	50 II	450

2. Intermediate standard I (100 µM): Prepare by transferring 100 µL of the stock methylmalonic acid into a 10 mL volumetric flask and diluting with methanol. Intermediate standard I is stable for 1 year from date of preparation when stored at –10 to –20°C.

3. Intermediate standard II (5 µM): Prepare by transferring 500 µL of the intermediate standard I stock into a 10 mL volumetric flask and diluting with type I water. Preparation is well mixed and aliquots were frozen in 500 µL volumes. Intermediate standard II is stable for 1 year from date of preparation when stored at –10 to –20°C. Thawed aliquots in use should be stored at 4–8°C and are stable for 1 month (*see* **Note 2**).

4. Intermediate standard III (1 µM): Prepare by transferring 100 µl of the intermediate standard I stock into a 10 mL volumetric flask and diluting with type I water. Mix well and

freeze in 500 µl aliquots. Intermediate standard II is stable for 1 year from date of preparation when stored at –10 to –20°C. Thawed aliquots in use are stored at 4–8°C and are stable for 1 month (*see* **Note 2**).

2.3. Internal Standard and Quality Controls

1. Internal standard stock (10 mM): Dissolve 12.4 mg of methylmalonic acid-d3 (Cambridge Isotope Laboratories, Andover, MA) in methanol and quantitatively transfer to a 10 mL volumetric flask and adjust to a final volume of 10 mL with methanol. The stock solution is stable for 2 years from date of preparation when stored at –10 to –20°C.

2. Working internal intermediate standard (WIS I 100 µM): Prepare by transferring 100 µL of the stock internal standard to a 10 mL volumetric flask and diluting with methanol. Working internal intermediate standard I is stable for 1 year from date of preparation when stored at –10 to –20°C.

3. WIS intermediate II (0.8 µM): Prepare by transferring 400 µL WIS I to a 50 mL volumetric flask and diluting with type I water. Aliquot in 2 mL volumes and store at –20°C. Frozen aliquots are stable for 1 year. Thawed aliquots in use are stored at 4–8°C and are stable for 1 month (*see* **Note 2**).

4. Quality controls: Prepare as described in **Table 36.2** using stock standards of a different lot from that used to prepare calibrators. Intermediate standard for quality control (QC), 60 µM is prepared by transferring 60 µL of stock methylmalonic acid for QC to a 10 mL volumetric flask and diluting with methanol.

Table 36.2
Preparation of in-house controls

Control	60 µM Intermediate std (µL)	Final volume with dialyzed plasma (mL)
Threshold (75 nM)	25	20
Positive (300 nM)	100	20

5. Threshold QC (75 nM): Prepare by transferring 25 µL of intermediate standard for QC to 20 mL of dialyzed plasma. Frozen aliquots are stable for 1 year.

6. Positive QC (300 nM): Prepare by transferring 100 µL of intermediate standard for QC to 20 mL of dialyzed plasma. Frozen aliquots are stable for 1 year.

7. Negative control: Dialyzed plasma is used as the negative control and is confirmed to be free of methylmalonic acid prior to use.

2.4. Tune Solution

Tune mix (40 μM): Add 40 μl of 10 mM methylmalonic stock standard to 10 mL of 50% Methanol: 50% of type 1 water which contains 0.1% formic acid. The tune solution is stable for 1 month when stored at −10 to −20°C (*see* **Note 3**).

2.5. Supplies

1. Amicon Centrifree YM30 filter units #4104 (Ionpure Technology, Lowell, MA)

2. Waters Symmetry C18 2.1 × 100 mm, 3.5 μm analytical column (Waters Corp, Milford, MA)

3. Waters Symmetry C18 2.1 × 10 mm, 3.5 μm guard column cartridge (Waters Corp, Milford, MA)

4. Waters 2.1 × 10 mm Guard column holder (Waters Corp, Milford, MA)

5. Max recovery auto sampler vials (Waters Corp, Milford, MA)

2.6. Equipment

1. Waters Acquity LC-MS-MS System (Waters Corp, Milford, MA)

2. Sorvall RC3C centrifuge

3. Methods

3.1. Stepwise Procedure

1. Label a 13 × 100 mm glass tube for each of four standards. Into the tubes labeled as standards add the amounts of methylmalonic intermediate standards and dialyzed plasma as described in **Table 36.1**.

2. After addition, vortex each tube to mix the contents.

3. Label a centrifree filter unit for each calibrator, control, and sample.

4. Pipette 200 μL of each calibrator, control, or sample into the top portion of filter unit. Tap unit to make sure sample is fully in the unit (*see* **Note 4**).

5. Pipette 50 μL of the WIS II to each filter unit. Tap unit to make sure volume is fully in unit. Mix well, using vortex mixer.

6. Centrifuge at 1,800 × *g* for 40 min at 22°C.

7. Transfer 100 μL of filtrate from bottom portion of filter unit to labeled maximum recovery autosampler vials, which contain 10 μL of 4% formic acid. Mix well.

3.2. Instrument Operating Conditions

1. LC Parameters

   Isocratic separation using mobile phase obtained from a mixture from pump A2 and B1 (*see* **Note 5**).

Flow:	0.4 mL/min
A2:	85 % – type 1 water with 0.1% formic acid
B1:	15 % – methanol with 0.1% formic acid
Autosampler temp:	10°C
Column temp:	30°C
Max press:	6,000 bar
Min press:	0
Run time:	3.0 min
Injection:	10 µL, partial loop with needle overfill

2. Tandem Quadrupole Detector (TQD)

   Atmospheric Pressure Ionization Source Settings

Capillary	2.5 kV
Extractor	2.0 V
Temperature	150°C
Desolvation gas flow	800 L/h
Desolvation temperature	400°C
Cone voltage	20 V

3. Analyzer Settings

MMA parent ion	116.9 amu
MMA daughter ion	72.9 amu
MMA-d3 parent ion	119.9 amu
MMA-d3 daughter ion	75.9 amu
Collision energy	9 eV
Collision gas flow	0.15 mL/min
Dwell time	0.5 sec
ESI mode	**Negative mode**
Repeats	1
Span	0
RT window	0–3 min

3.3. Data Analysis

1. The data are analyzed using QuanLynx software (Waters Corp, Milford, MA). Methylmalonic acid is quantified using a 4-point calibration curve, which plot response factors calculated from the ratio of the multiple reaction monitoring (MRM) peak area of methylmalonic acid to the peak area of

its deuterated internal standard. The concentration of calibrators used is listed in **Table 36.1**. The concentration of internal standard d3-MMA in each sample is 160 nM.

2. Representative LC-MS-MS chromatogram is shown in **Fig. 36.1**.

3. Linearity of the method is 25–3,000 nM.

4. Typical intra- and inter-assay variations are <10% CV.

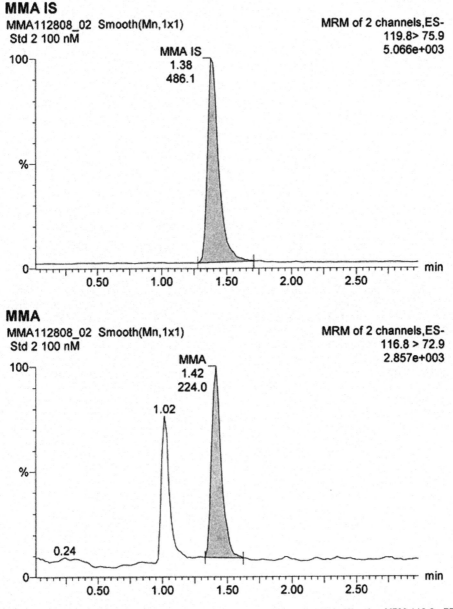

MMA IS

MMA112808_02 Smooth(Mn,1x1)
Std 2 100 nM

MRM of 2 channels,ES-
119.8> 75.9
5.066e+003

MMA IS
1.38
486.1

MMA

MMA112808_02 Smooth(Mn,1x1)
Std 2 100 nM

MRM of 2 channels,ES-
116.8 > 72.9
2.857e+003

MMA
1.42
224.0

1.02

0.24

Fig. 36.1. LC-MS-MS chromatogram for methylmalonic acid internal standard (MMA-d3) using MRM 119.8>75.9 and methylmalonic acid (MMA) using MRM 116.8>72.9.

4. Notes

1. Validate dialyzed plasma before use. Examine for the absence of methylmalonic acid and for adequate internal standard recovery. Plasma to be used for calibrator preparation should exhibit minimal matrix associated ion suppression.

2. Intermediate standards II, III, and working internal standard II should be prepared in type 1 water to avoid any protein precipitation caused by methanol. Micro precipitates may affect the centrifugal filtration efficiency negatively.

3. Analytes manually tuned once a month to ensure optimal detection parameters are in use. The mass spectrometer is calibrated every 6 months as suggested by the manufacturer.

4. Air bubbles trapped between the filtration membrane and the sample can impede the filtration process. Care should be taken to examine each filter unit after sample has been added to detect for this occurrence. Tap gently until bubble is dislodged when required.

5. Succinic acid has the same molecular mass as methylmalonic acid. Due to this similarity, succinic acid must be chromatographically resolved from methylmalonic acid.

References

1. Elin, R.J. and Winter, W. E. (2001) Methylmalonic acid: A test whose time has come? *Arch Pathol Lab Med* **125**, 824–7

2. Allen, R.H., Stabler, S.P., Savage, D.G., and Lindenbaum, J. (1990) Diagnosis of cobalamin deficiency I: Usefulness of serum methylmalonic acid and total homocysteine concentrations. *Am J Hematol* **34**, 90–8

3. Kushnir, M.M. and Komaromy-Hiller, G. (2000) Optimization and performance of a rapid gas chromatography-mass spectrometry analysis of methylmalonic acid determination in serum and plasma. *J Chrom B* **741**, 231–41

4. Rasmussen, K. (1989) Solid-phase sample extraction for rapid determination of methylmalonic acid in serum and urine by a stable-isotope-dilution method. *Clin Chem* **35**, 260–4

5. Straczek, J., Felden, F., Dousset, B., Gueant, J.L., and Belleville, F. (1993) Quantification of methylmalonic acid in serum measured by capillary gas chromatography-mass spectrometry as tert.-butyldimethylsilyl derivatives. *J Chrom* **620**, 1–7

6. Magera, M.J., Helgeson, J.K., Matern, D., and Rinaldo, P. (2000) Methylmalonic acid measurement in plasma and urine by stable-isotope dilution and electrospray tandem mass spectrometry. *Clin Chem* **46**, 1804–10

7. Schmedes, A., and Brandslund, I. (2006) Analysis of methylmalonic acid in plasma by liquid chromatography-tandem mass spectrometry. *Clin Chem* **52**, 754–7

8. Kushnir, M.M., Komaromy-Hiller, G., Shushan, B., Urry, F.M., and Roberts, W.L. (2001) Analysis of dicarboxylic acids by tandem mass spectrometry. high-throughput quantitative measurement of methylmalonic acid in serum, plasma and urine. *Clin Chem* **47**, 1993–02

9. Blom, H.J., Rooij, A.V., and Hogeveen, M. (2007) A simple high-throughput method for the determination of plasma methylmalonic acid by liquid chromatography-tandem mass spectrometry. *Clin Chem Lab Med* **45**, 645–50.

Chapter 37

Monitoring of Mycophenolate Acid in Serum or Plasma Using LC Tandem Mass Spectrometry

Catherine A. Hammett-Stabler, Diane Ciuffetti Geis, James C. Ritchie, and Christine Papadea

Abstract

Therapeutic drug management of patients receiving mycophenolic mofetil or mycophenolate sodium using mycophenolic acid (MPA) concentrations is controversial. Considered to be less toxic compared to many of the other drugs used in immunosuppression regimens, MPA is not tolerated by all patients. For these patients, monitoring is useful in achieving desired therapeutic targets, reducing adverse effects, and individualizing dosing. We describe an LC-MS-MS that permits the measurement of MPA and 7-O-glucuronide mycophenolic acid (MPAG) in serum or plasma.

Key words: Immunosuppressants, mycophenolate mofetil, mycophenolate acid, liquid chromatography tandem mass spectrometry.

---

1. Introduction

The greatest role for mycophenolic acid continues to lie in the prevention of organ transplant rejection. It is, however, gaining use in the treatment of various immune-mediated disorders such as lupus nephritis, psoriasis, and idiopathic thrombocytopenic purpura (1–3). Administered as an inactive prodrug, mycophenolate mofetil, or as mycophenolate sodium, MPA inhibits the rate-limiting enzyme involved in the production of guanosine monophosphate, inosine monophosphate dehydrogenase, and in doing so, inhibits DNA synthesis and proliferation of B and T lymphocytes. The adverse side effects of MPA include diarrhea, nausea, vomiting, and abdominal pain. Leukopenia, anemia, and an increased susceptibility to some infections also occur; however, these events are considered less toxic compared to those exhibited by other drugs used for immunosuppression, and as a result, MPA has replaced azathioprine

U. Garg, C.A. Hammett-Stabler (eds.), *Clinical Applications of Mass Spectrometry*, Methods in Molecular Biology 603, DOI 10.1007/978-1-60761-459-3_37, © Humana Press, a part of Springer Science+Business Media, LLC 2010

and is usually found as part of a regimen that includes a calcineurin inhibitor and prednisolone. The addition of MPA to immunosuppressive regimens usually permits a decrease in the dosing of the more toxic calcineurin inhibitors or prednisolone (4–7).

The active drug undergoes hepatic transformation to yield several metabolites. 7-O-glucuronide mycophenolic acid (MPAG) predominates, and although pharmacologically inactive, it is nevertheless important because it is reconverted back to MPA during extensive enterohepatic recirculation. This phenomenon gives rise to two peak plasma concentrations for MPA: the first 1–2.5 h and the second 4–12 h after dosing. A minor metabolite, mycophenolate acylglucuronide is thought to contribute to the adverse gastrointestinal events reported.

Therapeutic drug management by monitoring MPA concentrations has been controversial (8). Initially, monitoring was considered unnecessary, but with wider use and a greater understanding of its pharmacogenetics and pharmacokinetics, monitoring specific populations appear to provide benefit (9–10). Although, as discussed earlier, the drug is considered less toxic compared to others used in immunosuppression, many patients have difficulty tolerating the effects encountered. For some of these patients, monitoring is useful in achieving desired therapeutic targets, reducing adverse effects, and individualizing dosing. The use of LC-MS-MS-based methods rather than those that are immunoassay-based provides the ability to distinguish between the active drug, and when necessary the glucuronide and acylglucuronide metabolites.

2. Materials

2.1. Samples

Serum or plasma is used. Samples should be collected using a gel-free tube (royal blue, plain red, or EDTA anticoagulated). Specimens are typically collected as troughs just before the dose is given. The serum or plasma must be separated from cells immediately (see **Note 1**). Once separated from the red cells, samples are stable for up to 6 weeks when stored at 4–8°C and up to 11 months when stored at –20°C.

2.2. Reagents

1. Methanol, HPLC grade
2. Acetonitrile, HPLC grade
3. Water, HPLC grade
4. Fresh frozen plasma: Obtain outdated fresh frozen plasma from local blood center. Centrifuge at $5000 \times g$ for 10 min to remove particulates. Transfer 250 mL to a beaker and add 0.5 g of potassium oxalate monohydrate and 0.6 g of sodium fluoride as preservatives.

5. 0.1 M Zinc sulfate solution: In a 1 L volumetric, place 26.1 g of zinc sulfate heptahydrate. Bring to volume using HPLC-grade water. Stable for 1 year when stored at 2–8°C.

6. Precipitating/Internal Standard Solution (1.5 μg/mL mycophenolic acid carboxy butoxy ether, in acetonitrile):

   a. Prepare a fresh 1 mg/mL stock solution of mycophenolic acid carboxy butoxy ether (MPAC) (Roche Palo Alto LLC, Palo Alto, CA) in methanol (*see* **Note 2**).

   b. Add 1.5 mL of 1 mg/mL MPAC stock solution to ∼ 450 mL of acetonitrile in a 1 L volumetric flask. Mix well and bring to volume with acetonitrile. Stable for 1 year when stored at 2–8°C.

7. Mycophenolic acid (MPA, Sigma-Aldrich, St. Louis, MO) stock solution, 1 mg/mL: Quantitatively prepare a 1 mg/mL stock solution of mycophenolic acid in methanol. Stable for 6 months when stored at –80°C.

8. Mycophenolic acid glucuronide (MPAG, Analytical Services International Ltd, Caterham, UK) stock solution: Quantitatively prepare a 5 mg/mL stock solution of MPAG in methanol. Stable for 6 months when stored at –80°C.

9. 1 M Ammonium Acetate: In a 500 mL volumetric, place 38.5 g of ammonium acetate. Bring to volume using HPLC-grade water. Stable for 1 year when stored at 2–8°C.

10. Mobile Phase A and Purge Solvent (2 mM ammonium acetate in 0.1% formic acid): Mix 8 mL of 1 M ammonium acetate and 4 mL of 98% formic acid to 2.5 L of HPLC-grade water. Mix well and bring to 4 L with water. Mix again and store at room temperature. Stable for 6 months.

11. Mobile Phase B and Wash Solvent (2 mM ammonium acetate in 0.1% formic acid-methanol): Mix 8 mL of 1 M ammonium acetate and 4 mL of 98% formic acid with 2.5 L of methanol. Mix well and bring to 4 L with methanol. Mix again and store at room temperature. Stable for 6 months.

2.3. Calibrators

1. Add approximately 9 mL of fresh frozen plasma (FFP) to five 10 mL volumetric flasks. Add the appropriate amount of MPA and MPAG stock solutions to each flask as indicated in **Table 37.1** (*see* **Note 3**).

2. Fill to volume with FFP and mix well.

3. Aliquot 1 mL volumes into labeled polystyrene tubes and store at –80°C until used. Stable for 1 year.

4. Once thawed, calibrators are stable for 1 week when stored between uses at 2–8°C.

Table 37.1
Preparation of calibrators

Calibrator	MPA stock (μL)	MPAG stock (μL)	Plasma volume (mL)	MPA (μg/mL) concentration	MPAG (μg/mL) concentration
A	0	0	10.0	0	0
B	2.5	6	9.9915	0.25	3
C	10	20	9.9700	1	10
D	40	100	9.8600	4	50
E	160	400	9.4400	16	200

2.4. Controls

1. Add ~9 mL of fresh frozen plasma to three 10 mL volumetric flasks. Add the appropriate amount of MPA and MPAG stock solutions to each of the flasks as indicated in **Table 37.2** (*see* **Note 4**).

2. Fill to volume with FFP and mix well.

3. Aliquot 1 mL volumes into labeled polystyrene tubes and store at –80°C until used. Stable for 1 year.

4. Once thawed, controls are stable for 1 week when stored between uses at 2 and 8°C.

Table 37.2
Preparation of controls

Control	MPA stock (μL)	MPAG stock (μL)	Plasma volume (mL)	MPA (μg/mL) concentration	MPAG (μg/mL) concentration
1	20	50	9.93	2	25
2	50	150	9.80	5	75
3	100	300	9.60	10	150

2.5. Equipment and Supplies

1. 2795 Separations Module equipped with column heater assembly and column switching valve (Waters Corp. Milford, MA)

2. Quattro Micro Triple Quad Mass Spectrometer (Waters Corp. Milford, MA)

3. MassLynx V4.1 Software (Waters Corp. Milford, MA)

4. Atlantis dC18 5 uM 2.1 × 20 mm Guard Column (Waters Corp. Milford, MA)

5. Vacuum Pump

6. Polystyrene Microfuge tubes

7. Centrifuge

3. Method

3.1. Stepwise Procedure

1. Label a Microfuge tube for each calibrator, control, and sample and pipet 200 µL of 0.1 M zinc sulfate solution into each.

2. Into the corresponding tube, pipet 50 µL of each calibrator, control, or sample.

3. Add 500 µL of precipitating solution/internal standard.

4. Cap and vortex for 20 sec. Inspect each tube to verify material is thoroughly mixed.

5. Centrifuge for 2 min at $13,000 \times g$.

6. Transfer ~700 µL of each supernatant into the appropriately labeled vial and cap. Do not transfer any of the pellets. Place vials into the autoinjector.

7. Inject 5 uL of sample.

3.2. Chromatography and Mass Spectrometry Conditions

1. Chromatography conditions are given in **Table 37.3**.

2. The electrospray ion source is operated in positive ion mode. Instrument settings optimized for the Quattro Micro Triple Quad Mass Spectrometer are given in **Table 37.4**.

Table 37.3
Chromatography conditions

Column temperature 50°C

Flow rate 1 mL/min

Gradient conditions

Time	Mobile phase A	Mobile phase B
0	90	10
5	40	60
6.2	40	60
6.3	90	10
8.0	90	10

Table 37.4
MS-MS tune settings

Capillary (kV)	1.25
Cone (V)	14.00
Extractor (V)	2.50
RF lens (V)	0.4
Source temperature (°C)	130
Desolvation temperature (°C)	350
Cone gas flow	Off
Desolvation gas flow (L/hour)	800
LM1 resolution	12.5
LM 2 resolution	8.5
HM 1 resolution	12.5
HM 2 resolution	8.5

3.3. Analysis

1. Analyze data using MassLynx software. The qualifying ions are given in **Table 37.5** and representative ion chromatograms for MPA and MPAC are seen in **Fig. 37.1** Generate standard curves based on linear regression of the analyte concentration (x) to analyte to internal standard peak-area ratio (*y*). These curves are used to determine the concentrations of the controls and unknown samples and should have a correlation coefficient (r^2)>0.99.

2. The method is linear from approximately 0.125–16.0 µg/mL for MPA and 1.5–200 µg/mL for MPAG. Samples in which the drug concentrations exceed the upper limit of quantitation are diluted with drug-free plasma and retested.

Table 37.5
Retention times, precursor, and primary plus secondary product ions for MPA, MPAG, and MPAC

Compound	Retention time (min)	Precursor ion (*m/z*)	Product ions (primary/secondary) (*m/z*)
MPA	5.24	338.05	206.9/320.8
MPAG	4.08	513.95	320.7/206.9
MPAC	5.60	438.05	302.85/194.95

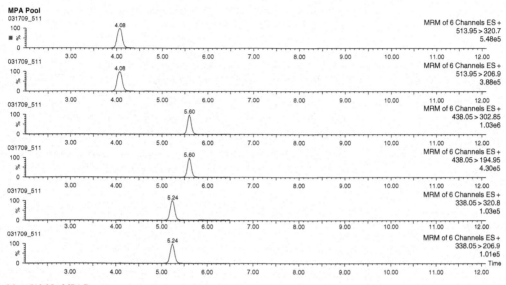

Mass 513.95 = MPAG

Mass 438.05 = MPAC

Mass 338.05 = MPA

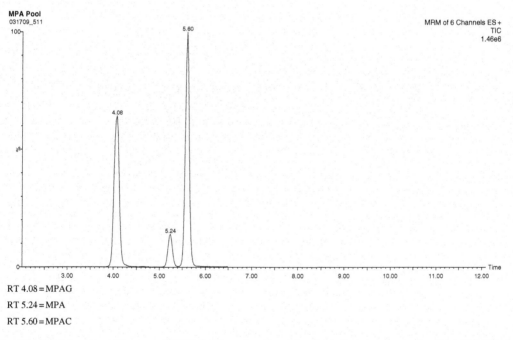

RT 4.08 = MPAG

RT 5.24 = MPA

RT 5.60 = MPAC

Fig. 37.1. Total ion chromatogram and MRMs for mycophenolic acid method. Shown are mycophenolic acid (MPA) at retention time 5.24 min, mycophenolic acid glucuronide (MPAG) at 4.08 min, and the internal standard, mycophenolic acid carboxy butoxy ether (MPAC) at 5.60 min.

3. Typical intra- and inter-assay imprecision for MPA obtained with this methods is <10%.

4. Quality control: The run is considered acceptable if calculated concentrations of drugs in the controls are within +/− two standard deviations of target values.

5. Interpretation: Therapeutic and toxic concentrations vary with dosing. For a patient receiving 2 g mycophenolic mofetil/day, trough mycophenolic acid plasma concentrations are typically between 1.0 and 3.5 μg/mL. For this dose, toxicity is reported for concentrations above 5 μg/mL.

4. Notes

1. Since EDTA anticoagulated whole blood is the sample of choice for cyclosporine, tacrolimus, and sirolimus monitoring, the use of EDTA plasma for MPA reduces blood collection. When the patient is receiving mycophenolic mofetil, the sample should be processed within 30 min of collection as mycophenolic mofetil undergoes temperature-dependent degradation to MPA (1).

2. Deuterated mycophenolic acid (Toronto Research Chemicals, Inc, North York, Ontario, CA) can also be used for the internal standard.

3. Each new lot of calibrators must be validated prior to placing into use. The samples should be first run as unknowns in a routine analysis. If results are within 10% of target values, prepare 20 extractions of each and use the results to determine the set points for each level of new calibrator. Also, compare calibration curves obtained with the new lot to the current lot in use. These runs should also include controls and patient samples. Results between the new lot and current lot must be within +/− 10%.

4. Avoid using the same MPA and MPAG stock solutions for preparation of calibrators and controls. When not possible, avoid preparing calibrators and controls simultaneously. Validate each new control lot by running as unknowns and comparing results with target concentration. Results should be within +/− 10%. Run the new lot twice a day for 10 days to derive a mean and CV.

References

1. Butch, A.W. (2007) Introduction to immunosuppressive drug monitoring. In: Hammett-Stabler, C.A. and Dasgupta A. Therapeutic Drug Monitoring: A Concise Guide, 3rd ed, . 129–161.

2. Navaneethan, S.D., Viswanathan, G., and Strippoli, G.F. (2008) Treatment options of proliferative lupus nephritis: an update of clinical trial evidence. Drugs 68, 2095–104.

3. Orvis, A.K., Wesson, S.K., Breza, T.S. Jr., Church, A.A., Mitchell, C.L., and Watkins, S.W. (2009) Mycophenolate mofetil in dermatology. J Am Acad Dermatol **60**, 183–99.

4. Tanchanco, R., Krishnamurthi, V., Winans, C., Wee, A., Duclos, A., Nurko, S., Fatica, R., Lard, M., and Poggio, E.D. (2008) Beneficial outcomes of a steroid-free regimen with thymoglobulin induction in pancreas-kidney transplantation. Transplant Proc **40**, 1551–4.

5. Becker, T., Foltys, D., Bilbao, I., D'Amico, D., Colledan, M., Bernardos, A., Beckebaum, S., Isoniemi, H., Pirenne, J., and Jaray, J. (2008) Patient outcomes in two steroid-free regimens using tacrolimus monotherapy after daclizumab induction and tacrolimus with mycophenolate mofetil in liver transplantation. Transplantation **86**, 1689–94.

6. Haller, M. and Oberbauer, R. (2009) Calcineurin inhibitor minimization, withdrawal and avoidance protocols after kidney transplantation. Transpl Int **22**, 69–77.

7. Zuckermann, A.O. and Aliabadi, A.Z. (2009) Calcineurin-inhibitor minimization protocols in heart transplantation. Transpl Int **22**, 78–99.

8. Barraclough, K.A., Staatz, C.E., Isbel, N.M., and Johnson, D.W. (2009) Therapeutic monitoring of mycophenolate in transplantation: Is it justified? Curr Drug Metabl **10**, 179–87.

9. Wavamunno, M.D. and Chapman, J.R. (2008) Individualization of immunosuppression: concepts and rationale. Curr Opin Organ Transplan **13**, 604–8.

10. Kuypers. D.R. (2008) Influence of interactions between immunosuppressive drugs on therapeutic drug monitoring. Ann Transplant **13**, 11–8.

Chapter 38

Quantitation of Nicotine, Its Metabolites, and Other Related Alkaloids in Urine, Serum, and Plasma Using LC-MS-MS

Bingfang Yue, Mark M. Kushnir, Francis M. Urry, and Alan L. Rockwood

Abstract

We describe a method for the quantitative analysis of nicotine, cotinine, trans-3'-hydroxy cotinine, nornicotine, and anabasine in urine, serum, and plasma using liquid chromatography-tandem mass spectrometry. A mix of deuterium-labeled internal standards (IS) is added to a specimen aliquot. The aliquot is extracted using mixed-mode solid phase extraction and eluted into an autosampler vial for injection into an LC-MS-MS system. An Atlantis silica column is used for LC separation in hydrophilic interaction mode. Tandem mass spectrometry detection is performed in positive ion mode with electrospray ionization and two multiple reaction monitoring (MRM) transitions monitored for each analyte and IS.

Key words: Nicotine, cotinine, trans-3'-hydroxycotinine, nornicotine, anabasine, solid phase extraction, liquid chromatography, hydrophilic interaction chromatography, tandem mass spectrometry, electrospray, multiple reaction monitoring.

1. Introduction

Cigarette smoking and tobacco use in other forms are deleterious to human health (1–3). Chemicals found in and as by-products of tobacco combustion produce both acute and chronic effects on many physiologic systems. Many related compounds are known carcinogens and are linked to numerous other pathological processes involving the oral cavity, cardiovascular, and respiratory systems. Newborns of mothers who smoke have an increased frequency of intrauterine growth retardation and subsequent low birth weight. Second-hand smoke, referred to as environmental tobacco smoke (ETS), produces effects in non-smokers, primarily in the form of respiratory disorders. In addition to its direct effect

U. Garg, C.A. Hammett-Stabler (eds.), *Clinical Applications of Mass Spectrometry*, Methods in Molecular Biology 603, DOI 10.1007/978-1-60761-459-3_38, © Humana Press, a part of Springer Science+Business Media, LLC 2010

on health, tobacco use influences the pharmacokinetics and pharmacodynamics of many drugs and can be responsible for both lack of efficacy and toxicity.

Nicotine is the primary alkaloid in tobacco products responsible for the addictive characteristics observed. The compound and its metabolites cotinine (COT) and trans-3′-hydroxy cotinine (3HCOT) are used as biomarkers of nicotine exposure. Nornicotine (NRNC) is both a tobacco alkaloid and a nicotine metabolite. In contrast, anabasine (ANAB) is present only in tobacco and is not a nicotine metabolite. Thus the presence of ANAB in biological fluids indicate active tobacco use (4, 5). The clinical purpose for monitoring these compounds includes assessment of nicotine replacement therapies (NRT) and smoking cessation programs following myocardial infarction or prior to surgical procedures. Their simultaneous measurement enables the classification of the donor of a specimen into categories: (1) an unexposed, non-tobacco user; (2) passive exposure; (3) a tobacco user abstinent for two or more weeks; or (4) an active tobacco user (*see* **Note 1**).

It is now recognized that liquid chromatography (LC)-tandem mass spectrometry (MS-MS) is the methodology of choice for the analysis of these compounds in biological fluids, such as serum/plasma (6–12), urine (8, 13–16), saliva (17), teeth (18), and meconium (19). The majority of published methods use atmospheric pressure chemical ionization (APCI) (6, 7, 9, 14, 17) or electrospray ionization (ESI) (8, 10, 11, 16). The method we describe uses dual-mode solid phase extraction (SPE) for sample preparation, hydrophilic interaction chromatography (HILIC) for LC separation, and positive ESI and multiple reaction monitoring (MRM) for mass spectrometric detection to simultaneously determine nicotine and the four related analytes in serum, plasma, or urine.

2. Materials

2.1. Specimens

1. Serum or EDTA anticoagulated plasma. Samples are stable for 2 weeks when stored at 4°C.
2. Random urine collected without preservative. Samples are stable for 10 days when stored at 4°C.

2.2. Reagents

1. 0.1 M acetic acid.
2. 0.1 M sodium bicarbonate (pH 9.0).
3. 100 mM ammonium formate, pH 3.0. Adjust pH to 3.0 using formic acid (99% purity). Stable for 3 days at room temperature.

4. Sodium hydroxide (6.0 M).

5. Ammonium hydroxide (30%, assay as NH_3).

6. Autosampler needle wash: Mix methanol (high purity solvent grade), Nanopure water (*see* **Note 2**), sodium hydroxide in proportions of 80:20:0.1. Stable for 10 days at room temperature.

7. Pump seal wash: Mix methanol and Nanopure water in 50:50 proportions. Stable for 2 weeks.

8. Purge solvent: 100% acetonitrile.

9. SPE elution solvent: Prepare by mixing ethyl acetate, isopropanol, and ammonium hydroxide in proportions of 74:24:2. The elution solvent must be prepared freshly prior to its use.

10. Nicotine (NIC) and cotinine (COT) reference materials, 1 mg/mL (Cerilliant, Round Rock, TX). Stable for 3 years, stored in the freezer (−65 to −75°C).

11. Anabasine (ANAB), nornicotine (NRNC), and trans-3′-hydroxy cotinine (HCOT) reference materials (Toronto Research Chemicals, North York, ON, Canada). ANAB and NRNC are supplied in oil form, containing 100 mg of material each. 3HCOT is supplied in powder form, containing 10 mg of material.

12. Deuterated reference materials COT d3 and 3HCOT d3 in powder form, and NIC d3, ANAB d4, and NRNC d4 in oil form (Toronto Research Chemicals, North York, ON, Canada).

2.3. Calibrators

1. Stock standard solutions (ANAB, NRNC, and 3HCOT). Prepare individually in methanol at concentration of 1 mg/mL. Stable for 3 years at −70°C.

2. Combined working standard solution, 0.2 ng/μL of each analyte in methanol. Prepare by combining 20 μL of each of the five 1 mg/mL stock standard solutions and dilute to a final volume of 100 mL with methanol. Stable for 3 months at 4°C and 1 year at −70°C.

3. Working calibrators: Prepare fresh with every batch of samples according to Table 1. Urine calibrators are used when testing plasma and serum (*see* **Note 3**).

2.4. Internal Standard Solutions

1. Stock internal standard (IS) solutions (COT d3, 3HCOT d3, NIC d3, ANAB d4, and NRNC d4). Prepare individually in methanol at concentration of 1 mg/mL. Stable for 3 years at −70°C.

2. Combined IS working standard solution, 1 ng/μL in methanol. Prepare by combining 100 μL of each of the five IS stock standard solutions and diluting to a final volume of 100 mL with methanol. Stable for 3 months at 4°C and 1 year at −70°C

2.5. Quality Control Materials

1. Negative control: Drug-free urine is used as the negative control (*see* **Note 3**).

2. Low urine positive control (*see* **Note 4**): Prepare by adding combined standard solution to certified drug-free urine to achieve approximate concentrations of 5 ng/mL COT, 50 ng/mL 3HCOT, and 2 ng/mL for each NIC, NRNC, and ANAB. Stable at 4°C for 3 months and at –7°C for 1 year.

3. High urine control: Prepare by adding standard solutions to achieve approximate concentrations of 2000 ng/mL 3HCOT, 20 ng/mL ANAB, and 200 ng/mL for each NIC, COT, and NRNC. Stable at 4°C for 3 months and at –7°C for 1 year.

4. Plasma negative control: Use drug-free plasma as the negative control.

5. Low plasma control: Prepare by adding standard solution to certified drug-free plasma to achieve concentrations of 2 ng/mL for each analyte.

6. High plasma control: Prepare by adding standard solution to certified drug-free plasma to achieve concentrations of 20 ng/mL for each analyte (*see* **Note 5**).

2.6. Equipment and Supplies

1. A Quattro Micro LC-MS-MS system (Waters, Milford, MA), equipped with an Alliance® HT HPLC system. The HPLC system includes solvent delivery module, autosampler and column oven. The system is controlled with Micromass MassLynx software.

2. LC column: Atlantis HILIC Silica, 100 mm length, 2.1 mm i.d., 3μm d_p, 100 Å pore (#186002013) (Waters, Milford, MA).

3. Solid phase extraction column: Trace B® (35 mg/3 mL) (SPEware, San Pedro, CA).

4. 48-place positive pressure manifold (SPEware, San Pedro, CA).

5. Cerex sample concentrator: 48-place sample dryer (SPEware, San Pedro, CA)

6. Microman® positive displacement precision pipettes (10 μL, 100 μL, 250 μL) (Gilson, Middleton, WI).

3. Methods

3.1. Stepwise Procedure

1. Organize the batch; label 16 × 100 mm glass tubes and autosampler vials with appropriate identifiers for calibrators, controls, and specimens.

2. Allow specimens, controls, and working standard solutions to equilibrate at room temperature.

3. Prepare working calibrators by spiking 10, 25, 100, and 250 μL of the combined working standard solution into 1,000 μL of certified blank urine to obtain equivalent concentration levels of 2, 5, 20, and 50 ng/mL (*see* **Table 38.1**).

4. Aliquot 1 mL of NEG CONT, POS CONT U/P, specimens (serum, plasma, or urine).

5. Add into each tube 25 μL of the IS working standard solution. The equivalent IS concentration corresponding to 1 mL of specimen volume is 25 ng/mL (*see* **Notes 6** and **7**).

6. Add into each tube 2 mL 0.1 M sodium bicarbonate buffer, pH 9.0 ± 0.1 (*see* **Note 8**).

7. Mix well and centrifuge for 5 min at 4°C and 3,500 rpm (*see* **Note 9**).

8. Condition Trace-B SPE cartridges at about one drop per second with 3 mL of methanol, 2 mL of water, 1 mL of sodium bicarbonate (pH 9.0).

9. Load samples onto the SPE columns and adjust flow rate of about one drop per 5 sec.

10. Wash each column with 1 mL of water, 3 mL of 0.1 M acetic Acid, 3 mL of methanol.

11. Dry columns for 5 min using air under pressure of 20–25 psi.

12. Elute with 1 mL of elution solvent (IPA:EtOAc:NH$_4$OH, 74:24:2) into autosampler vials at gravity flow or approximately one drop per 5 sec.

Table 38.1
Sample preparation

Test tube	WS,[a] μL	Blank urine, mL	IS,[b] μL	Specimen,[c] mL
2 Cal	10	1	25	
5 Cal	25	1	25	
20 Cal	100	1	25	
50 Cal	250	1	25	
NEG CONT		1		
POS CONT U – 2 POS CONT U – 200 POS CONT P – 2 POS CONT P – 20			25	1
urine or serum/plasma			25	1

[a]WS, working standard solution;
[b]IS: working internal standard solution;
[c]or controls.

13. Evaporate elution solvent for 5 min at room temperature. Do not evaporate to full dryness (*see* **Note 10**).

14. Cap vials and transfer to the instrument.

3.2. LC Conditions

1. Mobile phase A: Ammonium formate 100 mM, pH 3.0; mobile phase B: 100% acetonitrile. Isocratic elution (A: 15%; B: 85%) at flow rate of 0.54 mL/min, run time 6 min (*see* **Note 11**). *See* **Fig. 38.1** for typical chromatograms.

2. Column temperature: 30°C.

3. Autosampler conditions: Sequential injection type; partial loop fill, four pre-run and four post-run needle washes, injection volume of 20 μL with air gap of 2 μL.

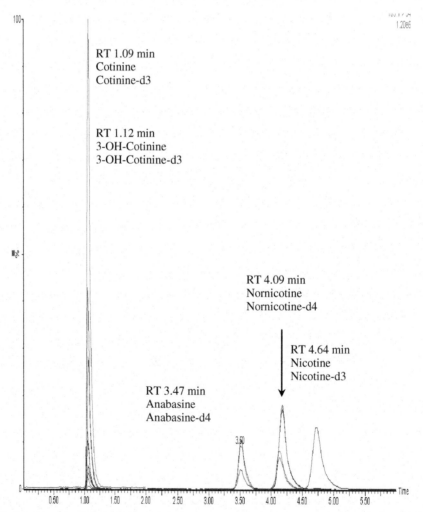

RT 1.09 min
Cotinine
Cotinine-d3

RT 1.12 min
3-OH-Cotinine
3-OH-Cotinine-d3

RT 4.09 min
Nornicotine
Nornicotine-d4

RT 4.64 min
Nicotine
Nicotine-d3

RT 3.47 min
Anabasine
Anabasine-d4

Fig. 38.1. Typical chromatogram of cotinine, trans-3'-hydroxy-cotinine, anabasine, and nicotine (two isomers, nicotine and anabasine are chromatographically resolved).

3.3. MS Settings

1. Source parameters: capillary: 1,000 V; extractor: 3 V; RF lens: 0.3 V; source temperature: 120°C; desolvation temperature: 400°C; cone gas flow: 25; desolvation gas flow: 800.

2. Analyzer parameters: LMI resolution: 13.0; HMI resolution: 13.0; ion energy 1: 0.1; entrance: –5; exit: 1; LM2 resolution: 13.0; LM2 resolution: 13.0; ion energy 2: 1.5. (*see* Note 12).

3. The MS-MS detection is in MRM and positive ion ESI mode with an ESCi probe. MRM transitions monitored are listed in Table 38.2 together with transition-specific parameters. For each compound, the quantitative MRM transitions are listed first, followed by the qualitative transitions (*see* Note 13 and 14).

Table 38.2
MRM transitions monitored (*m/z*)

Analyte	MRM	Cone	Collision	Analyte	MRM	Cone	Collision
NIC	163.1>130.1	28	20	NIC d3	166.1>130.1	28	20
	163.1>117.1	28	25		166.1>117.1	28	25
COT	177.1>98.1	35	20	COT d3	180.1>101.2	35	20
	177.1>70.1	35	30		180.1>73.1	35	30
3HCOT	193.1>80.0	35	25	3HCOT d3	196.1>80.1	35	25
	193.1>86.1	35	25		196.1>89.1	35	25
NRNC	149.0>130.0	25	15	NRNC d4	153.0>134.0	25	15
	149.0>117.0	25	23		153.0>121.0	25	23
ANAB	163.1>92.0	28	20	ANAB d4	167.0>96.1	28	20
	163.1>94.0	28	20		167.0>98.1	28	20

Note: Cone Voltage (V); Collisional energy (eV).

3.4. Data Analysis

The quantitative data analysis was performed using QuanLynx (Waters, Milford, MA). A linear calibration curve is generated using the peak-area ratio of MRM transitions for an analyte vs. the corresponding internal standard pair. Two calibration curves are produced for each analyte within every batch. The calculated concentrations by the two calibration curves should be within ±15% from 100% for each calibrator, control, or specimen. The qualitative criteria for the acceptance of the results in the batch include chromatographic peak shape, retention time, relative retention time, and the presence of peaks for each analyte and IS. Also for each analyte or IS, a ratio is calculated by dividing the peak area of the quantitative transition by that of the qualitative one, and should be with ±25% of the historical values or means established from the calibrators in the batch (20). The quantitative criteria include the correlation coefficient (*r*) ≥0.995 and the *y*-intercept ≤1 for the calibration curve, calculated concentrations in calibrators within ±20% of the target, calculated concentrations in negative control below the limit of detection for all analytes, and calculated concentrations in positive controls within ±20% of the target concentrations.

The analytical limit of detection (LOD, 1 ng/mL), lower limit of quantitation (LLOQ, 1 or 2 ng/mL), and upper limit of quantitation (ULOQ, 5,000 ng/mL) are the same for all analytes in serum, plasma, and urine. The total imprecision for each analyte at the LLOQ is below 10%.

4. Notes

1. In active tobacco users, concentrations of anabasine and nornicotine in serum/plasma are significantly lower than in urine.

2. Nanopure water refers to ASTM Type I reagent-grade water that is produced by a Barnstead NANOpure Infinity ultrapure water system. It was used in the method validation. Other grades of water from different sources need to be further evaluated.

3. Certified blank urine is used as blank matrix for preparing calibrators. Calibrators in urine matrix work well for serum/plasma specimen.

4. Standard solutions for preparing controls are separate preparations from the standard solutions used to prepare calibrators.

5. Concentrations of analytes in all newly prepared calibrators and controls need to be confirmed against previously used calibrators and controls.

6. It is necessary to carefully select the IS concentration-equivalent in 1 mL specimen to be used in the method, which is more critical for the lower part of the analytical measurement range.

7. For quantitative accuracy, it is important that the same amount of internal standard be added to each calibrator, control, and test specimen tube.

8. Composition of the loading buffer (pH 9.0), the wash solutions, and the elution solution are critical for cleanliness of the extract and recovery of the analytes.

9. This centrifugation prior to the SPE is especially important for serum/plasma specimens.

10. High concentration of ammonia in the final samples severely degrades the chromatographic separation in this method. The evaporation of the final extract is used to partially remove ammonia from the solvent before the chromatographic separation.

11. Hydrophilic interaction chromatography (HILIC) utilizes a polar stationary phase (e.g., silica, diol, etc.) and an aqueous-based mobile phase containing high percentage of acetonitrile. The LC method using HILIC column sufficiently retains the analytes and resolves the two isomers (anabasine and nicotine).

12. The MS settings in **Section 3.3** are listed as specified in the MassLynx software (Waters). The resolution settings on Q1 and Q3 correspond to unit resolution (0.7 Da peak width at 50% peak height).

13. The MRM transitions were carefully chosen to avoid potential interferences from other biogenic amines. The MRM transitions listed are not the most abundant, but were found to be free from interference in this method.

14. Use of individual stable-isotope labeled analogs (as internal standards for each of the targeted analytes) is critical for accurate quantification in this method.

References

1. (1988) The health consequences of smoking – nicotine addiction: a report of the Surgeon General. *USA Department of Health and Human Services.*

2. (1999) Achievements in public health, 1900–1999: Tobacco Use – United States, 1900–1999. *MMWR CDC Surveill.Summ.* **48**, 986–993.

3. (2004) The health consequences of smoking: a report of the Surgeon General. *USA Department of Health and Human Services.*

4. Jacob, P., Yu, L., Shulgin, A.T., and Benowitz, N.L. (1999) Minor tobacco alkaloids as biomarkers for tobacco use: comparison of users of cigarettes, smokeless tobacco, cigars, and pipes. *Am. J. Public Health* **89**, 731–736.

5. Jacob, P., Hatsukami, D., Severson, H., Hall, S., Yu, L., and Benowitz, N.L. (2002) Anabasine and anatabine as biomarkers for tobacco use during nicotine replacement therapy. *Cancer Epidemiol. Biomarkers Prev.* **11**, 1668–1673.

6. Xu, A.S., Peng, L.L., Havel, J.A., Petersen, M.E., Fiene, J.A., and Hulse, J.D. (1996) Determination of nicotine and cotinine in human plasma by liquid chromatography-tandem mass spectrometry with atmospheric-pressure chemical ionization interface. *J. Chromatogr. B Biomed. Appl.* **682**, 249–257.

7. Bernert, J.T. Jr., Turner, W.E., Pirkle, J.L., Sosnoff, C.S., Akins, J.R., Waldrep, M.K., Ann, Q., Covey, T.R., Whitfield, W.E., Gunter, E.W. et al. (1997) Development and validation of sensitive method for determination of serum cotinine in smokers and nonsmokers by liquid chromatography/atmospheric pressure ionization tandem mass spectrometry. *Clin. Chem.* **43**, 2281–2291.

8. Moyer, T.P., Charlson, J.R., Enger, R.J., Dale, L.C., Ebbert, J.O., Schroeder, D.R., and Hurt, R.D. (2002) Simultaneous analysis of nicotine, nicotine metabolites, and tobacco alkaloids in serum or urine by tandem mass spectrometry, with clinically relevant metabolic profiles. *Clin. Chem.* **48**, 1460–1471.

9. Stolker, A.L., Niesing, W., Hogendoorn, E.A., Bisoen Rambali, A., and Vleeming, W. (2003) Determination of nicotine and cotinine in rat plasma by liquid chromatography-tandem mass spectrometry. *J. Chromatogr. A* **1020**, 35–43.

10. Kellogg, M.D., Behaderovic, J., Bhalala, O., and Rifai, N. (2004) Rapid and simple tandem mass spectrometry method for determination of serum cotinine concentration. *Clin. Chem.* **50**, 2157–2159.

11. Xu, X., Iba, M.M., and Weisel, C.P. (2004) Simultaneous and sensitive measurement of anabasine, nicotine, and nicotine metabolites in human urine by liquid chromatography-tandem mass spectrometry. *Clin. Chem.* **50**, 2323–2330.

12. Kim, I., and Huestis, M.A. (2006) A validated method for the determination of nicotine, cotinine, trans-3'-hydroxycotinine, and norcotinine in human plasma using solid-phase extraction and liquid chromatography-atmospheric pressure chemical ionization-mass spectrometry. *J. Mass Spectrom.* **41**, 815–821.

13. Tuomi, T., Johnsson, T., and Reijula, K. (1999) Analysis of nicotine, 3-hydroxycotinine, cotinine, and caffeine in urine of passive smokers by HPLC-tandem mass spectrometry. *Clin. Chem.* **45**, 2164–2172.

14. Meger, M., Meger-Kossien, I., Schuler-Metz, A., Janket, D., and Scherer, G. (2002) Simultaneous determination of nicotine and eight nicotine metabolites in urine of smokers using liquid chromatography-tandem mass spectrometry. *J. Chromatogr. B Analyt. Technol. Biomed. Life Sci.* **778**, 251–261.

15. Heavner, D.L., Richardson, J.D., Morgan, W.T., and Ogden, M.W. (2005) Validation and application of a method for the determination of nicotine and five major metabolites in smokers' urine by solid-phase extraction and liquid chromatography-tandem mass spectrometry. *Biomed. Chromatogr.* **19**, 312–328.

16. Hoofnagle, A.N., Laha, T.J., Rainey, P.M., and Sadrzadeh, S.M. (2006) Specific detection of anabasine, nicotine, and nicotine metabolites in urine by liquid chromatography-tandem mass spectrometry. *Am. J. Clin. Pathol.* **126**, 880–887.

17. Bentley, M.C., Abrar, M., Kelk, M., Cook, J., and Phillips, K. (1999) Validation of an assay for the determination of cotinine and 3-hydroxycotinine in human saliva using automated solid-phase extraction and liquid chromatography with tandem mass spectrometric detection. *J. Chromatogr. B Biomed. Sci. Appl.* **723**, 185–194.

18. Marchei, E., Joya, X., Garcia-Algar, O., Vall, O., Pacifici, R., and Pichini, S. (2008) Ultrasensitive detection of nicotine and cotinine in teeth by high-performance liquid chromatography/tandem mass spectrometry. *Rapid Commun. Mass Spectrom.* **22**, 2609–2612.

19. Gray, T.R., Shakleya, D.M., and Huestis, M.A. (2008) Quantification of nicotine, cotinine, trans-3'-hydroxycotinine, nornicotine and norcotinine in human meconium by liquid chromatography/tandem mass spectrometry. *J. Chromatogr. B Analyt. Technol. Biomed. Life Sci.* **863**, 107–114.

20. Kushnir, M.M., Rockwood, A.L., Nelson, G.J., Yue, B., and Urry, F.M. (2005) Assessing analytical specificity in quantitative analysis using tandem mass spectrometry. *Clin. Biochem.* **38**, 319–327.

Chapter 39

Quantitation of Opioids in Blood and Urine Using Gas Chromatography-Mass Spectrometry (GC-MS)

Bruce A. Goldberger, Chris W. Chronister, and Michele L. Merves

Abstract

The opioid and 6-acetylmorphine assays utilize gas chromatography-mass spectrometry (GC-MS) for the analysis of morphine, codeine, hydromorphone, hydrocodone, and 6-acetylmorphine in blood and urine. The specimens are fortified with deuterated internal standard and a five-point calibration curve is constructed. Specimens are extracted by mixed-mode solid phase extraction. The morphine, codeine, hydromorphone, hydrocodone, and 6-acetylmorphine extracts are derivatized with *N*-methyl-bis(trifluoroacetamide) (MBTFA) producing trifluoroacetyl derivatives. The final extracts are then analyzed using selected ion monitoring GC-MS.

Key words: Opiate, opioid, morphine, codeine, hydromorphone, hydrocodone, 6-acetylmorphine, gas chromatography, mass spectrometry, solid phase extraction.

1. Introduction

Opium and its naturally occuring alkaloids, morphine and codeine, are obtained from the unripe pods of the opium poppy, *Papaver somniferum*. Many semisynthetic opioids have been derived over the years from both morphine (e.g., heroin) and codeine (e.g., hydrocodone and hydromorphone). Opioids are typically used as analgesics for the treatment of moderate-to-severe pain. Opioids are often abused for their central nervous system effects which include euphoria (1–3).

Usual routes of opioid administration include oral and intravenous; however, opioids can be snorted and smoked. The adverse effects of opioids include hypotension, respiratory depression, pulmonary edema, and coma (1–3).

U. Garg, C.A. Hammett-Stabler (eds.), *Clinical Applications of Mass Spectrometry*, Methods in Molecular Biology 603,
DOI 10.1007/978-1-60761-459-3_39, © Humana Press, a part of Springer Science+Business Media, LLC 2010

Morphine, codeine, hydromorphone, and hydrocodone are short-acting opioids with half-lives ranging from 3 to 6 h. Morphine is excreted principally in the urine as conjugated morphine. Codeine is metabolized to morphine and is excreted in urine as free and conjugated forms of codeine and morphine. Heroin has an extremely short half-life (~5 min). It is rapidly metabolized to 6-acetylmorphine (a unique heroin metabolite) and further metabolized to morphine. Hydrocodone is metabolized to a variety of metabolites including the pharmacologically active hydromorphone. Hydromorphone is excreted principally in the urine as conjugated hydromorphone (1–3).

The analysis of opioids includes immunoassay, as well as confirmation and quantitation by GC-MS. Solid phase extraction methods have been reported for the isolation of drug from blood and urine matrices. Derivatization of opioids is often essential in order to obtain satisfactory chromatographic performance (4–10).

The methods described below are validated for the analysis of morphine, codeine, hydromorphone, hydrocodone, and 6-acetylmorphine in blood and urine (*see* **Note 1**).

2. Materials

2.1. Chemical, Reagents, and Buffers

1. Blood, drug-free, prepared from human whole blood purchased from a blood bank and pretested to confirm the absence of analyte or interfering substance.

2. *N*-Methyl-bis(trifluoroacetamide), MBTFA (United Chemical Technologies)

3. 0.1 M Acetate Buffer, pH 4: Pipet 5.7 mL of glacial acetic acid into a 1,000 mL volumetric flask filled with 800 mL of water. Pipet 16 mL of 1.0 M potassium hydroxide. Mix well and check pH and adjust if necessary to pH 4.0. Q.S. to 1,000 mL with water. Stable for 1 year at room temperature.

4. Methylene Chloride:Isopropanol:Ammonium Hydroxide Solution (78:20:2; v:v:v): To 20 mL of isopropanol, add 2 mL of ammonium hydroxide. Add 78 mL of methylene chloride. Mix well. Prepare fresh.

5. 0.1 M Phosphate Buffer, pH 6: Dissolve 13.61 g of potassium phosphate monobasic in 900 mL of water. Adjust the pH to 6.0 with 5.0 M potassium hydroxide. Q.S. to 1,000 mL with water. Stable for 1 year at 0–8°C.

6. 1.0 M Potassium Hydroxide: Dissolve 5.6 g of potassium hydroxide in 50 mL of water in a 100 mL volumetric flask. Mix well and Q.S. to 100 mL with water. Stable for 1 year at room temperature.

7. 5.0 M Potassium Hydroxide: Dissolve 70.13 g of potassium hydroxide in 100 mL of water in a 250 mL volumetric flask. Mix well and Q.S. to 250 mL with water. Stable for 1 year at room temperature.

8. 1.0 M Sodium Hydroxide: Dissolve 40 g of sodium hydroxide in 500 mL of water in a 1,000 mL volumetric flask. Mix well and Q.S. to 1,000 mL with water. Stable for 1 year at room temperature.

2.2. Preparation of Calibrators (see Note 2)

2.2.1. Opioid Assay (Morphine, Codeine, Hydromorphone, and Hydrocodone)

1. Opioid Internal Standard Solution (10 µg/mL): Add the contents of 100 µg/mL D3-morphine, D3-codeine, D3-hydromorphone, and D3-hydrocodone vials (Cerilliant Corporation) to a 10 mL volumetric flask and bring to volume with methanol. Store at ≤−20°C.

2. Opioid Standard Stock Solution (100 µg/mL): Add the contents of 1.0 mg/mL morphine, codeine, hydromorphone, and hydrocodone vials (Cerilliant Corporation) to a 10 mL volumetric flask and bring to volume with methanol. Store at ≤−20°C.

3. Opioid Standard Solution (10 µg/mL): Dilute 1.0 mL of the 100 µg/mL opioid standard stock solution to 10 mL with methanol in a 10 mL volumetric flask. Store at ≤−20°C.

4. Opioid Standard Solution (1.0 µg/mL): Dilute 1.0 mL of the 10 µg/mL opioid standard solution to 10 mL with methanol in a 10 mL volumetric flask. Store at ≤−20°C.

5. Aqueous Opioid Calibrators: Add standard solutions to 2.0 mL of drug-free blood according to **Table 39.1**.

Table 39.1
Preparation of opioid calibration curve

Calibrator	Calibrator concentration (ng/mL)	Volume of 1.0 µg/mL standard solution	Volume of 10 µg/mL standard solution
1	50	100 µL	–
2	100	200 µL	–
3	250	–	50 µL
4	500	–	100 µL
5	1,000	–	200 µL

2.2.2. 6-Acetylmorphine Assay

1. 6-Acetylmorphine Internal Standard Solution (10 µg/mL): Add the contents of 100 µg/mL D3-6-acetylmorphine vial (Cerilliant Corporation) to a 10 mL volumetric flask and bring to volume with acetonitrile. Store at ≤−20°C.

2. 6-Acetylmorphine Standard Stock Solution (10 μg/mL): Add the contents of 100 μg/mL 6-acetylmorphine vial (Cerilliant Corporation) to a 10 mL volumetric flask and bring to volume with acetonitrile. Store at ≤−20°C.

3. 6-Acetylmorphine Standard Solution (1.0 μg/mL): Dilute 1.0 mL of the 10 μg/mL 6-acetylmorphine standard stock solution to 10 mL with acetonitrile in a 10 mL volumetric flask. Store at ≤−20°C.

4. Aqueous 6-Acetylmorphine Calibrators: Add standard solutions to 1.0 mL of drug-free blood according to **Table 39.2** .

Table 39.2
Preparation of 6-acetylmorphine calibration curve

Calibrator	Calibrator concentration (ng/mL)	Volume of 1.0 μg/mL standard solution
1	10	10
2	25	25
3	50	50
4	100	100
5	250	250

2.3. Preparation of Control Samples (see Note 2)

1. Negative (drug-free) Control: Prepare using human whole blood. Blood is pretested to confirm the absence of an opioid or interfering substance.

2.3.1. Opioid Assay (Morphine, Codeine, Hydromorphone, and Hydrocodone)

1. Opioid Control Stock Solution (100 μg/mL): Add the contents of 1.0 mg/mL morphine, codeine, hydromorphone, and hydrocodone vials (Cerilliant Corporation) to a 10 mL volumetric flask and bring to volume with methanol. Store at ≤−20°C.

2. Opioid Control Solution (10 μg/mL): Dilute 1.0 mL of the 100 μg/mL opioid control stock solution to 10 mL with methanol in a 10 mL volumetric flask. Store at ≤−20°C.

3. Aqueous Opioid Controls: Add control solution to 2.0 mL of drug-free blood according to **Table 39.3**.

Table 39.3
Preparation of opioid controls

Control	Quality Control concentration (ng/mL)	Volume of 10 μg/mL control solution
Low	250	50 μL
High	750	150 μL

2.3.2. 6-Acetylmorphine Assay

1. 6-Acetylmorphine Control Stock Solution (10 μg/mL): Add the contents of 100 μg/mL 6-acetylmorphine vial (Cerilliant Corporation) to a 10 mL volumetric flask and bring to volume with acetonitrile. Store at ≤−20°C.

2. 6-Acetylmorphine Control Solution (1.0 μg/mL): Dilute 1.0 mL of the 10 μg/mL 6-acetylmorphine control stock solution to 10 mL with acetonitrile in a 10 mL volumetric flask. Store at ≤−20°C.

3. Aqueous 6-Acetylmorphine Control: Add 6-acetylmorphine control solution to 1.0 mL of drug-free blood according to **Table 39.4**.

Table 39.4
Preparation of 6-acetylmorphine controls

Control	Quality Control concentration (ng/mL)	Volume of 1.0 μg/mL control solution
Low	25	25 μL
High	125	125 μL

2.4. Supplies

1. Clean Screen Extraction Columns (United Chemical Technologies)

2. Autosampler vials

3. Disposable glass culture tubes

4. Volumetric pipet with disposable tips

2.5. Equipment

1. 6890 Gas Chromatograph (Agilent Technologies, Inc.) or equivalent

2. 5973 Mass Selective Detector (Agilent Technologies, Inc.) or equivalent

3. 7673 Automatic Liquid Sampler (Agilent Technologies, Inc.) or equivalent

4. ChemStation with DrugQuant Software (Agilent Technologies, Inc.) or equivalent

5. Extraction manifold

6. Heat Block

7. Vortex Mixer

8. Centrifuge

9. Caliper Life Sciences TurboVap connected to nitrogen gas

3. Methods

3.1. Stepwise Procedure for Opioid Assay (Morphine, Codeine, Hydromorphone, and Hydrocodone)

1. Label culture tubes for each calibrator, control, and specimen to be analyzed and add 2.0 mL of the appropriate specimen to each corresponding tube.

2. Add 50 μL of the 10 μg/mL opioid internal standard solution to all culture tubes except the negative control.

3. Add 2.0 mL of 0.1 M acetate buffer (pH 4).

4. Vortex all specimens.

5. Add 1.0 mL of 0.1 M phosphate buffer (pH 6)

6. Add 200 μL of 1.0 M sodium hydroxide and vortex.

7. Centrifuge at ~1,500 × g for 5 min.

8. Place columns into the extraction manifold.

9. Prewash the columns with 3.0 mL of the methylene chloride:isopropanol:ammonium hydroxide solution.

10. Pass 3.0 mL of methanol through each column. Do not permit the columns to dry.

11. Pass 3.0 mL of water through each column. Do not permit the columns to dry.

12. Pass 2.0 mL of 0.1 M phosphate buffer (pH 6) through each column. Do not permit the columns to dry.

13. Pour specimen into column. Slowly draw specimen through column (at least 2 min) under low vacuum.

14. Pass 2.0 mL of water through each column.

15. Pass 2.0 mL of 0.1 M acetate buffer (pH 4) through each column.

16. Pass 3.0 mL of methanol through each column.

17. Dry column with full vacuum.

18. Turn off vacuum. Dry tips. Place labeled disposable culture tubes into column reservoir.

19. Add 3.0 mL of the methylene chloride:isopropanol:ammonium hydroxide solution and collect in disposable culture tubes.

20. Evaporate to dryness at 60 ± 5°C with a stream of nitrogen.

21. Add 50 μL of MBTFA. Cap tightly and vortex.

22. Place tubes in heating block for 30 min at 60 ± 5°C.

23. Transfer extract to autosampler vial and submit for GC-MS analysis.

3.2. Stepwise Procedure for 6-Acetylmorphine Assay

1. Label culture tubes for each calibrator, control, and specimen to be analyzed and add 1.0 mL of the appropriate specimen to each corresponding tube.

2. Add 10 μL of the 10 μg/mL 6-acetylmorphine internal standard solution to all culture tubes except the negative control.

3. Add 3.0 mL of 0.1 M phosphate buffer (pH 6).

4. Add 200 μL of 1.0 M sodium hydroxide and vortex.

5. Centrifuge at ~1,500 × *g* for 5 min.

6. Place columns into the extraction manifold.

7. Prewash the columns with 3.0 mL of the methylene chloride:isopropanol:ammonium hydroxide solution.

8. Pass 3.0 mL of methanol through each column. Do not permit the columns to dry.

9. Pass 3.0 mL of water through each column. Do not permit the columns to dry.

10. Pass 2.0 mL of 0.1 M phosphate buffer (pH 6) through each column. Do not permit the columns to dry.

11. Pour specimen into column. Slowly draw specimen through column (at least 2 min) under low vacuum.

12. Pass 2.0 mL of water through each column.

13. Pass 2.0 mL of 0.1 M acetate buffer (pH 4) through each column.

14. Pass 3.0 mL of methanol through each column.

15. Dry column with full vacuum.

16. Turn off vacuum. Dry tips. Place labeled disposable culture tubes into column reservoir.

17. Add 3.0 mL of the methylene chloride:isopropanol:ammonium hydroxide solution and collect in disposable culture tubes.

18. Evaporate to dryness at 60 ± 5°C with a stream of nitrogen.

19. Add 50 μL of MBTFA. Cap tightly and vortex.

20. Place tubes in heating block for 30 min at 60 ± 5°C.

21. Transfer extract to autosampler vial and submit for GC-MS analysis.

3.3. Instrument Operating Conditions

1. The GC-MS operating parameters are presented in **Tables 39.5, 39.6, 39.7**, and **39.8**.

2. The capillary column phase used for the opioid and 6-acetylmorphine assays is either 100% methylsiloxane or 5% phenyl–95% methylsiloxane.

Table 39.5
GC operating conditions for the opioid assay

Initial oven temp	150°C
Initial time	0.5 min
Ramp 1	25°C/min
Final temp.	320°C
Final time	3.00 min
Total run time	10.30 min
Injector temp	250°C
Detector temp.	290°C
Purge time	0.5 min
Column flow	1.0 mL/min

Table 39.6
GC operating conditions for the 6-acetylmorphine assay

Initial oven temp	140°C
Initial time	0.5 min
Ramp 1	25°C/min
Final temp	310°C
Final time	2.50 min
Total run time	9.80 min
Injector temp	265°C
Detector temp	290°C
Purge time	0.5 min
Column flow	1.0 mL/min

Table 39.7
Quantitative and qualifier ions for the opioid assay

Analyte	Quantitative ion (*m/z*)	Qualifier ions (*m/z*)
Morphine	364	477, 478
D3-Morphine	367	480
Codeine	282	395, 396

(continued)

Table 39.7 (continued)

Analyte	Quantitative ion (*m/z*)	Qualifier ions (*m/z*)
D3-Codeine	285	398
Hydromorphone	381	325, 310
D3-Hydromorphone	384	328
Hydrocodone	299	300, 270
D3-Hydrocodone	302	287

Table 39.8
Quantitative and qualifier ions for the 6-acetylmorphine assay

Analyte	Quantitative ion (*m/z*)	Qualifier ions (*m/z*)
6-Acetylmorphine	364	423, 364
D3-6-Acetylmorphine	367	426, 314

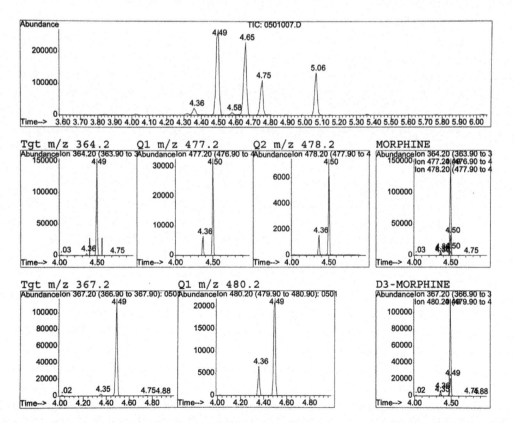

Fig. 39.1. Typical Agilent ChemStation report for opioids. **Top panel** is the total ion chromatogram; **middle and bottom panels** are selected ion chromatograms for morphine (retention time 4.49 min) and deuterated morphine (retention time 4.49 min), respectively. Codeine (retention time 4.65 min), hydromorphone (retention time 4.75 min), and hydrocodone (retention time 5.06 min) are also illustrated in the top panel. The retention time for each deuterated internal standard is similar to its native drug.

3. Set-up of the autosampler should include the exchange of solvent in both autosampler wash bottles with fresh solvent (bottle 1 – acetonitrile; bottle 2 – ethyl acetate).

4. A daily autotune must be performed with perfluorotributyla-mine (PFTBA) as the tuning compound prior to each GC-MS run.

3.4. Data Analysis

1. The review of the data requires the following information: retention times obtained from the selected ion chromato-grams; ion abundance, and ion peak ratios obtained from the selected ion chromatograms; and quantitative ion ratios between the native drug and its corresponding deuterated internal standard.

2. Typical Agilent ChemStation reports for the opioid and 6-acetylmorphine assays are illustrated in **Figs. 39.1** and **39.2**.

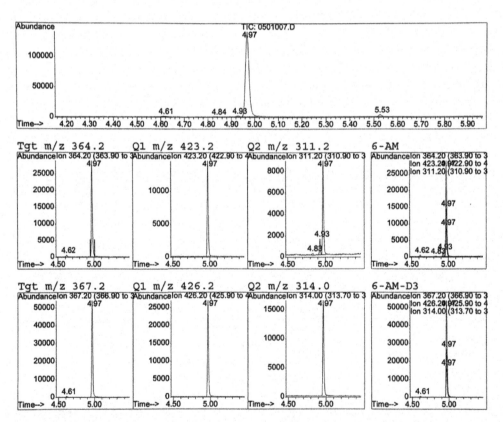

Fig. 39.2. Typical Agilent ChemStation report for 6-acetylmorphine. **Top panel** is the total ion chromatogram; **middle and bottom panels** are selected ion chromatograms for 6-acetylmorphine (retention time 4.97 min) and deuterated 6-acetylmorphine (retention time 4.97 min), respectively.

Table 39.9
Reporting criteria for opioids in blood

Opioid concentration (ng/mL)	Reported result
12.5–<25	Trace
25–<50	<50 ng/mL
50–1200	Report concentration
>1,200 ng/mL	Reanalyze "diluted" specimen

Table 39.10
Reporting criteria for 6-acetylmorphine in blood

Concentration (ng/mL)	Reported result
2.5–<5.0	Trace
5.0–<10	<10 ng/mL
10–300	Report concentration
>300	Reanalyze "diluted" specimen

3. In order for a result to be reported as "Positive," the following criteria must be satisfied: (1) the retention times of the ion peaks must be within ± 1% of the corresponding ions of the intermediate calibrator; (2) the ion peak ratio for the specimen must be within ± 20% of corresponding ion peak ratio of the intermediate calibrator; and (3) the correlation coefficient for the calibration curve must be 0.99 or greater if linear regression is used. Failure to meet one of the above criteria requires that the specimen be reported as "None Detected."

4. The limit of detection for the opioid assay is 12.5 ng/mL and the range of linearity is 50–1,000 ng/mL. The limit of detection for the 6-acetylmorphine assay is 2.5 ng/mL and the range of linearity is 10–250 ng/mL. The reporting criteria for blood are presented in **Tables 39.9** and **39.10**. Urine results are qualitative only and are reported as "Positive" when the concentration is greater than the limit of detection.

5. The intra- and inter-assay variability (%CV) is <10% for all analytes.

4. Notes

1. The opioid and 6-acetylmorphine assays can be applied to other specimens including those obtained at autopsy.

2. Separate sources of opioid analyte must be used when preparing standard and control solutions.

References

1. Stout, P.R. and Farrell, L.J. (2003) Opioids: Effects on human performance and behavior. *Forensic Sci Rev*, **15**, 29–59. Review.

2. Gutstein, H.B. and Akil, H. (2006) Opioid analgesics. In Brunton, L.L., Lazo, J.S. and Parker, K.L. (eds), *Goodman & Gilman's the Pharmacological Basis of Therapeutics*, 11th edition. McGraw-Hill, New York, pp. 547–90.

3. Lewis, L.S. (2006) Opioids. In Flomenbaum, N.E., Goldfrank, L.R., Hoffman, R.S., Howland, M.A., Lewin, N.A. and Nelson, L.S. (eds), *Goldfrank's Toxicologic Emergencies*, 8th edition. McGraw-Hill, New York, pp. 590–613.

4. Clean Screen Extraction Column Application Manual, United Chemical Technologies (2009).

5. Saady, J.J., Narasimhachari, N. and Blanke, R.V. (1982) Rapid, simultaneous quantification of morphine, codeine, and hydromorphone by GC/MS. *J Anal Toxicol*, **6**, 235–7.

6. Goldberger, B.A., Darwin, W.D., Grant, T.M., Allen, A.C., Caplan, Y.H. and Cone, E.J. (1993) Measurement of heroin and its metabolites by isotope-dilution electron-impact mass spectrometry. *Clin Chem.* **39**, 670–5.

7. Goldberger, B.A. and Cone, E.J. (1994) Confirmatory tests for drugs in the workplace by gas chromatography-mass spectrometry. *J Chromatogr A.* **674**, 73–86. Review.

8. Broussard, L.A., Presley, L.C., Pittman, T., Clouette, R. and Wimbish, G.H. (1997) Simultaneous identification and quantitation of codeine, morphine, hydrocodone, and hydromorphone in urine as trimethylsilyl and oxime derivatives by gas chromatography-mass spectrometry. *Clin Chem.* **43**, 1029–32.

9. Cremese, M., Wu, A.H., Cassella, G., O'Connor, E., Rymut, K. and Hill, DW. (1998) Improved GC/MS analysis of opiates with use of oxime-TMS derivatives. *J Forensic Sci.* **43**, 1220–4.

10. Meatherall, R. (2005) GC-MS quantitation of codeine, morphine, 6-acetylmorphine, hydrocodone, hydromorphone, oxycodone, and oxymorphone in blood. *J Anal Toxicol.* **29**, 301–8.

Chapter 40

Quantitation of Morphine, Codeine, Hydrocodone, Hydromorphone, Oxycodone, Oxymorphone, and 6-Monoacetylmorphine (6-MAM) in Urine, Blood, Serum, or Plasma Using Liquid Chromatography with Tandem Mass Spectrometry Detection

Tim Dahn, Josh Gunn, Scott Kriger, and Andrea R. Terrell

Abstract

Opiates and opioids currently rank among the most commonly prescribed pain medications. We describe two liquid chromatography tandem mass spectrometry (LC-MS-MS) methods for the quantification of morphine, codeine, hydrocodone, hydromorphone, oxycodone, oxymorphone, and 6-monoacetylmorphine (6-MAM). In the first, urine samples are pretreated by acidifying with sodium acetate containing appropriate deuterated internal standards and hydrolyzed with β-glucuronidase. Samples are cooled, diluted with water, vortexed, centrifuged, and a portion is transferred to an autosampler vial for analysis. The second method allows for the measurement of the compounds in blood, serum, or plasma specimens. Analysis of these samples involves pretreatment with acetonitrile containing deuterated internal standards to deproteinize the sample, which is subsequently vortexed and centrifuged. A portion of the organic layer is transferred to a clean test tube, dried under nitrogen, and reconstituted with water for analysis. Quantitation of analytes is accomplished using a commercially available single-point calibrator (urine samples) or an in-house prepared six-point standard curve (blood samples).

Key words: Morphine, codeine, hydrocodone, hydromorphone, oxycodone, oxymorphone, 6-monoacetylmorphine, 6-MAM, opiates, Opioids.

1. Introduction

Opioids are a class of drugs commonly prescribed for the treatment of moderate-to-severe pain. Morphine, the prototypical opioid, is a naturally occurring substance originally isolated from the milky substance in the seedpod of the Asian poppy, *Papaver*

U. Garg, C.A. Hammett-Stabler (eds.), *Clinical Applications of Mass Spectrometry*, Methods in Molecular Biology 603, DOI 10.1007/978-1-60761-459-3_40, © Humana Press, a part of Springer Science+Business Media, LLC 2010

somniferum (1). However, the majority of opioids are now synthesized. The interaction between an opioid and specific receptors located in the central nervous system, peripheral organs, and the immune response system lead to the observed effects (2). The primary receptors mediating opioid effects are the μ and the Κ receptors. Combined, these two receptors are responsible for analgesia, respiratory depression, euphoria, physical tolerance, and dependence. The euphoria resulting from usage of this class of drugs often leads to abuse.

The Drug Abuse Warning Network (DAWN) publishes data collected from hospital emergency departments across the United States. Between 2004 and 2006, there was a 43% rise in the number of emergency department visits related to non-medical use of opioids (3). The term non-medical implies use of a drug not as intended or prescribed. In many cases, drug is obtained with a legitimate prescription, but then diverted to another individual for monetary gain. As a consequence of widespread diversion, many physicians who prescribe opioids for pain control now rely on drug testing to monitor patient compliance with the prescribed drug regimen.

Urine is the specimen of choice for confirming compliance, detecting non-compliance, and detecting illicit drug use. Opioid measurement in blood specimens is appropriate in investigations of impairment or overdose. Blood may be used as well for the purpose of compliance monitoring; however, due to the short half-life of many of the opioids, negative results must be interpreted with caution.

We have developed two methods for confirmation and quantitation of seven natural and synthetic opiates. Each method involves very simple sample preparation processes, as compared to the sample preparation required for analysis using GC-MS. Many of the opioids are extensively metabolized by conjugation with glucuronic acid, thus hydrolysis is performed when total drug determination is desired. In urine specimens, measurement of total drug is preferred because the purpose of testing is to detect use of a drug. Hydrolysis is not performed for determination of opioids in blood specimens as each metabolite of the drug, including conjugated species, has variable physiological activities and should be measured individually.

2. Materials

2.1. Samples

1. Random urine collection
2. Whole blood, serum, or plasma collected into a non-gel barrier tube.

2.2. Reagents and Buffers

1) β-glucuronidase: 1,00,000 units/mL (Campbell Scientific)

2) 0.1 M Sodium Acetate buffer, pH 4.0

3) 0.1% Formic Acid in water

4) 0.1% Formic Acid in acetonitrile

5) 0.9% Saline

6) Synthetic drug-free urine: Combine 55.7 g potassium phosphate dibasic, 11.0 g sodium phosphate monobasic, 150 g sodium chloride, 3 g creatinine, 250 g urea, 0.05 g metanil yellow, and 0.10 g sodium azide in approximately 10 L of water. Adjust pH to 7.0 with potassium hydroxide or hydrochloric acid, as appropriate. Bring to a final volume of 20 L. Stable for 1 year when stored at 2–8°C (*see* **Note 1**).

7) Negative blood: Gently mix 500 mL of expired, packed-red cells with 550 mL of 0.9% saline. Add sodium fluoride to a concentration of 5 mg/mL. Prior to use, analyze to validate that the material is free of the drugs of interest. Stable for 1 year when stored at 2–8°C.

8) Drug stocks, 1.0 mg/mL in methanol for hydromorphone, morphine, oxymorphone, oxycodone, codeine, hydrocodone, and 6-monoacetylmorphine (Cerilliant, Inc., TX)

2.3. Quality Control (QC) Material (see Note 3)

1) Urine High QC pool: Prepare by adding sufficient drug stock solutions to the synthetic drug-free urine to achieve target concentrations of ~375 ng/mL for hydromorphone, morphine, oxymorphone, oxycodone, codeine, hydrocodone, and 62 ng/mL for 6-monoacetylmorphine.

2) Urine Low QC pool: Prepare by adding drug stock solutions to the synthetic drug-free urine to achieve target concentrations of ~ 120 ng/mL for hydromorphone, morphine, oxymorphone, oxycodone, codeine, hydrocodone, and 20 ng/mL for 6-monoacetylmorphine.

3) Urine Hydrolysis Control: Prepare by diluting morphine 3β-D-glucuronide, 1 mg/mL (Cerilliant, Inc.) to a concentration of ~250 ng/mL (as free morphine). Specifically, 80 μL of a 1 mg/mL solution is brought to 200 mL with drug-free urine to achieve a concentration of 400 ng/mL. Morphine accounts for 61.8% of the mass of the glucuronide conjugate, thus the final concentration of morphine, following hydrolysis, is ~250 ng/mL.

4) Urine Negative Control: Synthetic drug-free urine.

5) Blood QC stock solution: Prepared from hydromorphone, morphine, oxymorphone, oxycodone, codeine, hydrocodone, and 6-monoacetylmorphine methanol stocks, all 1.0 mg/mL (Cerilliant, Inc.). A dilution containing 100 μg/mL of codeine, morphine, hydrocodone, oxycodone,

and oxymorphone, and 50 µg/mL of hydromorphone is prepared in methanol. A separate dilution containing 10 µg/mL of 6-MAM is also prepared in Methanol. 6-MAM has a shorter shelf life as compared to the other opioids, and thus needs to be prepared on a more frequent basis.

6) Blood High QC working material: Prepare by diluting drug stock solutions to final concentrations of 400 ng/mL for codeine, morphine, hydrocodone, oxycodone, and oxymorphone, 200 ng/mL for hydromorphone, and 100 ng/mL for 6-MAM. The diluent for the control working material is negative blood.

7) Blood Low QC working material: Low quality control material is prepared from the stock solution to a final concentration of 100 ng/mL for codeine, morphine, hydrocodone, oxycodone, and oxymorphone, and 50 ng/mL for hydromorphone. The low QC does not contain 6-MAM.

8) Blood Negative Control: Validated drug-free blood.

2.4. Standards and Calibrators

1) Urine Positive Calibrator: Prepare by diluting the drug stock solutions with drug-free urine to concentrations of 150 ng/mL for each analyte except 6-MAM, which is prepared at a concentration of 10 ng/mL. Stable for 1 year when stored at 2–8°C.

2) Stock Blood Standards: Prepare a single solution containing all 7 analytes by diluting the drug stock solutions with acetonitrile to concentrations of 100 µg/mL for hydrocodone, codeine, morphine, oxycodone, and oxymorphone, and 25 µg/mL for hydromorphone and 6-MAM. Stable for 1 year when stored at −20°C.

3) Working Blood Standards: Combine 200 µL of the stock standard solution with 1.8 mL of drug-free blood to give a working standard of 1,000 ng/mL (hydrocodone, codeine, morphine, oxycodone, oxymorphone) and 250 ng/mL (hydromorphone, 6-MAM). Mix the working standard on a rotator for a minimum of 5 min at room temperature. This solution is prepared fresh on each day of use.

4) Blood calibrators: Prepare according to **Table 40.1**. This solution is prepared fresh on each day of use.

2.5. Internal Standards

1) Stock internal standards: Hydromorphone-d3, morphine-d3, oxymorphone-d3, oxycodone-d6, codeine-d3, hydrocodone-d3, and 6-monoacetylmorphine-d6, all 1.0 mg/mL (Cerilliant, Inc.).

2) Internal standard working solution for urine testing: Prepare by diluting stock internal standard to concentrations of 500 ng/mL using ultrapure water. Stable for 1 year when stored at 2–8°C.

Table 40.1
Preparation of calibrators for blood opioids. The higher concentration applies to morphine, codeine, oxycodone, oxymorphone and hydrocodone, and the lower concentration applies to 6-MAM and hydromorphone

Standard concentration (ng/mL)	Volume of working standard(μL)	Volume of negative blood (μL)
1,000/250	500	0
500/125	250	250
250/62.5	125	375
100/25	50	450
50/12.5	25	475
10/2.5	5	495

    3) Internal standard working solution, blood: Prepare by diluting the deuterated stock internal standards with acetonitrile. The final concentration is 200 ng/mL for all deuterated compounds except oxycodone, which is included at 100 ng/mL (*see* **Note 2**). The internal standard solution is maintained at −20°C.

2.6. Supplies

    1) Autosampler vials

    2) 11 mm Blue cut septa snap caps (MicroLiter, Inc.)

    3) 12 × 75 mm glass culture tubes

    4) Waters HSS T3 50 × 2.1 mm column, 1.7 μm particle size

    5) Thermo Aquasil C18 50 × 2.1 mm column, 3.0 μm particle size

    6) 60°C heat block

3. Methods

3.1. Urine Samples Preparation

    1) Add 250 μL of patient/donor sample, quality control samples (high QC, low QC, hydrolysis QC, and negative QC), and calibrator to appropriately labeled 12 × 75 mm glass tubes. Patient samples that are being analyzed at a dilution are brought to a volume of 250 μL using synthetic drug-free urine.

2) Add 500 µL of 0.1 M sodium acetate, pH 4.0 to each tube.

3) Add 20 µL of β-glucuronidase to each tube (*see* **Note 4**).

4) Add 75 µL internal standard working solution to each tube and vortex for 10–15 sec.

5) Place tubes in a 60°C heat block for 120 min. Cover samples with parafilm during hydrolysis.

6) Cool samples to room temperature.

7) Add 850 µL of ultrapure water and vortex for 10–15 sec.

8) Centrifuge for 5 min at $3,000 \times g$.

9) Transfer 200 µL to appropriately labeled autosampler vial and cap the vial.

3.2. Blood Samples Preparation

1) Add 500 µL of patient/donor sample, quality control samples (high QC, low QC, and negative QC), and calibrator to appropriately labeled 12 × 75 mm glass tubes. Patient samples that are being analyzed at a dilution are brought to a volume of 500 µL using negative blood.

2) Add 1.0 mL of cold internal standard solution to each tube and vortex for 20 sec.

3) Centrifuge for 10 min at $3,000 \times g$.

4) Transfer 100 µL to a new 12 × 75 mm glass tube.

5) Dry under nitrogen gas.

6) Reconstitute with 200 µL of ultrapure water.

7) Vortex for 10–15 sec.

8) Transfer to appropriately labeled autosampler vial and cap the vial.

3.3. Instrument Operating Conditions

Urine samples are analyzed on a Waters® Acquity TQD UPLC®-MS-MS system. Operating conditions for the Acquity® system are found in **Table 40.2**. Mass spectrometer operating conditions applicable to the entire method are found in **Table 40.3**. Specific transitions, cone voltages, and collision energies, all analyte specific parameters are shown in **Table 40.4**.

Blood specimens are analyzed on an API Sciex 3200™ Qtrap® MS-MS with an Agilent 1200® HPLC. Operating conditions for the HPLC system are found in **Table 40.5**. Mass spectrometer operating conditions applicable to the entire method are found in **Table 40.6**. Specific transitions, cone voltages, collision energies, all analyte specific parameters are shown in **Table 40.7**.

3.4. Data Analysis

For Waters® Acquity TQD UPLC®-MS-MS system, all aspects of instrument operation and data acquisition were controlled using MassLynx® software. Automated data processing was performed using QuanLynx® software. Both MassLynx® and QuanLynx®

Table 40.2
UPLC® method for urine opioids

Run time	Approximately 4.00 min		
Solvent A	1% Formic acid in ultrapure water		
Solvent B	1% Formic acid in acetonitrile		
Weak wash solvent	95% Water:5% acetonitrile		
Strong wash solvent	45% Acetonitrile: 45% Isopropanol: 10% Acetone		
Column temperature	35°C		
Sample temperature	20°C		
Gradient Table			
Time (minutes)	Flow rate	%A	%B
0.00	0.600	97.0	3.0
0.10	0.600	97.0	3.0
2.50	0.600	80.0	20.0
2.55	0.600	1.0	99.0
3.00	0.600	1.0	99.0
3.01	0.600	97.0	3.0

Table 40.3
Mass spectrometer operating parameters for urine opioids

Source (electrospray ionization, positive mode)	Settings
Capillary voltage (kV)	0.80
Extractor voltage (V)	4.00
RF Voltage (V)	0.10
Source temperature (°C)	150
Desolvation temperature (°C)	450
Cone gas flow (L/hr)	100
Desolvation gas flow (L/hr)	900
Collision gas flow (L/hr)	0.13

Table 40.4
Mass spectrometer method for urine opioids

Compound name	Quantification transition (T1)	Qualifier transition (T2)	Cone voltage (T1/T2)	Collision energy (T1/T2)
Hydromorphone	286.20 >157.10	286.20 > 185.20	40/50	50/35
Morphine	286.15 >152.10	286.15 > 57.90	42/42	58/30
Oxymorphone	301.80 >227.10	301.80 > 198.20	45/45	30/45
6-MAM	328.3 >165.20	328.30 > 211.20	45/45	45/28
Oxycodone	316.20 >241.10	316.20 > 256.20	40/40	28/25
Codeine	300.20 >152.10	300.20 > 115.20	45/45	68/65
Hydrocodone	300.30 >199.20	300.30 > 128.10	40/40	35/65
Hydromorphone IS	289.20 > 157.10	N/A	40	50
Morphine IS	289.09 > 152.60	N/A	39	48
Oxymorphone IS	304.80 >230.10	N/A	45	30
6-MAM IS	334.30 >165.20	N/A	45	45
Oxycodone IS	322.27 >262.20	N/A	42	28
Codeine IS	303.20 >152.10	N/A	45	68
Hydrocodone IS	303.30 >199.20	N/A	40	35

are products of Waters® Corporation. The quantitation of each drug was determined by reference to a single-point calibrator using multiple reaction monitoring and ratios of analyte to internal standard for the quantifying transition. For the target analyte, a qualifying transition is also monitored. The result is considered acceptable if the ratios for both the quantifying and the qualifying transition are within 20% of the calibration. Results from analysis of quality control specimens must fall within +/− 20% of target values, calculated by repeated analyses of each QC material. A total ion chromatogram is shown in **Fig. 40.1**. The lower limit of quantitation for the method is 10 ng/mL for all analytes, with a limit of detection of 5 ng/mL. The upper limit of quantitation is 10,000 ng/mL. Samples in which the analyte concentration exceeds the upper limit of quantitation are typically reported as ≥10,000 ng/mL. If an accurate quantitation is desired, samples may be diluted with synthetic drug-free urine and re-analyzed.

Table 40.5
HPLC method for blood opioids

Run time	Approximately 7.00 minutes			
Solvent A	1% Formic acid in ultrapure water			
Solvent B	1% Formic acid in acetonitrile			
Weak wash solvent	95% Water: 5% Acetonitrile			
Strong wash solvent	45% Acetonitrile: 45% Isopropanol: 10% Acetone			
Column temperature	20°C			
Sample temperature	20°C			
Gradient Table				
Time (seconds)	Flow rate (mL/min)	Function	%A	%B
30	0.500	Step	98.0	2.0
240	0.500	Ramp	60.0	40.0
241	0.500	Step	0.0	100.0
271	0.800	Step	0.0	100.0
272	0.600	Step	98.0	2.0
392	0.500	Step	98.0	2.0

Table 40.6
Mass spectrometer operating parameters for blood opioids

Source (electrospray ionization, positive mode)	Settings
Ion spray voltage (V)	2,000
Desolvation temperature (°C)	600
Curtain gas	30
Desolvation gas	70
Collision gas flow setting	High

For API Sciex 3200^{TM} Qtrap® MS-MS, data acquisition and data processing were controlled using Analyst^{TM} software (Applied Biosystems). Injection of specimens onto the HPLC is controlled by Aria^{TM} software (Thermo Fisher Scientific). The quantitation of each drug was determined by reference to a six-point calibration curve using multiple reaction monitoring and

Table 40.7
Mass spectrometer method for blood opioids

Compound name	Quantification transition (T1)	Qualifier transition (T2)	Cone voltage (T1/T2)	Collision energy (T1/T2)
Hydromorphone	286.05 > 185.20	286.05 > 157.20	56/56	41/55
Morphine	286.05 > 152.10	286.05 > 128.10	66/66	77/75
Oxymorphone	302.04 >227.20	302.04 >198.20	46/46	35/57
6-MAM	328.01 >165.20	328.01 >211.20	56/56	49/33
Oxycodone	316.01 >241.20	316.01 >256.30	41/41	35/33
Codeine	300.06 >115.10	300.06 >152.00	56/56	91/77
Hydrocodone	300.05 >199.20	300.05 >128.20	66/66	37/77
Hydromorphone IS	288.91 >157.20	N/A	61	55
Morphine IS	288.91 >152.20	N/A	61	75
Oxymorphone IS	304.86 >230.20	N/A	46	39
6-MAM IS	334.09 >165.20	N/A	51	49
Oxycodone IS	319.04 >244.30	N/A	41	39
Codeine IS	302.82 >152.20	N/A	51	83
Hydrocodone IS	302.83 >199.20	N/A	61	43

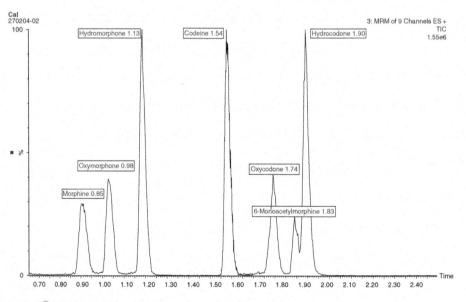

Fig. 40.1. UPLC®-MS-MS total ion chromatogram for urine opioids method. Shown is the 150 ng/mL calibrator (contains 6-MAM at 10 ng/mL).

ratios of analyte to internal standard for the quantifying transition. For the target analyte, a qualifying transition is also monitored. The result is considered acceptable if the ratios for both the quantifying and the qualifying transition are within 20% of the mean of the calibrators. Results from analysis of quality control specimens must fall within +/− 20% of target values calculated by repeated analysis of each QC material. A total ion chromatogram is shown in **Fig. 40.2**. The linearity/limits of quantitation are defined by the upper and lower calibrator points as shown in **Table 40.1**. Quantitation is provided for all samples, thus samples in which the analyte concentration exceeds the upper limit of quantitation are diluted with negative drug and re-analyzed.

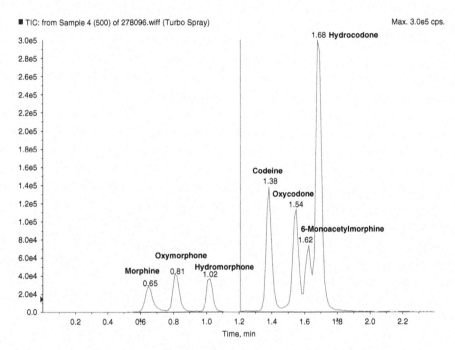

Fig. 40.2. HPLC-MS-MS total ion chromatogram for blood opioids method. Shown is the high calibrator, which contains 1,000 ng/mL of codeine, morphine, hydrocodone, oxycodone, and oxymorphone, and 250 ng/mL of hydromorphone and 6-MAM.

4. Notes

1. The use of synthetic drug-free urine instead of negative human urine was validated for use in this procedure.

2. We found that the deuterated analog of oxycodone contains a small amount of oxymorphone as a contaminant. If prepared at 200 ng/mL, the oxymorphone concentration in samples and in the negative control is above our limit of detection.

3. All positive quality control material is prepared in large volumes, aliquoted into single-use aliquots, and stored at −20°C.

4. β-glucuronidase should be gently mixed by inversion for 10–15 sec prior to use.

References

1. Jickells, S and Negrusz, A, eds: *Clarke's Analytical Forensic Toxicology*, Pharmaceutical Press, 2008.

2. Shaw, LM, et al., eds: *The Clinical Toxicology Laboratory: Contemporary Practice of Poisoning Evaluation*, American Association for Clinical Chemistry Press, 2001.

3. Drug Abuse Warning Network, 2006: National Estimates of Drug-Related Emergency Department Visits. http://DAWNinfo.samhsa.gov. August 2008.

Chapter 41

Urine Organic Acid Analysis for Inherited Metabolic Disease Using Gas Chromatography-Mass Spectrometry

Patricia M. Jones and Michael J. Bennett

Abstract

Urine organic acid analysis is an essential component of the workup of the patient suspected to have an inborn error of metabolism (IEM). Urine contains several hundred different organic acids, which arise from a multitude of different sources including both normal and abnormal metabolism. They may also arise from drugs and drug metabolism or from xenobiotics and dietary supplements. In addition to the diagnosis of inborn errors of metabolism, the identification of organic acids in a urine sample has a wide range of potential applications, including toxicology and poisonings. The method described below extracts the acidic fraction from urine samples, derivatizes the extracted compounds, and identifies intermediate metabolites by GC-MS. The method utilizes electron impact ionization gas chromatography-mass spectrometry (GC-MS) with total ion collection.

Key words: Organic acids, inborn errors of metabolism, gas chromatography-mass spectrometry.

1. Introduction

Normal human intermediary metabolism functions precisely and fluidly with enzymatic pathways designed to pass metabolites smoothly between enzymes, between pathways, and often even between organelles, cells, and cellular compartments. These pathways are essential for normal growth and development. Most generate either energy or metabolic building blocks for this purpose. Generally there are multiple feedback fail-safes in the metabolic processes to allow a continuous supply of necessary intermediates and a rerouting of excess metabolites. When an inborn error of metabolism (IEM) blocks one or more pathways, the body funnels excess intermediates through alternate pathways and in the process creates "abnormal" metabolites and metabolic profiles.

U. Garg, C.A. Hammett-Stabler (eds.), *Clinical Applications of Mass Spectrometry*, Methods in Molecular Biology 603, DOI 10.1007/978-1-60761-459-3_41, © Humana Press, a part of Springer Science+Business Media, LLC 2010

There are three main biochemical mechanisms that result in abnormal metabolic profiles in IEMs. First, normal metabolites of the pathway that are upstream of the blockage can accumulate. A notable example is the accumulation of glutarate in the genetic glutaric acidemias. Second, normal metabolites of other upstream pathways that cannot now feed into the blocked pathway may also accumulate. An example is with the accumulation of orotic acid and uracil in a number of urea cycle defects. Third, abnormal metabolites are formed when the excess intermediates are channeled through pathways they don't usually use. Examples include the accumulation of alloisoleucine in maple syrup urine disease and methylcitrate in both propionic and methylmalonic acidemias. Detecting the patterns of excretion of these metabolic intermediates in urine forms the basis for diagnosing IEM by urine organic acid analysis (1–6).

In this specific method for urine organic acid analysis, the compounds of interest are first extracted from acidified urine using an ethyl acetate liquid phase procedure followed by trimethylsilyl (TMS) derivatization to make the organic acids more volatile. TMS derivatization also allows for the creation of recognizable fragments from the compounds upon electron impact ionization in the mass spectrometer source. The fragments are then scanned while traveling through the single quadrupole mass spectrometer, and each compound results in a specific pattern of fragment ions, the mass spectrum. Data presentation is in the form of a total ion chromatogram (TIC). The mass spectra are then identified by library search and identifications are confirmed by operator-comparison against chromatographic retention time and spectral match.

2. Materials

2.1. Samples

Randomly collected urine, stored frozen until analysis (*see* **Note 1, 2,** and **3**).

2.2. Reagents

1. 6.0 M Hydrochloric acid (HCl)
2. Methanol (HPLC grade)
3. Ethyl acetate (HPLC grade)
4. Anhydrous sodium sulfate
5. N,O-bis(trimethylsilyl)trifluoroacetamide (BTSFA) with trimethylchlorosilane (TMCS), 99:1 (Supelco, Bellefonte, PA)
6. Internal standard: Prepare 1 mg/mL 2-phenylbutyric acid internal standard solution by mixing 50 mg of (S)-(+)-2-phenylbutyric acid (Sigma, St. Louis) with 50 mL of methyl alcohol. Stable for 6 months at 2–8°C.

Table 41.1
Organic acids for preparing stock solution (all from Sigma, St. Louis)

Compound	Weight (mg)
Lactic acid	9
β-Hydroxybutyric acid	12.6
Methylmalonic acid	11.8
Ethylmalonic acid	13.2
Fumaric acid	11.6
6-Methyluracil	63.1
Glutaric acid	13.2
3-Methylglutaric acid	14.6
Adipic acid	14.6
N-Acetyl – L – aspartic acid	17.5
Suberic acid	17.4
Orotic acid	15.6
Sebacic acid	20.2

7. Organic acid standard: Prepare according to **Table 41.1** by adding compounds to a 1 L volumetric flask. Add 50 mL of methyl alcohol to the flask and mix until all compounds are dissolved. Fill flask to volume with water, mix and aliquot into 1.5 mL eppendorf tubes. All organic acids are in a final concentration of 100 μmol/L except 6-methyluracil, which has a final concentration of 500 μmol/L. Solution is stable for 5 years at −70°C.

2.3. Quality Control Samples

1. Negative control: Deionized water
2. Positive control: Organic acid standard prepared as above but as a separate lot

2.4. Equipment and Supplies

1. Hewlett-Packard 5890 Series II gas chromatograph with a 5972 Series quadrupole mass spectrometer.
2. HP-5MS capillary column [30 m x 0.25 mm (i.d.)] coated with a 0.25 μm film of cross-linked 5% PH ME Siloxane (Agilent Technologies, Santa Clara, CA).
3. 16 × 100 mm (10 mL) glass tubes and Teflon-lined caps.
4. Autosampler vials (12 × 32 mm; crimp caps) with 0.2 mL limited volume inserts.

5. Heat block and evaporator.

6. Glass Pasteur transfer pipets.

7. Vortex.

3. Methods

3.1. Stepwise Procedure

1. Analyze all urine samples for creatinine in order to determine the volume of urine to use for the organic acid analysis. Calculate the volume of urine to use for each sample (*see* **Note 1, 2**, and **3**).

2. Label three 16 × 100 mm (10 mL) glass screw-cap tubes for each patient sample, a positive control, and a negative control.

3. Pipet 25 µL of 1 mg/mL 2-phenylbutyric acid internal standard solution into the first tube of each set of three.

4. Pipet the calculated volume of urine for each patient sample and 1 mL for each control into the correspondingly labeled tube containing internal standard.

5. *QS* all tubes containing urine to approximately 3 mL with deionized water. For sample volumes 3 mL or greater, do not add water (*see* **Note 4**).

6. Add 125 µL of 6 M HCl to all urine samples

7. Vortex for 15 sec.

8. Add 3 mL of ethyl acetate to each sample to extract the organic acids. Cap tightly and vortex for 30 sec.

9. Centrifuge the tubes at 2,000 × *g* for 2 min to separate the phases (*see* **Note 5**).

10. Remove the top ethyl acetate later to the second labeled tube in the set of three and then repeat steps 9–10 (*see* **Note 6**).

11. Remove the ethyl acetate layer from the second extraction and combine it with the first extracted layer in the second test tube. Discard the bottom layer and the first test tube.

12. Add approximately 0.5–1 g anhydrous sodium sulfate to the combined ethyl acetate layers in each tube (*see* **Note 7**). Cap and vortex for 15 sec.

13. Centrifuge at 3,000 × *g* for 2 min. Pour off the ethyl acetate sample into the clean third tube (*see* **Note 8**).

14. Dry the samples down under a gentle stream of nitrogen in a 37°C heat block.

15. Add 100 µL of BSTFA to each tube. Cap tightly and derivatize for at least 30 min (up to 2 h) at 70–80°C.

16. Transfer each sample to a 0.2 mL conical insert in a 1.5 mL autosampler vial and cap tightly.

17. Inject 1 μL of the sample.

3.2. GC-MS Operating Conditions

1. Gas chromatography: 1 μL sample is injected via a split/splitless injector in splitless mode at 250°C with helium carrier gas at a flow rate of 54.6 mL/min and pressure of 9.33 psi. The GC oven is programmed as follows: Initial temperature of 80°C for 5 min; ramp 3.8°C/min to 140°C; ramp 2.3°C/min to 200°C; ramp 6.6°C/min to 290°C, then held for 6 min. The program is 67 min in length.

2. The MS is set in SCAN mode to detect all ion fragments entering the mass spectrometer through a mass range of m/z 50–600. The solvent front can be eliminated by scanning from 7 to 67 min.

3.3. Data Analysis and Interpretation

1. For each patient and control sample, print the total ion chromatogram (TIC). Using the MS library, identify peaks, which are significantly above background on the TIC and label (*see* **Note 9**).

2. Also search for, and when present note, the compounds listed in **Table 41.2**. These are considered significant when present in any quantity. The presence of some of these compounds, regardless of relative amount, is considered pathognomonic.

Table 41.2
Compounds to be looked for in every patient sample

Compound	Ions		Range
4-Hydroxy-*n*-butyric Acid[1]	204	233	13–17 min
Tiglylglycine[2]	170	214	23–28 min
3-Hydroxyglutaric acid[3]	259	217	23–28 min
Hexanoylglycine[4]	158	230	25–30 min
Succinylacetone[5]	157	169	26–31 min
Orotic acid[6]	254	357	30–35 min

[1]Present in succinic semialdehyde dehydrogenase deficiency. May co-elute with urea and has similar spectrum to 3-hydroxyisobutyrate.
[2]Present in disorders of mitochondrial oxidative phosphorylation and in propionic acidemia. Has similar spectrum to 3-methylcrotonylglycine.
[3]Present in glutaric acidemia type 1 and short-chain 3-hydroxyacyl-CoA dehydrogenase deficiency. Small amounts are pathognomonic. May co-elute with 2-ketoglutaric acid.
[4]Present in medium-chain acyl-CoA dehydrogenase deficiency and glutaric acidemia type 2. May co-elute with 4-hydroxyphenylacetic acid.
[5]Present in tyrosinemia type 1. Any amount is diagnostic.
[6]Present in a number of urea cycle defects. Co-elutes with cis-aconitic acid.

3. The patterns are interpreted based upon a combination of compounds present and relative concentrations (7–9). **Figures 41.1, 41.2,** and **41.3** demonstrate representative organic acid profiles.

4. Specific steps may be eliminated or accelerated when needed to provide a stat result for a critically ill infant (*see* **Note 10**).

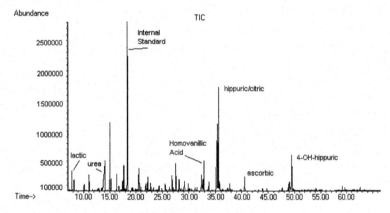

Fig. 41.1. Total ion chromatogram (TIC) of a urine organic acid profile from a sample demonstrating some of the compounds normally seen in a urine sample.

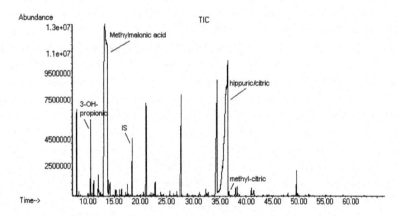

Fig. 41.2. TIC of an organic acid profile from an individual with methylmalonic aciduria, demonstrating a significant peak of methylmalonic acid, as well as peaks of methylcitrate and 3-hydroxypropionic acid.

4. Notes

1. A random urine sample collected without preservatives is used for testing. The samples should be stored frozen until analysis, since many of the compounds are volatile or labile. Before sample extraction begins, an aliquot should be first analyzed

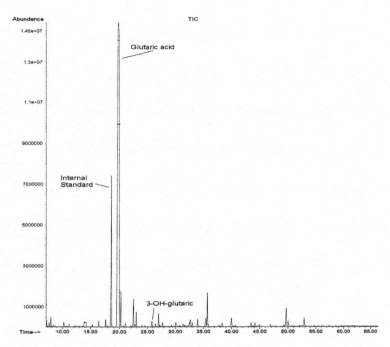

Fig. 41.3. TIC of an organic acid profile from an individual with glutaric acidemia, type 1, showing glutaric acid and 3-OH-glutaric acid.

for creatinine in order to determine the volume of urine to extract. This assures that relatively equal concentrations of urine are evaluated for every patient sample and facilitates interpretations. The method is used to identify compounds found in the urine sample and is not a quantitative method.

2. Numerous creatinine assays are available. The goal is to avoid over-loading or under-loading the column or detector. In general, if the sample being extracted contains ∼0.1 mg creatinine, the resulting chromatogram will not be over-loaded or under-loaded. The calculation used is: 10/creatinine concentration (mg/dL) = urine volume (mL). Sample volume must not be <0.5 mL or >5 mL. Typically, 0.5 mL of urine is extracted for samples with a creatinine concentration >20 mg/dL, while 5 mL is used for those with low creatinine concentrations of <2 mg/dL.

3. If the creatinine concentration is not available before extraction, use the sample's appearance to estimate the volume for testing. For those appearing dilute, i.e., clear, pale yellow-to-colorless, use 5 mL. For more concentrated urines, i.e., dark, thick urines, use less volume, but always stay between 0.5 and 5 mL.

4. Bring all tubes with less than 3 mL up to 3 mL total volume with water. This helps dilute the urea in the samples and reduces potential urea interference.

5. The phases in most samples will separate without centrifugation if the tubes are left to sit about 10 min.

6. Carefully remove the ethyl acetate layer with a glass Pasteur pipet, leaving behind the ethyl acetate close to the water layer. Avoid transferring any of the water phase. If necessary, leave some of the ethyl acetate phase behind.

7. The amount of anhydrous sodium sulfate does not need to be exact. The purpose of this step is to remove water, which may have been transferred or remains in the sample. A small mound of sodium sulfate on the end of a spatula is usually sufficient.

8. Occasionally, it is necessary to remove the ethyl acetate from the top of the sodium sulfate. However, the sodium sulfate will usually stay in the bottom of the tube and the ethyl acetate sample can simply be poured off.

9. Identification is accomplished by comparing retention times and the mass spectrum of the peak with those of a library. There are many libraries of mass spectra available. The two most commonly used libraries for organic acids are Wiley and the NIST, both of which can be obtained commercially at www.hdscience.com. The instrument computer can be programmed to print a report of the matches it finds, from 100% match down to 1% match, or any variation of % matches that the operator wishes. The report does not show the mass spectra, however, and the computer may be less than accurate in its assessment of how well the spectra match. We have found that the best way to make sure that all the necessary peaks are identified correctly is to have a human operator review and compare the mass spectra against the known spectra. It is also useful to maintain a hard copy database of most of the compounds that are known to be associated with IEMs.

10. Several steps may be eliminated or accelerated if a result is needed stat for a critically sick infant. The extraction can begin without measuring creatinine (*see* **Note 3**). A single ethyl acetate extraction (Steps 11–12) can be done. The nitrogen stream can be increased to accelerate the dry-down (Step 14). Care must be taken however not to "blast" the sample out of the tube. The derivatization time can be shortened to 30 min (step 16).

References

1. Chalmer, R.A., Lawson, A.M. (1982) *Organic Acids in Man: The Analytical Chemistry, Biochemistry and Diagnosis of the Organic Acidurias.* Chapman & Hall, London.

2. Goodman, S.I., Markey, S.P. (1981) Diagnosis of organic acidemias by gas chromatography-mass spectrometry. *Laboratory and Research Methods in Biology and Medicine,* Volume 6. Alan R. Liss, Inc., New York.

3. Sweetman, L. (1991) Organic acid analysis, in *Techniques in Diagnostic Human Biochemical Genetics: A Laboratory Manual.* (Hommes, F.A., ed), Wiley-Liss, New York, pp 143–176.

4. Hoffmann, G., Feyh, P. (2004) Organic acid analysis, in *Physician's Guide to the Laboratory Diagnosis of Metabolic Diseases,* second edition (Blau, N., Duran, M., Blaskovics, M.E., Gibson, K.M., eds), Springer, New York, pp. 27–44.

5. Bennett, M. (2006) Recommendations for the measurement of urine organic acids, in *Laboratory Medicine Practice Guidelines: Maternal-Fetal Risk Assessment and Reference Values in Pregnancy* (Sherwin, J.E., Lockitch, G., Rosenthal, P., et al, authors) National Academy Clinical Biochemistry, Washington, DC, pp. 59–62.

6. American College of Medical Genetics. (2006) F: Clinical biochemical genetics, in *Standards and Guidelines for Clinical Genetics Laboratories.* http://www.acmg.net/Pages/ACMG_Activities/stds-2002/f.htm

7. Rinaldo, P., Hahn, S., Matern, D. (2006) Inborn errors of amino acid, organic acid, and fatty acid metabolism, in *Tietz Textbook of Clinical Chemistry,* Fourth Edition (Burtis, C.A., Ashwood, E.R., Bruns, D.E., eds), Elsevier Saunders, St. Louis, MO, pp 2207–2247.

8. Ozand, P.T., Generoso G.G, (1991) Organic acidurias: A review. Part 2. *J. Child. Neurol,* **6**: 288–303.

9. Kumps, A., Duez, P., Mardens, Y. (2002) Metabolic, nutritional, iatrogenic, and artifactual sources of urine organic acids: A comprehensive table. *Clin. Chem,* **48(5)**: 708–717.

Chapter 42

Identification of Urine Organic Acids for the Detection of Inborn Errors of Metabolism Using Urease and Gas Chromatography-Mass Spectrometry (GC-MS)

Stanley F. Lo, Velta Young, and William J. Rhead

Abstract

A patient suspected of an inborn error of metabolism will commonly have urine organic acid analysis performed as part of their workup. The traditional urine organic acid method involves extraction of the acidic fraction from urine samples using an organic solvent, derivatization of extracted compounds, and identification using gas chromatography-mass spectrometry (GC-MS). Unfortunately, the extraction step results in the loss of many neutral and positively charged compounds, which may be of interest to metabolic physicians and biochemical geneticists. By replacing the traditional extraction step with an enzymatic treatment of the sample with urease, an abundance of organic molecules are available for separation and quantitation by GC-MS. The urease method is a useful adjunct to newborn screening follow-up and it has the additional benefit of being able to identify many classes of biochemical compounds, such as amino acids, acylglycines, neurotransmitters, and carbohydrates. The method below describes the urease treatment, derivatization, and the organic acids, and other biochemical metabolites that can be identified.

Key words: Organic acids, inborn errors of metabolism, gas chromatography-mass spectrometry.

1. Introduction

The identification of urinary organic acids is a key component in the diagnosis and treatment of inborn errors of metabolism. The addition of expanded newborn screening using tandem mass spectrometry has significantly increased the number of infants identified with fatty acid oxidation and organic acid disorders thus, increasing the value in determining the presence of urinary organic acids for both confirmation and treatment of disease. The Shoemaker and Elliott urease method (1) differs from the classical organic acid methods (2–4) by taking advantage of urease to remove urea in urine. The

U. Garg, C.A. Hammett-Stabler (eds.), *Clinical Applications of Mass Spectrometry*, Methods in Molecular Biology 603,
DOI 10.1007/978-1-60761-459-3_42, © Humana Press, a part of Springer Science+Business Media, LLC 2010

elimination of extracting from the acidic fraction of urine, as done in the classical method, is replaced by the enzymatic removal of urea and enables specimens to have an abundance of biochemical metabolites to derivatize and detect using GC-MS. While the method described here is limited to organic acids, it can easily be used to detect amino acids, carbohydrates, acylglycines, neurotransmitters, purines and pyrimidines, vitamins, and peroxisomal metabolites, and dipeptides. The method can also be adapted to measure metabolites in plasma, cerebrospinal fluid, and amniotic fluid.

2. Materials

2.1. Samples

1. Random urine. Preferred specimen volume is 3–10 mL.
2. Store in −20°C freezer. Samples that will not be analyzed within 2 weeks should be stored at −70°C. Retained specimen samples are stored in the −70°C freezer for up to 2 years and as space permits.

2.2. Reagents and Solutions

1. Dichloromethane (Biotech grade 99.9%)
2. Triethylammonium trifluoroacetic acid (TEA/TFA) (*see* **Note 1**): Prepare an equimolar mixture of triethylamine (TEA, Sigma) and trifluoroacetic acid (TFA, Sigma) in dichloromethane as follows. This is necessary to insure derivatization of organic acid metabolites for the organic acids assay.
   a. Calculate the necessary volumes of each compound, so that after mixing an equimolar mixture will result.
   b. Determine the volume of dichloromethane needed by adding the two calculated volumes and multiply this value by 50.
   c. Add this volume of dichloromethane to a 1 L round bottom flask with a stir bar. Place in an ice bath on a magnetic stir plate and chill for 10 min. After 10 min, slowly add the TEA. Begin stirring and chill the mixture for 5 min (*see* **Note 2**).
   d. Using a glass disposable pipet, add TFA drop-wise to the stirring solution (*see* **Note 3**).
   e. Once the TFA is completely added, allow the reaction to occur for 10 min by stirring the solution.
   f. Use a Rotovap apparatus to remove the dichloromethane (*see* **Note 4**).
   g. Once the dichloromethane has been removed, transfer the TEA/TFA mixture to a glass container that can be tightly sealed. Label appropriately. Mixture is stable at room temperature. Do not refrigerate.
3. Urease (from Jackbeans), 80 mg vials (Calzyme Laboratories) store at −20°C.

4. Aqueous urease solution, 100 mg/mL (~150 U): Prepare fresh just prior to use. Remove one vial of ~80 mg urease from freezer. Add 0.8 mL of type 1 water and mix thoroughly. Let sit for 5 min.

5. Acetonitrile (ACS grade, 99.9%)

6. Acetone (HPLC grade, 99.9%): Cool at time of use, 0°C or colder

7. N-methyl-N-trifluoroacetamide (MSTFA) (Pierce Biotechnology)

2.3. Standards and Calibrators

Calibration is performed by separating the analytes of interest into groups (**Table 42.1**). Stock solutions of 0.01 M (10 mL) for each analyte are prepared and then combined with other analytes to create group solutions. A five-point standard curve is prepared. In general, calibration samples of 4, 20, 50, 100, and 200 nmol are prepared for each analyte in a group by adding 200 µL of each

Table 42.1
Calibration groups. All reagents are from Sigma/Aldrich/Fluka unless otherwise noted. Spain refers to the University of Madrid, Spain

Group 1 and 2	Group 3
Lactic acid	Alanine
Glycolic acid	2-Aminoisobutyric acid
α-Hydroxybutyric acid	β-Alanine
Malic acid	Leucine
α-Ketoglutaric acid	Isoleucine
p-Hydroxyphenylacetic acid	Proline
Fucose	Threonine
Xylose	Methionine
Fructose	Phenylalanine
Mannose	Asparagine
Citric acid	Glutamine
Galactose	Histidine
L-2-Hydroxyglutaric acid	Lysine
	Tyrosine
	Tryptophan
	Cystine
	L-Citrulline
Group 4	**Group 5**
Adipic acid	Pyruvic acid
Ascorbic acid	Oxalic acid
Ethylmalonic	Methylmalonic acid
Fumaric	Octanoic acid
Glutaric	Succinic acid
Hexanoic	Methylsuccinic acid
Hippuric	Suberic acid

(continued)

Table 42.1 (continued)

3-Hydroxyisovaleric acid	Phenylpyruvic acid
4-Hydroxybutyric acid	Homovanillic acid
Glucuronic acid	Homogentisic acid
Malonic acid	Sebacic acid
D-2-Hydroxyglutaric acid	Uric acid
3-Hydroxy-3-methylglutaric acid	
Group 6	**Group 7**
Propionyl glycine (Spain)	Glycine
Butyroyl glycine (Spain)	Valine
Isovaleryl glycine (Spain)	Ethanolamine
Tiglyl glycine(Spain)	Serine
Suberyl glycine (Spain)	Homoserine
Hexanoyl glycine (Spain)	Pyroglutamic acid
3-Methylcrotonyl glycine (Spain)	Aspartic acid
Phenylpropionyl glycine (Spain)	Hydroxyproline
	γ-Aminobutyric acid
	Homocysteine
	n-Acetylaspartic acid
	Ornithine
	3-Methylhistidine
	Hydroxylysine
	Cystathionine
	Homocystine
Group 8	**Group 9**
Sarcosine	Maleic acid
Ketoleucine	Glycerol
Ketovaline	Creatinine
Ketoisoleucine	4-Hydroxy-3-methoxyphenyl glycol
3-Hydroxybutyric acid	Formiminoglutamic acid
Pipecolic acid	Vanillylmandelic acid
Orotic acid	Metanephrine
Mannitol	Pantothenic acid
Xanthine	5-Hydroxyindoleacetic acid
Glutamic acid	5-Aminolevulinic acid
Succinylacetone	Glyceric acid
	Sialic acid
Group 10	**Group 11**
Acetoacetic acid	2-Methylcitric acid (Spain)
3-Hydroxypropionic acid (Spain)	2-Hydroxydecanedioic acid
3-Hydroxyisobutyric acid (Spain)	2-Hydroxy-3-methylbutyric acid
5-Hydroxyhexanoic acid (Spain)	Arginine
3-Methylglutaconic acid (Spain)	Alloisoleucine
3-Hydroxyglutaric acid (Spain)	Argininosuccinic acid
3-Hydroxy-2-methylvaleric acid (Spain)	3-Methylglutaric acid
Uridine	Uracil
Thymidine	Cis-aconitic acid
Deoxyuridine	Mevalonic acid

All reagents are from Sigma/Aldrich/Fluka, unless otherwise noted. Spain refers to the University of Madrid, Spain.

stock solution to a 10 mL volumetric flask. Then different volumes of the group solution (20, 100, 250, 500, and 1,000 µL) are processed for analysis (*see* **Note 5**).

2.4. Internal Standards and Quality Control

1. *Internal Standard Stock Solution*: d_3-lactate (C/D/N Isotopes), $^{13}C_3$-pyruvate (Isotec Sigma), d_3-methylmalonic acid (C/D/N Isotopes), d_3-serine (Cambridge Isotopes), d_5-phenylalanine(C/D/N Isotopes), $^{15}N_2$-orotic acid (Cambridge Isotopes), d_4-sebacic acid (C/D/N Isotopes), $^{13}C_6$-glucose (Isotec Sigma), d_6-inositol (C/D/N Isotopes), d_5-tryptophan (C/D/N Isotopes), and 500 nmol d_3-creatine (C/D/N Isotopes) are used as internal standards. Prepare 0.01 M stock solutions for each internal standard in water, and if needed, methanol. Add ammonia gas if needed for some standards ($^{15}N_2$-orotic acid). Stable for 2 years at $-70°C$.

2. *Internal Standard working solution*: Place 400 µL of each 0.01 M stock solution in a 10 mL volumetric flask as described in **Table 42.2**. 250 µL of this composite standard mix is added to each reactivial. Each sample reactivial contains 100 nmol of ten internal standards and 500 nmol of creatine. A set of reactivials with internal standards added is frozen at $-70°C$ until ready for use and is stable for 1 year at $-70°C$.

Table 42.2
Amounts needed to make a 10 mL solution of internal standard and controls

Chemical	F.W.	Percent of purity	Amount needed (g)	Water soluble?
d_3-Lactate	115.08	99.7	0.01154	yes
$^{13}C_3$-Pyruvate	113.02	99.0	0.01142	yes
d_3-Methylmalonic acid	121.11	99.7	0.01215	yes
d_3-Serine	108.11	98.0	0.01103	yes
d_5-Phenylalanine	170.22	99.4	0.01712	+ NH_3 gas
$^{15}N_2$-Orotic acid	158.08	98.0	0.01613	+ NH_3 gas
d_4-Sebacic acid	206.27	99.4	0.02075	+ methanol
$^{13}C_6$-Glucose	186.10	99.0	0.01880	yes
d_6-Inositol	186.19	99.4	0.01873	yes
d_5-Tryptophan	209.26	98.8	0.02118	+ NH_3 gas
d_3-Creatine	152.17	99.7	0.07631	at 70°C

2.5. Supplies

1. GC column: (Agilent) fused silica capillary column, DB-5, with dimensions of 25 m $\times$ 0.32 mm $\times$ 0.5 µm film thickness

2. Reactivials 2 mL (Supelco)

3. Red rubber plug septa (Wheaton)

4. Teflon septa (Supelco)

5. Disposable 3 mL glycerol-free syringe

6. Anotop 0.22 µm syringe filters (Whatman)

7. Borosilicate transfer pipets, Pasteur type

8. Disposable 12 $\times$ 75 mm glass test tubes

9. 2.0 mL Microcentrifuge tubes

10. Disposable 1 mL syringe

11. Disposable 21-gauge syringe needles

12. Sample vials 1.8 mL (Wheaton)

13. Aluminum seals (Wheaton)

14. Vial inserts (Supelco)

15. Vacutainer® blood collection tube, used to hold ammonium hydroxide to generate ammonia gas to aid dissolution of standards in water.

16. Blood collection set 21-gauge (butterfly) (Becton Dickenson) used to transfer ammonia gas from the collection tube to the volumetric flask holding a standard in water.

17. Cryovials 1.2 mL (Nalgene)

18. Carbon dioxide gas (CO_2)

2.6. Equipment

1. A gas chromatograph/mass spectrometer system (GC-MS; 6890/5973) with autosampler and operated in electron impact mode.

2. Refractometer, TS Meter (Leica)

3. Reacti-Vap Evaporator III (Pierce) with nitrogen source

4. Reacti-Therm III Heating Module (Pierce)

5. Reacti-Block T-1 (Pierce)

6. Rotovapor, Model RE 111 (Buchii, Brinkmann)

3. Method

3.1. Sample Preparation

1. Determine the specific gravity of each urine sample to be analyzed.

1. The volume of urine used for analysis is determined by its specific gravity:

     a. specific gravity ≤ 1.010 1.0 mL

     b. specific gravity >1.010 but ≤ 1.020 0.5 mL

     c. specific gravity >1.020 0.25 mL + 0.25 mL H_2O

2. Using a 3 mL glycerol-free syringe, filter urine through a 0.2-micron aqueous syringe filter into a clean glass test tube.

3. Label a reactivial for each sample.

4. Transfer appropriate urine volume to the appropriately labeled reactivial containing 250 µL of the working internal standard mixture. Cap vial with a rubber septum plug and mix by vortexing.

5. Add CO_2 into the vial with a needle connected to a CO_2 line through the septum with the cap loose and after 10 seconds tighten the cap and quickly remove the needle.

6. Add the 50 µL of the urease solution to the sample. Mix well by vortexing, reintroduce CO_2 blanket, and reseal.

7. Incubate sample at 37°C for 30 min. Add additional CO_2 at 15 min without loosening the cap to maintain pressure.

8. After 30-min incubation, add additional 50 µL of urease solution. Mix well by vortexing, reintroduce CO_2 blanket, and incubate at 37°C for an additional 15 min.

9. Chill sample in an ice bath for 5 min.

10. Change existing septum to a Teflon-lined rubber septum.

11. Add cold acetone to ~50%, mix well by vortexing, and incubate in ice bath for 15 min. (*see* **Note 6**) Use new pipet tip for each acetone addition.

12. Transfer sample using a glass transfer pipet to a 2.0 mL plastic conical centrifuge tube and centrifuge for 5 min at $13,600 \times g$.

13. Pour supernatant into a clean 2 mL reactivial.

14. Add TEA/TFA and mix well by vortexing:
     a. Use 20 µL of TEA/TFA, if 1.0 mL of urine (or standard) was used

     b. Use 40 µL of TEA/TFA, if <0.5 mL of urine was used.

15. Add acetonitrile to 2 mm below the top of reactivial using a glass transfer pipet, and vortex.

16. In a 70°C heating block, evaporate the sample under a nitrogen stream for 10 min.

17. Add 0.5 mL acetonitrile, vortex briefly, and evaporate under nitrogen. Continue adding acetonitrile until a constant volume (to dryness) is obtained (*see* **Note 7**).

18. Add dichloromethane in 1.0 mL portions to the sample, and vortex or hand mix carefully (*see* **Note 8**).

19. Evaporate under nitrogen and check after 3 min. Repeat this process two more times until the sample is anhydrous and reduced to constant volume.

20. Fill the vial with nitrogen and cap it tightly.

21. Add 250 µL of MSTFA and mix by vortexing (*see* **Note 9**).

22. Tighten cap and incubate at 70°C for 1 h 15 min (*see* **Note 10**).

23. After incubation, cool, and transfer to a 2 mL plastic conical centrifuge tube with a glass transfer pipet. Centrifuge for 5 min at 12,000 rpm ($13,600 \times g$) at 4°C.

24. Transfer supernatant to autosampler vials containing a glass insert. Carefully cover sample with N_2 before capping. Inject 0.5 µL on GC-MS for analysis (*see* **Note 11**).

3.2. Creatinine Determination

Two methods are used to determine the creatinine concentration in the sample urine. One is to use an automated method, for example, picric acid or enzymatic reaction-based. Alternatively, isotope dilution using d_3-creatine can be used. The sample preparation process quantitatively converts creatine to creatinine as long as the aqueous solutions remain in an alkaline environment. Our laboratory typically determines both values and utilizes the isotope-dilution method to provide results and the clinical analyzer result as a comparison check.

3.3. GC-MS Operating Conditions

See Table 42.3.

Table 42.3
GC-MS operating conditions

Front inlet mode	Splitless
Column pressure	4 psi
Injector temp.	200°C
Purge time on	0.75 min.
Detector temp.	280°C
Initial oven temp.	80°C
Initial time	1.0 min.
Temperature ramp 1	4°C/min.
Oven temp. 1	130°C
Temperature ramp 2	6°C/min.

(continued)

Table 42.3 (continued)

Oven temp. 2	200°C
Temperature ramp 3	12°C/min.
Oven temp. 3	285°C
Final time 3	10 min.
MS source temp.	230°C
MS mode	Electron impact at 70 eV, scan mode

3.4. Data Analysis

After testing of the sample on the GC-MS, the total ion chromatogram (TIC) is reviewed for each patient. Derivatized compounds are identified by their mass spectral fragmentation patterns. Metabolite identities are confirmed using mass spectra of compounds in our custom library. Other libraries used in the identification process include those from NIST, St. Louis University School of Medicine Metabolic Screening Laboratory, and proprietary libraries used with the permission of James Pitt and Lawrence Sweetman. A trained technologist is critical to the proper identification and search for potentially hidden metabolites within the TIC. Representative chromatograms from patients with known inborn errors of metabolism are retained for comparison. An example of the TIC of a normal sample is seen in **Fig. 42.1**, while **Figs. 42.2** and **42.3** are illustrative of various disorders identified using the urease method.

Additionally, each sample is reviewed to assure acceptability of the peak-area ratios of the quantifying ions of internal standards to the 332 ion of d_3-creatinine. Acceptable quality control limits are set at $+/-$ 3 standard deviations of the ion ratio mean.

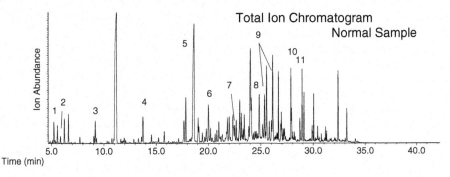

Fig. 42.1. Total ion chromatogram of a urine organic acid profile from a normal sample. 1. L-d_3-lactate, 2. $^{13}C_6$-pyruvate, 3. d_3-methylmalonic acid, 4. DL-d_3-serine, 5. d_3-creatine, 6. L-phenyl-d_5-alanine, 7. $^{15}N_2$-orotic acid, 8. d_4-sebacic acid, 9. D-$^{13}C_6$-glucose, 10. d_6-myoinositol, and 11. L-d_5-tryptophan.

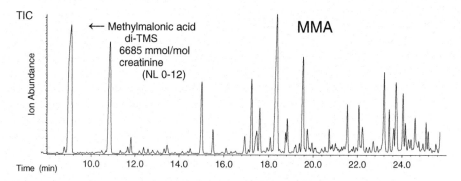

Fig. 42.2. Total ion chromatogram of a urine organic acid profile from a methylmalonic aciduria patient.

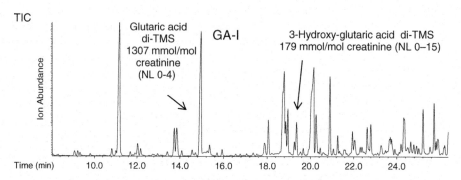

Fig. 42.3. Total ion chromatogram of a urine organic acid profile from a glutaric acidemia, type 1 patient.

4. Notes

1. The triethylammonium trifluoroacetic acid (TEA/TFA) reagent insures derivatization of the organic acid metabolites. Both chemicals are liquid, so densities are required in their calculations. CAUTION: both chemicals are extremely hazardous and should be handled with extreme care. Mixing the acid and base too quickly will produce extreme heat! In addition, dichloromethane is highly flammable. Read each chemical's MSDS before beginning this procedure, work in a chemical fume hood, and wear appropriate personal protective equipment.

2. The round bottom flask should be chosen so that the volume added to it is no more than 60–70% of the flask volume.

3. Beware of the reaction getting too hot too quickly. A minimal amount of smoke resulting from this reaction is expected. If the reaction starts to boil, smoke profusely, etc., STOP ADDING THE TFA IMMEDIATELY! Add more ice to the ice bath and allow the reaction to cool before adding more TFA.

4. The boiling point of dichloromethane is 40°C. Placing a warm bath under the RBF up to the neck should be all that is needed to remove the dichloromethane. If the house vacuum is used with this solution, a second cold trap in between the Rotovap and the vacuum source is required to pull the dichloromethane entering into the house vacuum. Make sure both the collection flask and the second cold trap are up to their neck in ice water.

5. The calibration range is increased for lactic acid, hippuric acid, and citric acid by either increasing the stock solution concentration and/or increasing the amount added to the 10 mL group solution. A single blank can be used for all standard curves. Each calibration group sample is added to a reactivial containing the 11 internal standards and then processed as other samples, starting with the addition of TEA/TFA.

6. If 1 mL of sample was used, add 1.4 mL of acetone. If 0.5 mL of sample was used, then add 0.73 mL of acetone. If 0.25 mL of sample was used, then add 0.5 mL of acetone. Use new pipet tip for each acetone addition.

7. This step may require 3–7 additions. When a precipitate forms (~1.0 ml level), mix carefully to avoid precipitate from coating the vial.

8. Allow the sample to cool before adding dichloromethane.

9. This is best done using a syringe. Withdraw MSTFA from the hypo-vial and add MSTFA into reactivial through the septum.

10. Check the sample after 30 min and vortex to remove any residue that may have remained on the sides of the vial. Observe gradual darkening of color. After 30 min the solution usually takes on a yellowish hue and continues to get darker, possibly turning reddish brown, or occasionally, purple color.

11. Sample preparation of internal standard and calibration samples can omit urease treatment.

References

1. Shoemaker, J.D., and Elliot, W.H. (1991) Automated Screening of Urine Samples for Carbohydrates, Organic and Amino Acids after Treatment with Urease. *Journal of Chromatography*, **562**: 125–138.

2. Goodman, S.I. and Markey, S.P. (1981) Diagnosis of Organic Acidemias by Gas Chromatography-Mass Spectrometry. *Laboratory and Research Methods in Biology and Medicine*, Volume 6. Alan R. Liss, Inc., New York.

3. Sweetman, L. (1991) Organic acid analysis, in *Techniques in Diagnostic Human Biochemical Genetics: A Laboratory Manual.* (Hommes, F.A., ed), Wiley-Liss, New York, pp. 143–176.

4. American College of Medical Genetics. (2006) F: Clinical biochemical genetics, in *Standards and Guidelines for Clinical Genetics Laboratories.* http://www.acmg.net/Pages/ACMG_Activities/stds-2002/f.htm

Chapter 43

Quantitation of Orotic Acid in Urine Using Isotope Dilution-Selected Ion Gas Chromatography-Mass Spectrometry

Jie Chen and Michael J. Bennett

Abstract

The measurement of urinary orotic acid excretion is an important test for establishing a diagnosis of hereditary orotic aciduria, a genetic defect of pyrimidine biosynthesis. Measurement of secondary urinary orotic acid elevation is also an important clinical test for the differential diagnosis of hyperammonemia due to some of the primary disorders of the urea cycle including ornithine transcarbamylase (OTC) deficiency, and the hyperornithinemia–hyperammonemia–homocitrullinemia (HHH) syndrome. Low levels of orotic acid are observed in carbamylphosphate synthetase (CPS) defects. This method utilizes a stable-isotope labeled internal standard (1, 3-^{15}N-orotic acid), which is added to the standards, controls, and patient samples prior to extraction. Interference from urea is removed by incubation of samples with urease and the orotic acid is derivatized by trimethylsilylation. Quantitation is made against an eight-point standard curve using specific selected ions from both the labeled and unlabeled orotic acid.

Key words: Hereditary orotic aciduria, urea cycle defects, hyperammonemia, isotope dilution mass spectrometry.

1. Introduction

Orotic acid (**Fig. 43.1**) is a metabolic intermediate in the pathway of *de novo* biosynthesis of pyrimidine bases (1). Carbamyl phosphate is an important precursor of this pathway and also, in liver, an important precursor of ureagenesis through the urea cycle (2).

Increased urinary excretion of orotic acid is seen in hereditary orotic aciduria, which results from a specific defect of the enzyme uridine 5-monophosphate (UMP) synthase, a metabolic disorder that presents with severe megaloblastic anemia and crystaluria due to the poor solubility of orotic acid (3). Carbamyl phosphate levels are increased in a number of disorders of the urea cycle, in

U. Garg, C.A. Hammett-Stabler (eds.), *Clinical Applications of Mass Spectrometry*, Methods in Molecular Biology 603, DOI 10.1007/978-1-60761-459-3_43, © Humana Press, a part of Springer Science+Business Media, LLC 2010

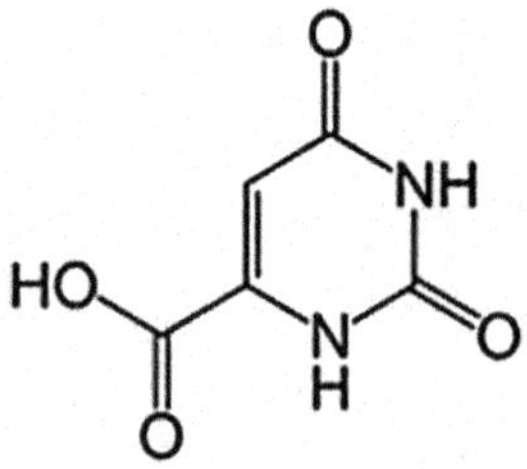

Fig. 43.1. Orotic acid molecule. For the internal standard, both nitrogen molecules are labeled with ^{15}N.

particular in OTC deficiency, the most commonly encountered urea cycle disorder (4). The increased levels of carbamyl phosphate drive pyrimidine biosynthesis as the pathway to urea synthesis is blocked. This metabolic redirection results in increased orotic acid excretion but concentrations are considerably less and more variable than those observed in hereditary orotic aciduria (5). Defects of carbamyl phosphate synthase (CPS) and related pathways lead to low concentrations of this intermediate, and subsequently low concentrations of orotic acid. CPS defects are clinically similar to OTC defects and the differential diagnosis of low or elevated orotic acid excretion is used as an important tool when evaluating hyperammonemia (6).

This assay which utilizes stable-isotope dilution–selected ion monitoring gas chromatography-mass spectrometry is capable of measuring the very low levels of orotic acid that are seen in CPS deficiency. A stable-isotope labeled internal standard of 1, 3-^{15}N-orotic acid is used and pairs of selected ions monitored for definitive identification and for quantitation (7).

2. Materials

2.1. Specimens

1. Randomly collected urine: Collect urine specimens without preservatives and store at $-20°C$, if analysis is not to be performed immediately. Complete testing within 14 days.

2.2. Reagents

1. Orotic acid stock standard, 1 mM: Prepare by dissolving 0.0039 g of mono potassium salt (Sigma-Aldrich, St. Louis, MO) in 20 mL of deionized water. Stable for 5 years when stored frozen at $-40°C$.

2. 1,3 ^{15}N orotic acid stock standard, 0.1 mM, (Cambridge Isotope Laboratories, Andover, MA). Stable for 5 years when stored at $-40°C$.

3. Urease, type C-3 (Sigma-Aldrich, St Louis, MO): 30 units/ 10 µl. Make fresh each time. Prepare appropriate amount for each batch of controls and samples.

4. Quality control samples: Prepare high and normal control materials by pooling previously analyzed samples. The high quality control material may be prepared by the addition of orotic acid to pooled normal urine. Aliquot and store at −40°C.

5. Bis(trimethylsilyl) trifluoroacetamide:trimethylchlorosilane 99:1 (Pierce Inc., Rockford, IL).

2.3. Equipment and Supplies

1. Instrument: Agilent 6890/5973 GC-MSD instrument with a HP-5MS column (30 m × 250 µm × 0.25 µm) cross-linked with 5% phenylmethylsilane (Agilent Technologies, Wilmington, DE).

3. Methods

3.1. Stepwise Procedure

1. Remove and thaw an aliquot of the frozen orotic stock standard solution and use to prepare seven standards spanning the analytical measuring range of orotic acid from 0.1 to 10.0 µM. Include a water blank.

2. Place 100 µl of each standard, control, and patient sample into a labeled microcentrifuge tube.

3. Remove and thaw an aliquot of the internal standard.

4. Add 10 µL of the 0.1 µmol/L internal standard to each standard, control, and patient sample and vortex well.

5. Add 30 units of urease to each tube, mix well, and incubate at 37°C for 30 min (*see* **Note 1**).

6. Add 1 ml of ethanol to all tubes to precipitate proteins. Vortex and centrifuge at 18,000 × g for 10 min.

7. Transfer the supernatant to labeled 13 x 100 mm screw-capped tubes and evaporate to dryness under a stream of nitrogen at room temperature.

8. Add 100 µl BSTFA: TMCS and 100 µl pyridine.

9. Vortex and seal tubes. Incubate at 70°C for 30 min (*see* **Note 2**).

10. Allow to cool and transfer 100 µl to GC vials. Seal vials.

11. Inject 1 µl into gas chromatograph.

3.2. Gas Chromatography Conditions

1. The gas chromatographic running conditions are provided in **Table 43.1**.

Table 43.1
Gas chromatography running conditions for orotic acid assay

Ramp	°C/min	Next °C	Hold (min)	Run time (min)
Initial		100	1.00	1.00
Ramp 1	10.00	240	0.00	15.00
Ramp 2	30.00	300	10.00	27.00

3.3. Data Acquisition and Analysis

1. Two pairs of ions are acquired for both the internal standard and for the standards and unknown orotic acid at m/z 254 and 357 for the standards and unknown and m/z 256 and 359 for the internal standard (5). The ratios of each pair are monitored to ensure that there is no contamination. The m/z 254 and 256 ions are used for quantification. **Figures 43.2** and **43.3** show selected ion chromatograms and mass spectra of unlabeled and labeled orotic acid.

2. Data are extracted in the time range 11.4−12.0 min for quantification and the areas under the respective peaks are integrated for the calculation.

3. The standards are used to generate a standard curve for which the slope k and intercept b are utilized in the calculation of control and unknown values.

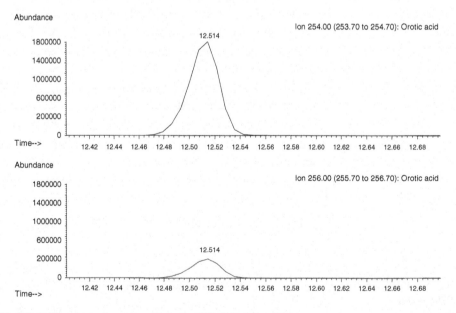

Fig. 43.2. Selected ion chromatograms unlabeled and labeled orotic acid.

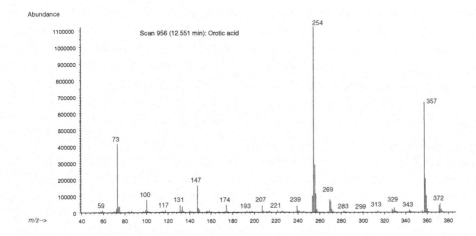

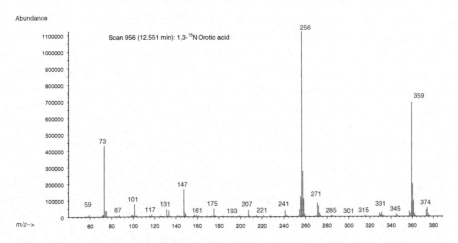

Fig. 43.3. Mass spectra of unlabeled and labeled orotic acid.

4. A correction factor is applied to the calculation to take into account the natural abundance of ^{13}C and ^{15}N in unlabeled orotic acid. This was calculated in the laboratory by replicate analysis (N = 15) of the signal from the 10 μM orotic acid standard to be 0.1 atoms percent of the m/z 254 signal.

5. The orotic concentration is calculated using the following equation:

$$\text{Orotic acid } (\mu M) = \frac{mz\,254/mz\,256 - 0.1 \times m/z\,254 - b}{K}$$

6. The final report is presented as μmol/mmol (mmol/mol) of urinary creatinine (*see* **Note 3**).

3.4. Quality Control, Proficiency Testing

1. Quality control samples are included with each analytical run. Results should be within acceptability criteria established by the laboratory.

2. There is no accredited proficiency program for urine orotic acid measurement. Proficiency testing is accomplished by biannually sending samples to a second laboratory that also measures orotic acid as a clinical service, and comparing the results.

3.5. Reference Values, Analytical Measurement Range, and Clinical Reportable Range

1. *Reference range*: The reference range determined from individual patients for whom there was no evidence of hereditary orotic aciduria or a urea cycle defect varies with age. 1–14 days: 2.27–4.13 mmol/mol creatinine; 15 days–12 months: 1.51–3.09 mmol/mol creatinine; 1–10 years 1.12–2.52 mmol/mol creatinine; and over 10 years of age 0.5–0.98 mmol/mol creatinine. Concentrations below 0.5 mmol/mol creatinine raise a suspicion of CPS deficiency (*see* **Note 4**).

2. *Analytical measurement range*: This was determined to be 0.1–100 mM, using both aqueous standards and serial dilution of a high urine value.

3. *Clinical reportable range*: This was determined to be 0.1–100 mM. Values below 0.1 should be reported <0.1 and values >100 should be re-evaluated by serial dilution.

4. Notes

1. This step removes urea, a significant interference, from the sample.

2. These steps should be performed under a fume hood.

3. Correction to the urine creatinine concentration is important to adjust for variable urine concentrations. Creatinine is measured using a validated method.

4. Urine samples with orotic acid concentrations less than 0.5 mmol/mol creatinine are very rare and occur in only a few conditions, such as CPS deficiency.

References

1. Nyhan WL. (2009) Purine and pyrimidine metabolism. In: Sarafoglou K, Hoffmann GF, Roth KS (eds.) Pediatric Endocrinology and Inborn Errors of Metabolism. New York: McGraw-Hill, 757–786.

2. Summar ML. (2009) Urea cycle disorders. In: Sarafoglou K, Hoffmann GF, Roth KS (eds.) Pediatric Endocrinology and Inborn Errors of Metabolism. New York: McGraw-Hill, 141–152.

3. Smith LJ, Sullivan M, Huguley C. (1961) Pyrimidine metabolism in man: IV. The enzymatic defect of orotic aciduria. J Clin Invest **40**, 656–662.

4. Brusilow SW, Horwich AL. (2001) Urea cycle enzymes. In: Scriver CR, Beaudet AL, Sly WS, Valle D (eds.) The Metabolic and Molecular Bases of Inherited Disease. New York: McGraw-Hill, 1909–1963.

5. Brusilow SW. (1991) Determination of urine orotate and orotidine and plasma ammonium. In: Hommes FA (ed.) Techniques in Diagnostic Human Biochemical Genetics. New York: Wiley-Liss, 345–357.

6. Cowan TM, Yu C. (2009) Laboratory investigation of inborn errors of metabolism. In: Sarafoglou K, Hoffmann GF, Roth KS (eds.) Pediatric Endocrinology and Inborn Errors of Metabolism. New York: McGraw-Hill, 857–874.

7. McCann MT, Thompson MM, Gueron IC, Tuchman M. (1995) Quantification of orotic acid in died filter paper urine samples by stable isotope dilution. Clin Chem 41, 739–743.

Chapter 44

Quantitation of Oxycodone in Blood and Urine Using Gas Chromatography-Mass Spectrometry (GC-MS)

Bruce A. Goldberger, Chris W. Chronister, and Michele L. Merves

Abstract

Oxycodone is a semisynthetic opioid analgesic used in pain management. In the following method, gas chromatography-mass spectrometry (GC-MS) is used to determine the presence and concentration of the drug in blood and urine. The specimens are fortified with deuterated internal standard and a five-point calibration curve is constructed. Specimens are extracted using mixed-mode solid phase extraction and derivatized with N,O-bis(trimethylsilyl)trifluoroacetamide (BSTFA) producing trimethylsilyl derivatives. The final extracts are then analyzed using selected ion monitoring GC-MS.

Key words: Opiate, opioid, oxycodone, gas chromatography, mass spectrometry, solid phase extraction.

1. Introduction

Oxycodone is an opioid analgesic medication indicated for the treatment of moderate-to-severe pain. It is a semisynthetic opioid derived from thebaine with potency similar to morphine, when administered parenterally. Oxycodone is typically administered orally and is often used in combination with nonopioid analgesics such as acetaminophen or aspirin. Oxycodone is also available in sustained-release formulations. Oxycodone is often abused for its central nervous system effects which include euphoria (1–3).

Oxycodone has a half-life of approximately 3–6 h. The metabolism of oxycodone includes *O*- and *N*-demethylation which form oxymorphone and noroxycodone, respectively. Oxymorphone is pharmacologically active and contributes to oxycodone's analgesic potency. Metabolites of oxycodone undergo glucuronidation prior to elimination (1–3).

U. Garg, C.A. Hammett-Stabler (eds.), *Clinical Applications of Mass Spectrometry*, Methods in Molecular Biology 603, DOI 10.1007/978-1-60761-459-3_44, © Humana Press, a part of Springer Science+Business Media, LLC 2010

The analysis of oxycodone includes immunoassay, as well as confirmation and quantitation by GC-MS. Solid phase extraction methods have been reported for the isolation of drug from blood and urine matrices. Derivatization of oxycodone is often essential in order to obtain satisfactory chromatographic performance (4–9).

The method described below is validated for the analysis of oxycodone in blood and urine (*see* **Note 1**).

2. Materials

2.1. Chemical, Reagents, and Buffers

1. Blood, drug-free, prepared from human whole blood purchased from a blood bank and pretested to confirm the absence of analyte or interfering substance.

2. N,O-Bis(trimethylsilyl)trifluoroacetamide with 10% trimethylchlorosilane (TMCS), BSTFA (United Chemical Technologies).

3. 1.0 M Acetic Acid: Add 28.6 mL of acetic acid to 400 mL of water. Dilute to 500 mL with deionized water. Stable for 1 year at room temperature.

4. 0.1M Acetate Buffer, pH 4: Pipet 5.7 mL glacial acetic acid into a 1,000 mL volumatric flask filled with 800 mL of water. Pipet 16 mL of 1.0 M potassium hydroxide. Mix well and check pH and adjust it necessary to pH 4.0. Q.S. to 1,000 mL with water. Stable for 1 year at room temperature.

5. Methylene Chloride:Isopropanol:Ammonium Hydroxide Solution (78:20:2; v:v:v): To 20 mL of isopropanol, add 2 mL of ammonium hydroxide. Add 78 mL of methylene chloride. Mix well. Prepare fresh.

6. 0.1 M Phosphate Buffer, pH 6: Dissolve 13.61 g of potassium phosphate monobasic in 900 mL of water. Adjust the pH to 6.0 with 5.0 M potassium hydroxide. Q.S. to 1,000 mL with water. Stable for 1 year at 0–8°C.

7. 5.0 M Potassium Hydroxide: Dissolve 70.13 g of potassium hydroxide in 100 mL of water in a 250 mL volumetric flask. Mix well and Q.S. to 250 mL with water. Stable for 1 year at room temperature.

8. 1.0 M Potassium Hydroxide: Dissolve 5.6 g of potassium hydroxide in 50 mL of water in a 100 mL volumetric flask. Mix well and Q.S. to 100 mL with water. Stable for 1 year at room temperature.

2.2. Preparation of Calibrators (see Note 2)

1. Oxycodone Internal Standard Solution (10 μg/mL): Add the contents of 100 μg/mL D6-oxycodone vial (Cerilliant Corporation) to a 10 mL volumetric flask and bring to volume with methanol. Store at $\leq-20°C$.

2. Oxycodone Standard Stock Solution (100 μg/mL): Add the contents of 1.0 mg/mL oxycodone vial (Cerilliant Corporation) to a 10 mL volumetric flask and bring to volume with methanol. Store at $\leq-20°C$.

3. Oxycodone Standard Solution (10 μg/mL): Dilute 1.0 mL of the 100 μg/mL oxycodone standard stock solution to 10 mL with methanol in a 10 mL volumetric flask. Store at $\leq-20°C$.

4. Oxycodone Standard Solution (1.0 μg/mL): Dilute 1.0 mL of the 10 μg/mL oxycodone standard solution to 10 mL with methanol in a 10 mL volumetric flask. Store at $\leq-20°C$.

5. Aqueous Oxycodone Calibrators: Add standard solutions to 1.0 mL of drug-free blood according to **Table 44.1**.

Table 44.1
Preparation of oxycodone calibration curve

Calibrator	Calibrator concentration (ng/mL)	Volume of 1.0 μg/mL standard solution	Volume of 10 μg/mL standard solution
1	50	50 μL	–
2	100	100 μL	–
3	250	–	25 μL
4	500	–	50 μL
5	1,000	–	100 μL

2.3. Preparation of Control Samples (see Note 2)

1. Negative (drug-free) Control: Prepare using human whole blood. Blood is pretested to confirm the absence of an opioid or interfering substance.

2. Oxycodone Control Stock Solution (100 μg/mL): Add the contents of 1.0 mg/mL oxycodone vial (Cerilliant Corporation) to a 10 mL volumetric flask and bring to volume with methanol. Store at $\leq-20°C$.

3. Oxycodone Control Solution (10 μg/mL): Dilute 1.0 mL of the 100 μg/mL oxycodone control stock solution to 10 mL with methanol in a 10 mL volumetric flask. Store at $\leq-20°C$.

4. Aqueous Oxycodone Control: Add oxycodone control solution to 1.0 mL of drug-free blood according to **Table 44.2**.

Table 44.2
Preparation of oxycodone controls

Control	Quality control concentration (ng/mL)	Volume of 10 µg/mL control solution
Low	250	25 µL
High	750	75 µL

2.4. Supplies

1. Clean Screen Extraction Columns (United Chemical Technologies)
2. Autosampler vials
3. Disposable glass culture tubes
4. Volumetric pipet with disposable tips

2.5. Equipment

1. 6890 Gas Chromatograph (Agilent Technologies, Inc.) or equivalent
2. 5973 Mass Selective Detector (Agilent Technologies, Inc.) or equivalent
3. 7673 Automatic Liquid Sampler (Agilent Technologies, Inc.) or equivalent
4. ChemStation with DrugQuant Software (Agilent Technologies, Inc.) or equivalent
5. Extraction manifold
6. Heat Block
7. Vortex Mixer
8. Centrifuge
9. Caliper Life Sciences TurboVap connected to nitrogen gas

3. Methods

3.1. Stepwise Procedure for Oxycodone Assay

1. Label culture tubes for each calibrator, control, and specimen to be analyzed and add 1.0 mL of the appropriate specimen to each corresponding tube.
2. Add 25 µL of the 10 µg/mL oxycodone internal standard solution to all culture tubes except the negative control.
3. Add 4.0 mL of 0.1 M phosphate buffer (pH 6).
4. Vortex all specimens.
5. Add 200 µL of 1.0 M potassium hydroxide and vortex.
6. Centrifuge at $\sim$1,500 $\times$ g for 5 min.

7. Place columns into the extraction manifold.

8. Prewash the columns with 3.0 mL of the methylene chloride:isopropanol:ammonium hydroxide solution.

9. Pass 3.0 mL of methanol through each column. Do not permit the columns to dry.

10. Pass 3.0 mL of water through each column. Do not permit the columns to dry.

11. Pass 1.0 mL of 0.1 M acetate buffer (pH 4) through each column. Do not permit the columns to dry.

12. Pour specimen into column. Slowly draw specimen through column (at least 2 min) under low vacuum.

13. Pass 3.0 mL of water through each column.

14. Pass 1.0 mL of 1.0 M acetic acid through each column.

15. Dry column with full vacuum.

16. Pass 2.0 mL of hexane through each column.

17. Pass 3.0 mL of methanol through each column.

18. Dry column with full vacuum.

19. Turn off vacuum. Dry tips. Place labeled disposable culture tubes into column reservoir.

20. Add 3.0 mL of the methylene chloride:isopropanol:ammonium hydroxide solution and collect in disposable culture tubes.

21. Evaporate to dryness at 40°C ± 5°C with a stream of nitrogen.

22. Add 50 µL of BSTFA and 50 µL of ethyl acetate. Cap tightly and vortex.

23. Place tubes in heating block for 30 min at 60°C ± 5°C.

24. Transfer extract to autosampler vial and submit for GC-MS analysis.

3.2. Instrument Operating Conditions

1. The GC-MS operating parameters are presented in **Tables 44.3** and **44.4**.

2. The capillary column phase used for the oxycodone assay is either 100% methylsiloxane or 5% phenyl–95% methylsiloxane.

3. Set-up of the autosampler should include the exchange of solvent in both autosampler wash bottles with fresh solvent (bottle 1 – acetonitrile; bottle 2 – ethyl acetate).

4. A daily autotune must be performed with perfluorotributylamine (PFTBA) as the tuning compound prior to each GC-MS run.

Table 44.3
GC operating conditions for the oxycodone assay

Initial oven temp	140°C
Initial time	0.5 min
Ramp 1	20°C/min
Final temp	320°C
Final time	2.00 min
Total run time	11.50 min
Injector temp	250°C
Detector temp	290°C
Purge time	0.5 min
Column flow	1.0 mL/min

Table 44.4
Quantitative and qualifier ions for the oxycodone assay

Analyte	Quantitative ion (*m/z*)	Qualifier ions (*m/z*)
Oxycodone	387	372, 388
D6- Oxycodone	393	378

3.3. Data Analysis

1. The review of the data requires the following information: retention times obtained from the selected ion chromatograms; ion abundance and ion peak ratios obtained from the selected ion chromatograms; and quantitative ion ratios between the native drug and its corresponding deuterated internal standard.

2. A typical Agilent ChemStation report for oxycodone is illustrated in **Fig. 44.1**.

3. In order for a result to be reported as "Positive," the following criteria must be satisfied: (1) the retention times of the ion peaks must be within ± 1% of the corresponding ions of the intermediate calibrator; (2) the ion peak ratio for the specimen must be within ± 20% of corresponding ion peak ratio of the intermediate calibrator; and (3) the correlation coefficient for the calibration curve must be 0.99 or greater if linear regression is used. Failure to meet one of the above criteria requires that the specimen be reported as "None Detected."

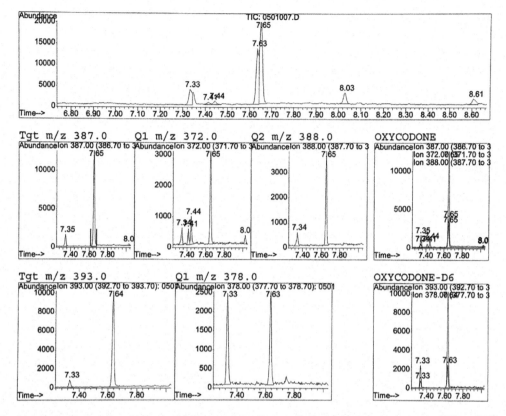

Fig. 44.1. Typical Agilent ChemStation report for oxycodone. *Top panel* is the total ion chromatogram; *middle and bottom panels* are selected ion chromatograms for oxycodone (retention time 7.65 min) and deuterated oxycodone (retention time 7.64 min), respectively.

4. The limit of detection for the oxycodone assay is 12.5 ng/mL and the range of linearity is 50–1,000 ng/mL. The reporting criteria for blood are presented in **Table 44.5**. Urine results are qualitative only and are reported as "Positive" when the concentration is greater than the limit of detection.

5. The intra- and inter-assay variability (%CV) is less than <10%.

Table 44.5
Reporting criteria for oxycodone in blood

Concentration (ng/mL)	Reported result
12.5<25	Trace
25–<50	<50 ng/mL
50–1200	Report concentration
>1,200 ng/mL	Reanalyze "diluted" specimen

4. Notes

1. The assay can be applied to other specimens including those obtained at autopsy.

2. Separate sources of oxycodone analyte must be used when preparing standard and control solutions.

References

1. Stout, P.R. and Farrell, L.J. (2003) Opioids: Effects on human performance and behavior. *Forensic Sci Rev*, **15**, 29–59. Review.

2. Gutstein, H.B. and Akil, H. (2006) Opioid Analgesics. In Brunton, L.L., Lazo, J.S. and Parker, K.L. (eds), *Goodman & Gilman's the Pharmacological Basis of Therapeutics*, 11th edition. McGraw-Hill, New York, pp. 547–590.

3. Lewis, L.S. (2006) Opioids. In Flomenbaum, N.E., Goldfrank, L.R., Hoffman, R.S., Howland, M.A., Lewin, N.A. and Nelson, L.S. (eds), *Goldfrank's Toxicologic Emergencies*, 8th edition. McGraw-Hill, New York, pp. 590–613.

4. Clean Screen Extraction Column Application Manual, United Chemical Technologies (2009).

5. Ropero-Miller, J.D., Lambing, M.K. and Winecker, R.E. (2002) Simultaneous quantitation of opioids in blood by GC-EI-MS analysis following deproteination, detautomerization of keto analytes, solid-phase extraction, and trimethylsilyl derivatization. *J Anal Toxicol*, **26**, 524–528.

6. Moore, K.A., Ramcharitar, V., Levine, B. and Fowler, D. (2003) Tentative identification of novel oxycodone metabolites in human urine. *J Anal Toxicol*, **27**, 346–352.

7. Meatherall, R. (2005) GC-MS quantitation of codeine, morphine, 6-acetylmorphine, hydrocodone, hydromorphone, oxycodone, and oxymorphone in blood. *J Anal Toxicol*, **29**, 301–308.

8. Moore, C., Rana, S. and Coulter, C. (2007) Determination of meperidine, tramadol and oxycodone in human oral fluid using solid phase extraction and gas chromatography-mass spectrometry. *J Chromatogr B Analyt Technol Biomed Life Sci*, **850**, 370–375.

9. McKinley, S., Snyder, J.J., Welsh, E., Kazarian, C.M., Jamerson, M.H. and Klette, K.L. (2007) Rapid quantification of urinary oxycodone and oxymorphone using fast gas chromatography-mass spectrometry. *J Anal Toxicol*, **31**, 434–441.

Chapter 45

Quantitation of Phencyclidine (PCP) in Urine and Blood Using Gas Chromatography-Mass Spectrometry (GC-MS)

Angela M. Ferguson and Uttam Garg

Abstract

Phencyclidine (PCP) is a cycloalkylamine and is classified as a dissociative anesthetic. In the1950s, PCP was tested as an intravenous anesthetic but due to its severe side effects, it was withdrawn from the clinical use. Since then PCP has become an illegal street drug making its laboratory analysis forensically essential. PCP can be detected in urine, serum, or plasma by immunoassays and quantified by gas or liquid chromatography mass spectrometry. In the method described here, a deuterated internal standard is added to the sample and the drug is extracted under alkaline conditions. Analysis is conducted using gas chromatography mass spectrometry (GC-MS). Quantitation of PCP is done by comparing the responses of unknown samples to the standards using selected ion monitoring.

Key words: Phencyclidine, PCP, mass spectrometry, gas chromatography, drugs of abuse.

1. Introduction

Phencylidine, a dissociative anesthetic and illegal drug, can be administered by smoking, nasal insufflation, intravenous injection, or by oral ingestion. Once investigated as intravenous anesthetic, now PCP is considered an illegal drug (1). It interacts with several neurotransmitter systems, but its major actions are through inhibition of N-methyl-D-aspartate (NMDA) receptor (2). Its effects are unpredictable and include feelings of euphoria, elevated blood pressure, tachycardia, agitation, anxiety, paranoia, delusions of grandeur, muscle rigidity, ataxia, lethargy, and coma. Feelings of superhuman strength coupled with a lack of pain intolerance can lead to self-induced trauma and can result in rhabdomyolysis and myoglobulinuric renal failure (3). Treatment of PCP intoxication is largely supportive and may include enhanced elimination by urine acidification (2, 3).

U. Garg, C.A. Hammett-Stabler (eds.), *Clinical Applications of Mass Spectrometry*, Methods in Molecular Biology 603,
DOI 10.1007/978-1-60761-459-3_45, © Humana Press, a part of Springer Science+Business Media, LLC 2010

Phencyclidine is included in the Substance Abuse and Mental Health Services Administration (SAMHSA) list of drugs that are screened for in workplace drug testing programs as well as in most other drug of abuse panels (4). The commonly used methods for analysis of PCP include immunoassay for screening and GC-MS for confirmation. Although immunoassays are rapid and sensitive, many drugs can produce false positive results. High concentrations of dextromethorphan (5), diphenhydramine (6), and thioridazine (7) are shown to produce false positive immunoassay results making confirmation using an alternative method essential (3). The most commonly used method for the confirmation of PCP is GC-MS, although newer methods using liquid chromatography and tandem mass spectrometry are starting to appear in the literature, especially for alternate matrices such as oral fluid, sweat, and hair (3, 8, 9). Also sample preparation procedures are less extensive in liquid chromatography and tandem mass spectrometry as compared to GC-MS. The drug from the samples is extracted by liquid–liquid or solid phase extraction. In this chapter, a GC-MS method involving liquid–liquid alkaline extraction is described. The quantification of the drug is done using selected ion monitoring.

2. Materials

2.1. Sample

Serum or plasma (heparin or EDTA) or randomly collected urine is an acceptable sample for this procedure. Samples are stable for 1 week when refrigerated and for 6 months when stored at −20°C.

2.2. Reagents and Buffers

1. Buffer salts mixture: Sodium carbonate: sodium bicarbonate: sodium chloride (1:1:12). The mixture is stable at room temperature for 1 year.

2. Extraction solvent: Methylene chloride: cyclohexanes: isopropanol (9:9:2). The extraction solvent is stable at room temperature for 1 year.

3. Extraction tubes: Add 1 g of buffer salt and 3 mL of extraction solvent to each extraction tube. The tubes are stable for 3 months.

4. Human drug-free urine and serum (UTAK Laboratories, Inc., Valencia, CA).

2.3. Standards and Calibrators

1. 1 mg/mL primary phencyclidine standard (Cerilliant Corporation, Round Rock, TX). The primary standard is stable for 1 year at −20°C.

2. 100 μg/mL secondary phencyclidine standard: Prepared by transferring 1.0 mL of primary phencyclidine standard to a 10 mL volumetric flask and diluting with methanol. The secondary standard is stable for 1 year at −20°C.

3. 10 μg/mL tertiary phencyclidine standard: Prepared by transferring 1.0 mL of secondary phencyclidine standard to a 10 mL volumetric flask and diluting with methanol. Tertiary standard is stable for 1 year at −20°C.

4. Working calibrators are made according to **Table 45.1** using 10 mL volumetric flasks. The calibrators are stable for 1 year when stored at −20°C.

Table 45.1
Preparation of calibrators

Calibrator	Drug-free urine (mL)	Tertiary standard (mL)	Secondary standard (mL)	Concentration (ng/mL)
1	9.98	0.02	0	20
2	9.90	0.10	0	100
3	9.95	0	0.05	500
4	9.90	0	0.10	1,000

2.4. Internal Standard and Quality Controls

1. 100 μg/mL primary internal standard: Phencyclidine d5 (Cerilliant Corporation, Round Rock, TX).

2. 4 μg/mL working internal standards: Prepared by transferring 1.0 mL of primary internal standard to a 25 mL volumetric flask and diluting with methanol. Working internal standard is stable for 1 year at −20°C.

3. Quality control samples: Commercial controls are Bio-Rad Liquichek™ Urine Toxicology Controls C3 and C4 (Bio-Rad Inc., Irvine, CA). The target values for these two quality control samples are established in-house.

2.5. Supplies

1. 16 × 100 mm glass tubes (Fisher Scientific, Fair Lawn, NJ).

2. Concentration vials (Fisher Scientific, Fair Lawn, NJ).

3. GC column used: Zebron ZB-1, 15 m × 0.25 mm × 0.25 μm (Phenomenex, Terrance, CA).

2.6. Equipment

1. A gas chromatograph/mass spectrometer system: GC-MS; 6890/5975 or 5890/5972 utilizing electron impact mode (Agilent Technologies, Wilmington, DE).

2. Zymark TurboVap® IV Evaporator (Zymark Corporation, Hopkinton, MA)

3. Methods

3.1. Stepwise Procedure

1. Prepare an unextracted standard by adding 40 μL of the 10 μg/mL tertiary standard and 100 μL of the working internal standard to a concentration tube. Set-aside until step 9.

2. Prepare and label extraction tubes for each calibrator, control, and sample.

3. Add 3 mL of deionized water to each tube.

4. Pipette 1 mL sample into appropriately labeled tube.

5. Add 100 μL of internal standard to each tube.

6. Cap and rock for 5 min.

7. Centrifuge at ~1,600 × *g* for 5 min.

8. Transfer the upper organic phase to the concentration tubes (*see* **Note 1**).

9. Evaporate the extract to dryness in a water bath at 45°C under nitrogen (*see* **Note 2**).

10. Reconstitute the residue with 100 μL of ethyl acetate and inject 1 μL on GC-MS for analysis.

3.2. Instrument Operating Conditions

The instrument's operating conditions are given in **Table 45.2**.

Table 45.2
GC operating conditions

Initial oven temp.	90°C
Initial time	1.0 min
Ramp 1	32°C/min
Temp. 2	170°C
Time	2.0 min
Ramp 2	20°C/min
Final temp.	280°C
Injector temp.	250°C

(continued)

Table 45.2 (continued)

Detector temp.	280°C
Purge time on	0.5 min.
Column pressure	5 psi
MS source temp.	230 °C
MS mode	Electron Impact at 70 eV, selected ion monitoring
MS tune	Autotune

3.3. Data Analysis

1. Representative GC-MS chromatogram of phencyclidine is shown in **Fig. 45.1**. GC-MS selected ion chromatograms for PCP and PCP-d5 are shown in **Fig. 45.2**. Electron impact ionization mass spectra of PCP are shown in the **Fig. 45.3** (*see* **Note 3**). Ions used for identification and quantification are listed in **Table 45.3**.

2. Analyze data using Target Software (Thru-Put Systems, Orlando, FL) or similar software. The quantifying ions (**Table 45.3**) are used to construct standard curves of the peak-area ratios (calibrator/internal standard pair) vs. concentration. These curves are then used to determine the concentrations of the controls and unknown samples (*see* **Note 4**).

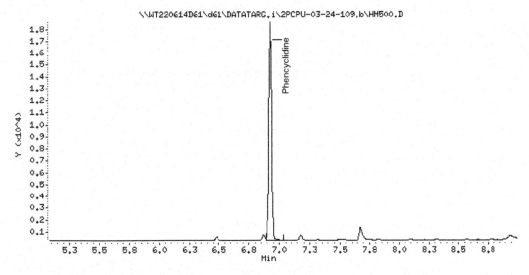

Fig. 45.1. GC-MS chromatogram of PCP (500 ng/mL). Deuterated internal standard co-elute with the compound.

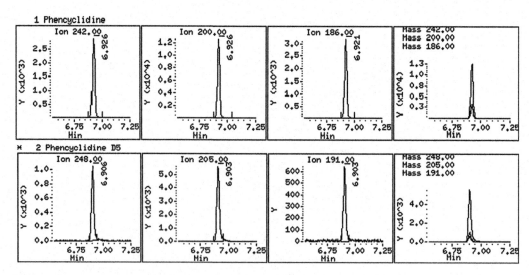

Fig. 45.2. Selected ion chromatograms of PCP and deuterated PCP.

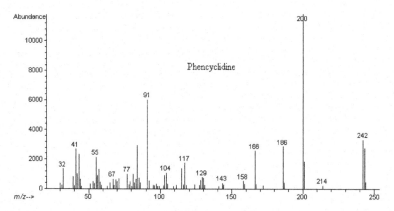

Fig. 45.3. Electron impact ionization mass spectrum of PCP.

Table 45.3
Quantitation and qualifying ions for cocaine and its metabolites

	Quantitation ion	Qualifier ions
Phencyclidine (PCP)	242	200, 186
Phencyclidine d5	248	205, 191

3. The linearity/limit of quantitation of the method is 20–1,000 ng/mL. Samples in which the drug concentrations exceed the upper limit of quantitation should be diluted with the appropriate sample matrix and retested.

4. Standard curves should have a correlation coefficient (r^2) >0.99.

5. Typical intra- and inter-assay imprecision is <10%.

6. Quality control: The run is considered acceptable if calculated concentrations of drugs in the controls are within +/− 20% of target values. Quantifying ion in the sample is considered acceptable if the ratios of qualifier ions to quantifying ion are within +/− 20% of the ion ratios for the calibrators.

4. Notes

1. Make sure no aqueous layer is transferred.

2. Do not over-dry. This will result in low recovery.

3. Electron impact ionization spectra are needed in the initial stages of method set up to establish retention times and later on, if there is a need for change in quantifying or qualifying ions. They are not needed for routine quantitaion.

4. The use of urine calibrators and controls for serum testing has been validated for this procedure. There is no matrix effect.

References

1. Baselt, R.C., *Phencyclidine*, in *Disposition of Toxic Drugs and Chemicals in Man*. 2002, Biomedical Publications: Foster City, pp. 827–830.

2. Fenton, J.J., *Drugs of Abuse*, in *Toxicology: A Case-oriented Approach*. 2002, CRC Press: Boca Raton, pp. 359–401.

3. Porter, W.H., *Clinical Toxicology*, in *Tietz Textbook of Clinical Chemistry and Molecular Diagnostics*, C.A. Burtis, E.R. Ashwood, and D.E. Bruns, Editors. 2006, Elsevier Saunders: St. Louis, MO, pp. 1287–1369.

4. Kwong, T.C., *Introduction to Drug of Abuse Testing*, in *Handbook of Drug Monitoring Methods: Therapeutics and Drugs of Abuse*, A. Dasgupta, Editor. 2008, Humana Press: Totowa, NJ, pp. 297–315.

5. Boeckx, R., False positive EMIT DAU PCP assay as a result of an overdose of dextromethorphan. *Clin Chem*, 1987, **33**: 974–975.

6. Levine, B. and M. Smith, Effects of diphenhydramine on immunoassay of phencyclidine in urine. *Clin Chem*, 1990, **36**: 1258.

7. Long, C., J. Crifasi, and D. Maginn, Interference of thioridazine (Mellaril) in identification of phencyclidine. *Clin Chem*, 1996, **42**: 1885–1886.

8. Kala, S.V., et al., Validation of analysis of amphetamines, opiates, phencyclidine, cocaine, and benzoylecgonine in oral fluids by liquid chromatography-tandem mass spectrometry. *J Anal Toxicol*, 2008, 32(8): 605–611.

9. Coulter, C., K. Crompton, and C. Moore, Detection of phencyclidine in human oral fluid using solid-phase extraction and liquid chromatography with tandem mass spectrometric detection. *J Chromatogr B Analyt Technol Biomed Life Sci*, 2008, **863**(1): 123–128.

Chapter 46

Quantitation of Sirolimus Using Liquid Chromatography-Tandem Mass Spectrometry (LC-MS-MS)

Magdalena Korecka and Leslie M. Shaw

Abstract

A multiple reaction monitoring positive ion HPLC method with tandem mass spectrometric detection (MS-MS) for determination of sirolimus in human blood samples is described. This method utilizes an online cleanup step that provides simple and rapid sample preparation with a switching valve technique. This procedure includes: instrumentation, API 3000 triple quadrupole with turbo-ion spray (Applied Biosystems, Foster City, CA); HPLC system (Agilent Technologies series 1100, Wilmington, DE); two position switching valve (Valco, Houston, TX); 10 mm guard cartridge (C_{18}) used as an extraction column (Perkin Elmer, Norwalk, CT); analytical column (Nova-Pak C_{18} column, 2.1 × 150 mm I.D., 4 μm, Waters Corp, Milford, MA) maintained at 65°C; extraction solution, ammonium acetate (30 mM, pH 5.2), flow rate 1.0 mL/min; eluting solution, methanol:30 mM ammonium acetate buffer (pH 5.2, 97:3 v/v), flow rate 0.8 mL/min with ~ 1/3 of the flow split post-column into the MS-MS; total run-time 3.5 min. Sample preparation is based on simple protein precipitation with a mixture of methanol and zinc sulfate (7:3, v/v) followed by online sample cleanup. This procedure provides a decreased sample preparation time by a factor of four compared to a method that uses an SPE column. The first and third quadrupoles were set to detect the ammonium adduct ion and a high mass fragment of sirolimus (m/z 931.8→864.6), and of an internal standard (ascomycin) (m/z 809.5→756.5). The lower limit of quantification of this method is 2.5 μg/L. The quantification of drug is made from standard curve using peak-area ratio of analyte vs. internal standard. Calibration curve is constructed using non-weighted linear through zero regression.

Key words: Sirolimus, mass spectrometry, online cleanup, switching valve.

1. Introduction

Sirolimus, a macrocyclic antibiotic, is a fermentation product of the actinomycete, *Streptomyces hygroscopicus* and was first isolated from soil samples collected on Rapa Nui (Easter Island) following a search for novel antifungal agents (1). Structurally, sirolimus is a

U. Garg, C.A. Hammett-Stabler (eds.), *Clinical Applications of Mass Spectrometry*, Methods in Molecular Biology 603, DOI 10.1007/978-1-60761-459-3_46, © Humana Press, a part of Springer Science+Business Media, LLC 2010

lipophilic macrocyclic lactone comprised of a 31-membered macrolide ring. It was shown to possess antifungal, antitumor, and immunosuppressive activity in animal-model studies and was subsequently shown effective in preventing acute rejection of renal transplants. Formal clinical trials were undertaken and in 1999, sirolimus was approved for this indication by the US FDA.

The complex of sirolimus and the intracellular immunophilin FK-BP12 modulates the immune response by combining with the specific cell-cycle regulatory protein mTOR and inhibiting its activation. This inhibition results in suppression of cytokine driven T-lymphocyte proliferation, inhibiting the progression from the G_1 to the S phase of the cell cycle (2).

Sirolimus is administered orally as a solution containing a combination of phosphatidyl choline, propylene glycol, monoglycerides, ethanol, soy fatty acids, ascorbyl palmitate, and Polysorbate 80 with a sirolimus concentration of 1 g/L. A 1 mg tablet formulation was also approved but this formulation is not bioequivalent to the oral solution. The two were shown to be clinically equivalent at a 2 mg dose, based on comparable rates of efficacy failure, graft loss, or death. Sirolimus is rapidly absorbed from the gastrointestinal tract with an average time to reach maximal concentration in whole blood of about 2 h (3). Sirolimus distributes extensively into blood cells as reflected by the average 36.5 blood-to-plasma ratio in renal transplant patients. Approximately 95% distributes into red blood cells, 3% in plasma, and 1% each in lymphocytes and granulocytes (4).

The relationship between sirolimus whole blood trough concentrations and efficacy and toxicity has been investigated in renal transplant patients who received concomitant full-dose CsA and corticosteroid therapy. According to these analyses the minimum effective sirolimus concentration below which there is a significant increase in risk for acute rejection is 4–5 µg/L (5). The threshold concentration of 13–15 µg/L was identified above which the risks for the concentration-related side effects, thrombocytopenia ($<100,000$ platelets/mm^3), leukopenia (<4000 leukocytes/mm), and hypertriglyceridemia (>300 mg/dL serum triglycerides) are increased (5). Therapeutic drug monitoring (TDM) of sirolimus is advocated by the drug's manufacturer, since it provides clinically useful prediction of risk for inadequate immunosuppression or possible adverse events.

According to the 2007 CAP survey (6), participating laboratories analyze sirolimus with either microparticle enzyme immunoassay (MEIA) (Abbot IMx) (70% participants) or HPLC with mass spectrometry detection (HPLC-MS) (30%). The reports from the Analytical Services International Ltd. (7) immunosuppressive drug measurement proficiency testing scheme show that 46% of participants use HPLC-MS methodology and 39.2% use immunoassay. The remainder of the participants uses an HPLC method with UV detection. This chapter will present an HPLC method with tandem mass spectrometry detection that is used in our clinical laboratory for sirolimus TDM. In the past few years, high-pressure liquid chromatography with mass spectrometry (HPLC-MS) has been popularly utilized in drug quantitation and pharmacokinetics studies and now is considered to be the gold standard analytical method in therapeutic drug monitoring. The main attraction of HPLC-MS is high selectivity and sensitivity. This technique permits the quantification of the main drug independently from its metabolites. Since the immunosuppressive agents are used in combined regimens a very important advantage of HPLC-MS quantitation methods is the possibility of simultaneous analysis of several compounds in one short run. For four out of five major immunosuppressants (CsA, tacrolimus, sirolimus, and everolimus) the desired specificity and sensitivity can be achieved without chromatographic separation in time, which shortens analysis time, increases throughput, and also makes a big impact on the laboratory budget. The method presented here for the analysis of sirolimus can easily be used for quantification of other immunosupressants in blood samples including tacrolimus, everolimus, and CsA (8–13) with mass transitions m/z 1220→1203 and m/z 1234→1217 for ammonium adducts of CsA and its internal standard CsD (14).

2. Materials

2.1. Samples

1. EDTA anticoagulated whole blood. Stable for up to 2 days when stored at room temperature protected from light (15), up to 14 days at 2–8°C, and up to 3 months at −40°C (1). Samples must be well mixed prior to analysis.

2.2. Supplies

1. Disposable 1.5 mL polypropylene, microcentrifuge tubes
2. Pipettes (with disposable tips) 20, 200, and 1,000 μL
3. Autosampler vials (11 mm, pp, blue snap top) with built-in 0.250 mL insert (Agilent Tech., Wilmington, DE)
4. Volumetric flasks

2.3. Equipment

1. Agilent HPLC system series 1100 (autosampler, two pumps, degasser; Agilent Technologies, Wilmington, DE) with 10-port switching valve (can be six port) (VICI Valco Instruments Co., Houston, TX)
2. Tandem MS: API 3000 (Applied Biosystems, Foster City, CA)

2.4. Chemicals

1. Sirolimus, (Wyeth-Ayerst, Princeton, NJ) (stored at −70°C)
2. Ascomycin (Sigma-Aldrich, St.Louis, MO)
3. Human drug-free EDTA anticoagulated whole blood (stored at −20°C)

2.5. Buffers

1. Extraction solution: 30 mM ammonium acetate buffer pH 5.2. Dissolve 3.32 g of anhydrous ammonium acetate in 1 L of water. Make fresh with each analytical run and adjust pH with acetic acid. Prior to use, filter using a 0.45 μm nylon membrane filter under vacuum.
2. Eluting solution: Mixture of methanol and 30 mM ammonium acetate buffer pH 5.2 (97:3, v/v). Make fresh with each analytical run and filter as described previously.
3. 0.2 M zinc sulfate: Prepare by dissolving 57.5 g of zinc sulfate in 1 L of water. Stable for 1 year when stored at room temperature.
4. Precipitation solution: Mix methanol and 0.2 M zinc sulfate (70:30, v/v). Prepare as needed. Stable for 1 month when stored at room temperature.

2.6. Standards

1. Sirolimus stock solution, 600 mg/L: Prepare by dissolving sirolimus in methanol. Stored at −70°C and covered with aluminum foil is stable for a year.
2. Sirolimus working solution, 5 mg/L: Prepare by diluting stock solution with methanol. Stored at −70°C and covered with aluminum foil is stable for a year.
3. Spiking standard solutions: Prepare by diluting sirolimus working solution with methanol as outlined in **Table 46.1**. This gives sirolimus spiking standard solutions of 125, 250, 625, 1,250, and 2,500 μg/L, which are used to prepare

Table 46.1
Preparation of calibrators. Dilutions are made with methanol

Calibrator (μg/L of blood)	Dilution of working solution (5 mg/L)
2.5	1:40
5.0	1:20
12.5	1:8
25.0	1:4
50.0	1:2

calibrators of the following concentrations in blood: 2.5, 5.0, 12.5, 25.0, and 50.0 µg/L. A blank sample is included into every run. Standard solutions are stored at −20°C and prepared as needed.

4. Internal standard stock solution, 100 mg/L: Prepare by dissolving ascomycin in methanol. This solution must be stored tightly sealed at −20°C and covered with aluminum foil. It is stable for up to a year.

5. Internal standard working solution, 0.5 mg/L: Prepare by diluting the stock solution with methanol. Store at +4°C (*see* **Note 1**).

2.7. Quality Control
Samples

1. Three quality control samples (QCs) are prepared to achieve sirolimus concentrations of 5, 10.0, and 20.0 µg/L. Prepare each by adding sirolimus working solution into 10 mL of drug-free EDTA whole blood as given in **Table 46.2**. Aliquot into 1.5 mL tubes and store at −70°C. Prepare as needed. All three levels of controls are run daily.

Table 46.2
Preparation of in-house controls

Quality control (µg/L)	Working solution (5 mg/L) (µL)	Drug-free whole human blood (mL)
5.0	10.0	9.990
10.0	20.0	9.980
20.0	40.0	9.960

3. Methods

3.1. Stepwise
Procedure

1. Prepare each concentration of calibrator by adding 0.005 mL of each sirolimus spiking standard solution to 0.25 mL of drug-free whole human blood and mix.

2. To each 0.25 mL sample or control add 0.005 mL of methanol.

3. To each calibrator, control, or sample add 1.0 mL of mixture containing precipitation reagent and internal standard 200:1 v/v (1.0 mL precipitation reagent + 0.005 mL of internal standard for each sample).

4. Vortex tubes vigorously for 20 sec. Inspect to make sure that all sediment has been removed from the bottom of the bullet (*see* **Note 2**).

5. Centrifuge the sample for 15 min at $14,000 \times g$

6. Transfer supernatant into autosampler vials. Make sure no particulate matter is transferred (*see* **Note 3**).

7. Place the vials into the autosampler rack.

3.2. Instrument Operating Conditions (see Notes 4 and 5).

This method is based on online cleanup of the injected blood sample extract (injection volume 40 µL) with subsequent introduction into the mass spectrometer by using a Valco switching valve. The set up for the two position switching valve is shown in **Fig. 46.1**. Extraction solution at a flow rate of 1.0 mL/min and an extraction column (Validated-C_{18}, 5 µm; Perkin Elmer, MA) are used for online cleanup. After 1.3 min of cleaning, the switching valve is activated and the sample extract is back flushed from the extraction column onto the analytical column (position B on **Fig. 46.1.**) (Nova-Pak C_{18} column, 2.1 × 150 mm I.D., 4 µm, Waters Corp, Milford, MA) maintained at 65°C. The flow rate of the eluting solution is 0.8 mL/min with an approximately one-third split of the post-column flow to the mass spectrometer.

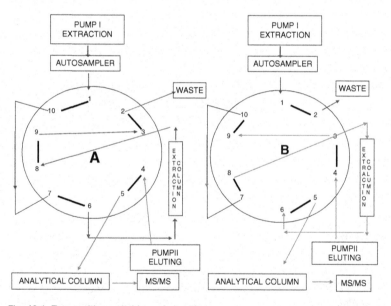

Fig. 46.1. Two position switching valve setting.

After 3.0 min of run the valve switches back the flow of solutions to the initial conditions. Total run time is 3.5 min. There is 0.2 min re-equilibration time before the next injection.

The API 3000 mass spectrometer, with a turbo-ion spray heated at 480°C is operated in the positive ion mode and MRM scan type. The first quadrupole is set to select the ammonium adducts $[M-NH_4]^+$ of sirolimus (*m/z* 931.8) and internal standard (*m/z* 809.5). The second quadrupole is used as a collision cell with

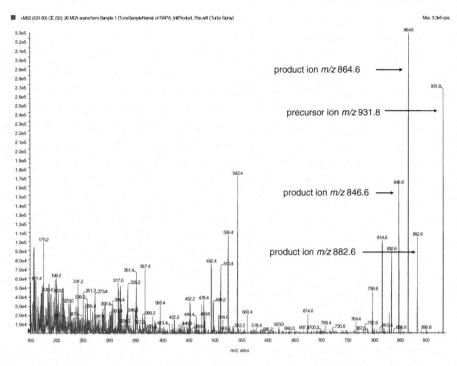

Fig. 46.2. Mass spectrum showing precursor (ammonium adduct) and product ions of sirolimus obtained during direct injection (flow 10 µL/min) of sirolimus (200 µg/L) in methanol with 3% of 20 mM ammonium acetate.

an exit potential of 55 V. The third quadrupole is set to detect the most abundant fragment ion of sirolimus, (m/z 864.4) (**Fig. 46.2**) and that of the internal standard, (m/z 756.5). Declustering and focusing potentials are set at 39 and 270 V, respectively.

3.3. Data Analysis

The data are collected and analyzed using Analyst 1.3 software (Applied Biosystems, Foster City, CA). The quantitation of drug is made from a standard curve using the peak-area ratio of analyte to that of the internal standard. The calibration curve is constructed using non-weighted linear through zero regression. The limit of quantification of the method is 2.5 µg/L and the linearity is up to 50 µg/L. Data from the validation study for this method and ongoing experience in the clinical laboratory show that the inter-day precision does not exceed 10%. Furthermore, no matrix effects or carryover have been observed.

4. Notes

1. The working solution of internal standard provides an internal standard concentration of 10 µg/L of blood.

2. Vortexing the sample after adding the precipitating solution is a crucial step. Vortex first for 10 sec, invert the tube for 2–3 sec, and vortex again for 10 sec. In some cases it may be necessary to sharply rap the tube on the countertop to dislodge material that fails to mix.

3. After transferring the supernatants into the vial inserts, check for the presence of air bubbles and any precipitated protein. Remove any air bubbles in the vials by firmly tapping the vials with fingers. If any particulate material has been transferred, recentrifuge and transfer to a clean vial.

4. The switching valve technique requires two separate pumps since two outlets are needed, for the extraction and eluting solution. A binary pump, since it has only one outlet, is not sufficient to perform the online cleanup.

5. Even small increases in pressure on the extraction column indicate contamination and the necessity of replacing the extraction column.

References

1. Napoli, K.L. and Taylor, P.J. (2001) From beach to bedside: history of the development of sirolimus. *Ther Drug Monit*, **23**, 559–86.

2. Shaw, L.M., Kaplan, B. and Brayman, K.L. (2000) Advances in therapeutic drug monitoring for immunosuppressants: a review of sirolimus. Introduction and overview. *Clin Ther*, **22 Suppl B**, B1–B13.

3. Zimmerman, J.J. and Kahan, B.D. (1997) Pharmacokinetics of sirolimus in stable renal transplant patients after multiple oral dose administration. *J Clin Pharmacol*, **37**, 405–15.

4. Yatscoff, R., LeGatt, D., Keenan, R. and Chackowsky, P. (1993) Blood distribution of rapamycin. *Transplantation*, **56**, 1202–6.

5. Kahan, B.D., Napoli, K.L., Kelly, P.A., Podbielski, J., Hussein, I., Urbauer, D.L., Katz, S.H. and Van Buren, C.T. (2000) Therapeutic drug monitoring of sirolimus: correlations with efficacy and toxicity. *Clin Transplant*, **14**, 97–109.

6. Klee, G. (2007) Surveys 2007. College of Ammerican Patologists.

7. Scheme, I.P.T. http://www.bioanalytics.co.uk.

8. Volosov, A., Napoli, K.L. and Soldin, S.J. (2001) Simultaneous simple and fast quantification of three major immunosuppressants by liquid chromatography – tandem mass-spectrometry. *Clin Biochem*, **34**, 285–90.

9. Streit, F., Armstrong, V.W. and Oellerich, M. (2002) Rapid liquid chromatography-tandem mass spectrometry routine method for simultaneous determination of sirolimus, everolimus, tacrolimus, and cyclosporin A in whole blood. *Clin Chem*, **48**, 955–8.

10. Christians, U., Jacobsen, W., Serkova, N., Benet, L.Z., Vidal, C., Sewing, K.F., Manns, M.P. and Kirchner, G.I. (2000) Automated, fast and sensitive quantification of drugs in blood by liquid chromatography-mass spectrometry with on-line extraction: immunosuppressants. *J Chromatogr B Biomed Sci Appl*, **748**, 41–53.

11. Wang, S., Magill, J.E. and Vicente, F.B. (2005) A fast and simple high-performance liquid chromatography/mass spectrometry method for simultaneous measurement of whole blood tacrolimus and sirolimus. *Arch Pathol Lab Med*, **129**, 661–5.

12. Kirchner, G.I., Vidal, C., Jacobsen, W., Franzke, A., Hallensleben, K., Christians, U. and Sewing, K.F. (1999) Simultaneous on-line extraction and analysis of sirolimus (rapamycin) and ciclosporin in blood by liquid chromatography-electrospray mass spectrometry. *J Chromatogr B Biomed Sci Appl*, **721**, 285–94.

13. Annesley, T.M. and Clayton, L. (2004) Simple extraction protocol for analysis of immunosuppressant drugs in whole blood. *Clin Chem*, **50**, 1845–8.

14. Keevil, B.G., Tierney, D.P., Cooper, D.P. and Morris, M.R. (2002) Rapid liquid chromatography-tandem mass spectrometry method for routine analysis of cyclosporin A over an extended concentration range. *Clin Chem*, **48**, 69–76.

15. Holt, D.W., Lee, T. and Johnston, A. (2000) Measurement of sirolimus in whole blood using high-performance liquid chromatography with ultraviolet detection. *Clin Ther*, **22 Suppl B**, B38–B48.

Chapter 47

Quantitation of Tacrolimus in Whole Blood Using High Performance Liquid Chromatography-Tandem Mass Spectrometry (HPLC-MS-MS)

Keri J. Donaldson and Leslie M. Shaw

Abstract

We describe a multiple reaction monitoring positive ion HPLC/tandem mass spectrometric method for quantification of tacrolimus in human whole blood with online extraction and cleanup. Included in this procedure: API 2000 triple quadrupole mass spectrometer with turbo-ion spray source (Applied Biosystems, Foster City, CA); 10-port diverter/switching valve (Valco, Houston, TX); HPLC system (Agilent Technologies series 1100, Wilmington, DE); 10 mm (C_{18}) guard cartridge (Perkin Elmer, Norwalk, CT) used as an extraction column; a Nova-Pak C18 analytical column (2.1 × 150 mm I.D., 4 μm, Waters Corp, Milford, MA); washing solution, methanol: 30 mM ammonium acetate pH 5.1 (80:20); eluting solution, methanol:30 mM ammonium acetate pH 5.1 (97:3); flow rate 0.8 mL/min; and a run-time of 2.8 min. The first and third quadrupoles were set to detect the ammonium adduct ion and a high mass fragment of tacrolimus (m/z 821.5→768.3), and of an internal standard (ascomycin) (m/z 901.8→834.4). The lower limit of quantification of this method is 3.75 mg/L. The concentration of drug is determined by comparing peak-area ratios for tacrolimus and internal standard to a standard curve constructed using non-weighted linear through zero regression.

Key words: Tacrolimus, mass spectrometry, liquid chromatography, high pressure liquid chromatography, therapeutic drug monitoring.

---

1. Introduction

Tacrolimus "Tsukuba macrolide immunosuppressant" (FK-506, Fujimycin, Prograf, Avagraf, Protopic) is a naturally occurring macrocyclic lactone isolated from *Streptomyces tsukubaensis* (discovered near Mt. Tsukuba, Ibaraki, Japan) in 1984 (1, 3). In vivo, tacrolimus reduces peptidyl–prolyl isomerase activity by binding to

U. Garg, C.A. Hammett-Stabler (eds.), *Clinical Applications of Mass Spectrometry*, Methods in Molecular Biology 603,
DOI 10.1007/978-1-60761-459-3_47, © Humana Press, a part of Springer Science+Business Media, LLC 2010

the immunophilin FKBP-12 (FK506 binding protein) creating a new complex. The newly formed FKBP12–FK506 complex then forms a pentameric complex with calmodulin and calcineurins A and B; which inhibits calcineurin's phosphatase activity. Transcription factors requiring dephosphorylation for nuclear transport are functionally inhibited by their cytoplasmic localization. The decreased expression of IL-2, IL-3, IL-4, IL-5, IFNγ, GM-CSF, and TNF-α causes immunosuppression (1, 4). The use of tacrolimus as an immunosuppressive agent in the setting of allogenic organ transplantation was first described in 1989 (5). In April 1994, the FDA approved the drug for use in liver transplantation; this has been extended to include kidney, heart, pancreas, lung, small bowel, trachea, skin, cornea, bone marrow, and limb transplants.

The pharmacokinetics of the absorption and elimination of tacrolimus are variable with peak blood or plasma concentrations being reached in 0.5–6 h with a mean bioavailability of the oral dose of 17–22% (range 4–89%). Tacrolimus is extensively bound to red blood cells with a mean blood-to-plasma ratio of about 15:1 (range 15:1–45:1). Albumin and α1-acid glycoprotein appear to primarily bind tacrolimus in plasma (PB 72–98.8%). Tacrolimus is >99.5% metabolized prior to elimination. The mean disposition half-life is 12 h (range 3.5–40.5 h) and the total body clearance based on blood concentration is approximately 0.06 L/h/kg (0.041+/− 0.036 L/h/kg), mainly in bile (6, 7). The elimination of tacrolimus is decreased in the presence of liver impairment and in the presence of several drugs. Because of this pharmacokinetic variability combined with a narrow therapeutic index of tacrolimus and the potential for several drug interactions, careful monitoring of tacrolimus blood concentrations is needed for the optimization of therapy and dosage regimen design, based upon therapeutic ranges (*see* **Table 47.1**).

In the 2009 CAP survey (2), participating laboratories analyze tacrolimus with either microparticle enzyme immunoassay (MEIA) (Abbot IMx, Abbot Architect, Siemens Dimension, Siva EMIT 2000) (88% participants) or high-pressure liquid chromatography with mass spectrometry detection (HPLC-MS) (12%) (2). This chapter will present an HPLC based method with tandem mass spectrometry detection that is used in our clinical laboratory for tacrolimus total drug monitoring. The appeal HPLC-MS in drug quantitation and pharmacokinetics studies is the methods' high selectivity and sensitivity. An added advantage of HPLC-MS is that it permits the quantification of the main drug independently from its metabolites, and it is possible to simultaneously analyze several compounds in one short run.

Table 47.1
Therapeutic range and critical values

Renal transplant patients	
Post-transplant interval	Therapeutic range*
0–3 months	8–10 ug/L
4–6 months	6–8 ug/L
6–12 months	5–7 ug/L
>12 months	4–6 ug/L
Heart transplant patients	
Post-transplant interval	Therapeutic range*
0–3 months	10–12 ug/L
3–6 months	10 ug/L
6–12 months	
>12 months	5–8 ug/L
Liver transplant patients	
Post-transplant interval	Therapeutic range*
1st week	10–20 ug/L
1–4 weeks	5–15 ug/L
>4 weeks	5–10 ug/L
Lung transplant patients	
Post-transplant interval	Therapeutic range*
0–12	8–12 ug/L
>12	6–8 ug/L
Bone marrow transplant patients	
Therapeutic range* 5–15 ug/L *-peripheral whole blood concentration	
Critical Value >20 ug/L	

2. Materials

2.1. Samples

1. EDTA anticoagulated whole blood. Samples must be well mixed prior to analysis. Samples should be stored at 28°C for 1 week after completion of analysis.

2.2. Supplies

1. Disposable 1.5 mL polypropylene, microcentrifuge tubes
2. Pipettes (with disposable tips) 20, 200, and 1,000 µL
3. Autosampler vials (11 mm, pp, blue snap top) with built-in 0.250 mL insert (Agilent Tech., Wilmington, DE)
4. Volumetric flasks

2.3. Equipment

1. API 2000 triple quadrupole mass spectrometer with turbo-ion spray (Applied Biosystems, Foster City, CA), with switching valve

2. Agilent HPLC system series 1100 (autosampler, two pumps, degasser; Agilent Technologies, Wilmington, DE) with 10-port switching valve (can be six port) (VICI Valco Instruments Co., Houston, TX)

3. Nova-Pak C18 (Perkin Elmer, Norwalk, CT) analytical and guard column

2.4. Reagents and Buffers

1. Ammonium Acetate (NH4-Ac) Solution, 30 mM, pH 5.2: Add 4.64 g of NH4-Ac to a 2,000 mL flask. While swirling, add deionized water up to the neck of the flask. Bring the pH of the solution to 5.2 by adding 2 mL of concentrated acetic acid. Fill the flask to the mark with deionized water and mix.

2. Mobile Phase #1 ("Extraction Solution"), 80% NH4-Ac, 20% Methanol: To a 2,000 mL flask, add 400 mL of MeOH and slowly add 1,600 mL of the above prepared NH4-Ac. solution. Cover and let cool. Filter with a 0.45 uM filter.

3. Mobile Phase #2 ("Elution Solution"), 3% NH4-Ac, 97% Methanol: To a 1,000 mL flask, add 30 mL of the above prepared NH4-Ac solution. Slowly add 970 mL methanol while mixing. Cover. No cooling is necessary. Filter with 0.45 uM filter.

4. Needle Wash Solution, 50% H_2O, 50%MeOH: Combine 500 mL of DI H_2O with 500 mL MeOH. Mix well, cover, and let cool.

5. Wash Solution (for Shutdown), 30% H_2O, 70% MeOH: Combine 300 mL of H_2O with 700 mL of MeOH. Prepare two 1,000 mL bottles – one for each of the #1 and #2 mobile phase positions. Cover and let cool. Filter as above.

6. 0.2 M Zinc Sulfate Solution. Add 15 g of $ZnSO4.7H_2O$ to a 250 mL flask. With mixing, add H_2O to the 250 mL mark.

7. Precipitation Solution. 70% MeOH, 30% 0.2 M $ZnSO_4$. Add 30 mL of the above Zinc Sulfate solution to a 100 mL flask. While mixing, fill to the mark with MeOH.

8. Let cool.

2.5. Standard Solutions

1. Tacrolimus Stock Standard (10 µg/mL in methanol): Dissolve 1 mg in 100 mL volumetric flask. Solution stable when stored at −20°C for up to 1 year.

2. Tacrolimus Standard #4 in Whole Blood (30.0 ng/mL): Add 150 uL of Tacrolimus Stock to a 50 mL flask. While mixing, add negative whole blood to the 50 mL mark of the flask and store at 4°C.

3. Std #3 (15.0 ng/mL): Dilute the #4 std X2 with whole blood as diluent and mix.

4. Std #2 (7.50 ng/mL): Dilute the #3 std X2 with whole blood as diluent and mix.

5. Std #1 (3.75 ng/mL): Dilute the #2 std X2 with whole blood as diluent. Mix solution with methanol (1:100). Stable for 6 months when stored at -20°C.

2.6. Working Calibrators and Internal Standard

1. Prepare working calibrators in whole blood, aliquot, and freeze at -20°C. Calibrators are stable for 6 months when stored frozen. The aliquot in use is stored refrigerated and is stable for 1 month.

2. Internal Standard Stock Solution: Ascomycin $-$ 1 mg/10 mL Methanol. Stable for up to 1 year when stored at -20°C.

3. Working Internal Standard solution (Ascomycin) 100 ng/mL: Add 100 uL of stock to 100 mL of 50/50 MeOH/H_2O. Store at 4°C. Stable for up to 1 year.

2.7. Quality Controls

1. Bio-Rad Lyphochek® Level 1, Level 2, and Level 3 Whole Blood Controls (Irvine, CA) – Reconstitute with 2 mL of type 1 distilled water. Controls are good for 1 week when stored refrigerated. Equilibrate for at least 15 min before mixing thoroughly.

2. Run all three levels of controls each day after calibration, alternating controls with each batch of specimens that are set up throughout the day.

3. Methods

3.1. Stepwise Procedure

1. Remove the following items from the refrigerator: 50 ml flask of Tacrolimus #4 Blood standard, tube of Tacrolimus internal std, Bio-Rad Lyphochek® Controls and Negative blood.

2. To 2 mL microcentrifuge tubes, add 800 uL of precipitation reagent.

3. Add 40 ul of working internal std to each.

4. Add 200 uL of well-mixed whole blood standard, control, or patient sample.

5. After every two blood additions, stop and vortex the two tubes vigorously for 30 sec making sure in the process that all sediment has been removed from the bottom of the bullet (*see* **Note 1**).

6. Preset the refrigerated microcentrifuge with the following settings: 8,000 $\times g$, 4°C, 10 min. Equilibrate to temperature before use.

7. Centrifuge using the above settings.

8. Using disposable plastic micropipettes transfer the supernatant to clear Agilent MS 1.5 mL vials without inserts. Make sure no particulate matter is transferred (*see* **Note 2**).

9. Place the vials in the PE 200 autosampler of LC-MS.

10. Before patient samples, run a negative blood, the three Bio-Rad Lyphochek® whole blood controls. Each batch should contain a control at the beginning and one at the end.

3.2. Instrument Operating Conditions/ Setup

1. LC-MS conditions: *See* **Table 47.2**
   Set the column heater temperature to 65°C

Table 47.2
HPLC MS-MS conditions

Flow gradient	extracting solvent 1.0–3.0 mL/min	eluting solvent 0.5–1.0 mL/min
Injection volume	100 μL	
Run time	3.3 min	
Detection	Tandem MS	
MRM monitoring	positive ionization mode	
Declustering potential	39 V	
Focusing potential	260 V	
Interface heater temperature	450°C	
Collision cell exit potential	55 V	
Nebuliser, curtain, collison gas	Nitrogen	

2. HPLC MS-MS conditions:
   HPLC column: 2 guard cartridges: Extraction Column – 4.6 × 12.5 mm, 5 μm, Agilent #820950-926 Analytical Column – 2.1 × 150 mm, 4 μm, Waters #WAT023655.

   Switching valve set up: (*see* **Fig. 47.1**)
   0.0 min position A
   2.0 min position B
   3.2 min position A

   connections: pump A (washing solvent) to 3 (*see* **Notes 3** and **4**)
   from 2 to column guard
   from column guard to 5
   from 4 to waste

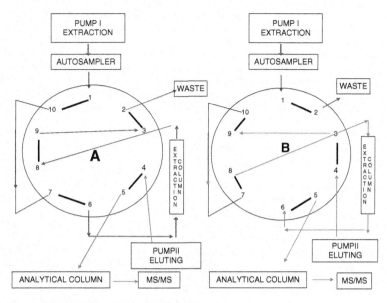

Fig. 47.1. Two position switching valve setting.

pump B (eluting solvent) to 6
from 6 to 7 – loop to 1
from 1 to MS-MS
9–8 – idle

Position A: 2–3	Position B:1–2
4–5	3–4
7–6	5–6
1–10	10–9
9–8	8–7

3.3. Data Analysis

1. Data are collected and analyzed using Analyst 1.3 software (Applied Biosystems, Foster City, CA).

2. Peak-area ratio for Tacrolimus (m/z 821.5 → 768.3) and internal standard (m/z 809.5 → 756.5) were used for quantification. Standard curve is constructed using non-weighted linear through zero regression. *See* **Fig. 47.2** for positive ion mass of ammonium adduct of Tacrolimus.

3. Retention time for internal standard and Tacrolimus is approximately 2.8 min.

4. Each run should be evaluated to assure that the internal standard peak area for all samples is within ± one factor of 10. The assay is linear from 1 to 30 ng/mL.

5. Results >30 ng/mL should be diluted with negative whole blood control and rerun. Multiply result by appropriate dilution factor. Results less than 1 should be reported as "<1".

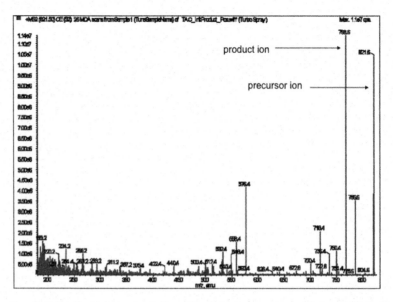

Fig. 47.2. Positive ion mass spectrum showing precursor (ammonium adduct) and product ions of TAC obtained on API 5000(Turbo Spray) during infusion of TAC (200 µg/L), in MeOH with 3% of 20 mmol/L ammonium acetate, flow rate 10 µL/min.

4. Notes

1. Vortexing the sample after adding the precipitating solution is a crucial step. Vortex first for 10 sec, invert the tube for 2–3 sec, and vortex again for 10 sec. In some cases it may be necessary to sharply rap the tube on the countertop to dislodge material that fails to mix.

2. After transferring the supernatants into the vial inserts, check for the presence of air bubbles and any precipitated protein. Remove any air bubbles in the vials by firmly tapping the vials with fingers. If any particulate material has been transferred, re-centrifuge and transfer to a clean vial.

3. The switching valve technique requires two separate pumps, since two outlets are needed, for the extraction and eluting solution. A binary pump, since it has only one outlet, is not sufficient to perform the online cleanup.

4. Even small increases in pressure on the extraction column indicate contamination and the necessity of replacing the extraction column.

References

1. Kino T, Hatanaka H, Hashimoto M, Nishiyama M, Goto T, Okuhara M, Kohsaka M, Aoki H, Imanaka H. (1987). FK-506, a novel immunosuppressant isolated from a Streptomyces. I. Fermentation, isolation, and physico-chemical and biological characteristics. *J Antibiot* 40: 1249–55.

2. Magnani B, et al. (2009). Surveys 2009. *College of Ammerican Patologists.*

3. Ponner B, Cvach B (Fujisawa Pharmaceutical Co.). (2005). *Protopic Update.*

4. Tamura K, Fujimura T, Iwasaki K, Sakuma S, Fujitsu T, Nakamura K, Shimomura K, Kuno T, Tanaka C, Kobayashi M. (1994). Interaction of tacrolimus(FK506) and its metabolites with FKBP and calcineurin. *Biochem. Biophys. Res. Commun*: 437–443.

5. Starzl TE, Fung J, Venkataramman R, Todo S, Demetris AJ, Jain A. (1989). FK506 for liver, kidney and pancreas transplantation. *Lancet* ii: 1000–1004.

6. Venkataramanan R, Swaminathan A, Prasad T, Jain A, Zuckerman S, Warty V, McMichael J, Lever J, Burckart G, Starzl T. (1995). Clinical pharmacokinetics of tacrolimus. *Clin Pharmacokinet* 29: 404–30.

7. Wallemacq P, Verbeeck R. (2001). Comparative clinical pharmacokinetics of tacrolimus in paediatric and adult patients. *Clin Pharmacokinet* 40: 283–295.

8. Volosov A, Napoli KL, Soldin SJ (2001). Simultaneous simple and fast quantification of three major immunosuppressants by liquid chromatography tandem mass-spectrometry. *Clin Biochem* 34: 285–290.

Chapter 48

Analysis of Testosterone in Serum Using Mass Spectrometry

Robert L. Fitzgerald, Terrance L. Griffin, and David A. Herold

Abstract

For either gas chromatography mass spectrometry (GC-MS) or liquid chromatography tandem mass spectrometry (LC-MS-MS) methods, the first step in the analysis is to add a deuterium-labeled internal standard such as testosterone-16,16,17-d$_3$. Testosterone in the sample is then isolated by liquid–liquid extraction and the extract is dried under a stream of nitrogen. For the GC-MS method we describe; the residue is transformed to the pentafluorobenzyl/trimethylsilyl derivative and is injected into the GC-MS, separated on a dimethypolysiloxane column, and ionized using electron capture negative chemical ionization (ECNCI). Quantification of testosterone in the samples is by selected ion monitoring, measuring peak ratios of testosterone relative to the deuterium-labeled internal standard. For the LC-MS-MS analysis of testosterone, the sample extract is reconstituted in mobile phase, injected on a C18 column, and quantified using multiple reaction monitoring of testosterone relative to the internal standard. There are no interferences from common steroids found in human serum. For both methods the run-to-run precision and accuracy is generally less than 6% and the methods are linear from 5 to 2000 ng/dL.

Key words: Testosterone, mass spectrometry, mass spectrometry/mass spectrometry, gas chromatography, liquid chromatography.

1. Introduction

Testosterone was first characterized in 1935 when Laqueur isolated about 10 mg of a compound that had a melting point of 154°C from 100 kg of steer testis tissue (1). From this point forward, scientists were interested in quantifying the amount of testosterone in various specimens for clinical purposes.

Today many laboratories use more sensitive immunoassay based methods, such as chemiluminescent and electrochemical techniques, combined with automation to perform testosterone measurements. These assays are useful for the diagnosis of hyper- and hypo-androgen states in men, but are often less than optimal

U. Garg, C.A. Hammett-Stabler (eds.), *Clinical Applications of Mass Spectrometry*, Methods in Molecular Biology 603, DOI 10.1007/978-1-60761-459-3_48, © Humana Press, a part of Springer Science+Business Media, LLC 2010

when analyzing specimens from women and infants (2, 3). The antigen-antibody reactions are relatively nonspecific compared to chromatography linked mass spectrometric techniques, and immunoassays often require extensive sample preparation in order to obtain reliable results. Even immunoassays that utilize an extraction step, followed by a chromatographic separation, and then immunoassay suffer from accuracy and precision problems (4, 5). A 2007 endocrine society position statement summarizes many of the limitations of using immunoassays to measure testosterone (6). In fact, it has been suggested that when testing specimens from women, guessing testosterone concentrations can be more accurate than using some immunoassays (7).

Mass spectrometry is poised to make major improvements in how clinical laboratories measure a variety of compounds, including testosterone. A recent comparison of GC-MS and LC-MS-MS showed fairly good accuracy and precision over a wide range of concentrations (8). Although mass spectrometry assigned values have some variability, the precision and accuracy in the mass spectrometry based measurements are far superior to those of the immunoassays. Below are some general considerations as well as detailed methods for GC-MS and LC-MS-MS analyses of testosterone.

Testosterone requires extraction and derivatization prior to GC-MS analysis. Some descriptions include complicated extraction protocols followed by column chromatography plus derivatization prior to injection. In our laboratory, we initially developed a simple ethyl acetate extraction and pentafluorobenzyl/trimethylsilyl derivatization for GC-MS analysis; however we have since implemented a simple organic extraction without derivatization for LC-MS-MS analysis.

Some of the more common testosterone derivatives are shown in **Fig. 48.1**. When derivatizing testosterone, one must be careful because the ketone at the 3-position exists as both keto and enol tautomers. The enol form contains active hydrogen that will bond with GC columns causing poor chromatography. This tautomerization can also cause multiple products to be formed during derivatization reactions. The t-butyldimethylsilyl derivative derived by reacting testosterone with N-methyl-[N-tert-butyltrimethylsilyl]-trifluoroacetamide (MTBSTFA) under anhydrous conditions forms a single stable t-butyl dimethylsilyl derivative (9). We developed the first GC-MS method capable of quantifying testosterone suitable for the range of concentrations expected for females using electron capture negative chemical ionization with a pentafluorobenzyl-trimethylsilyl derivative (2). We have used this derivative on thousands of patient samples. In our hands, it forms a derivative with excellent chromatographic and electron capturing properties and is stable for weeks. **Section 3** contains a description of the extraction and derivatization protocol used in our laboratory for GC-MS analysis.

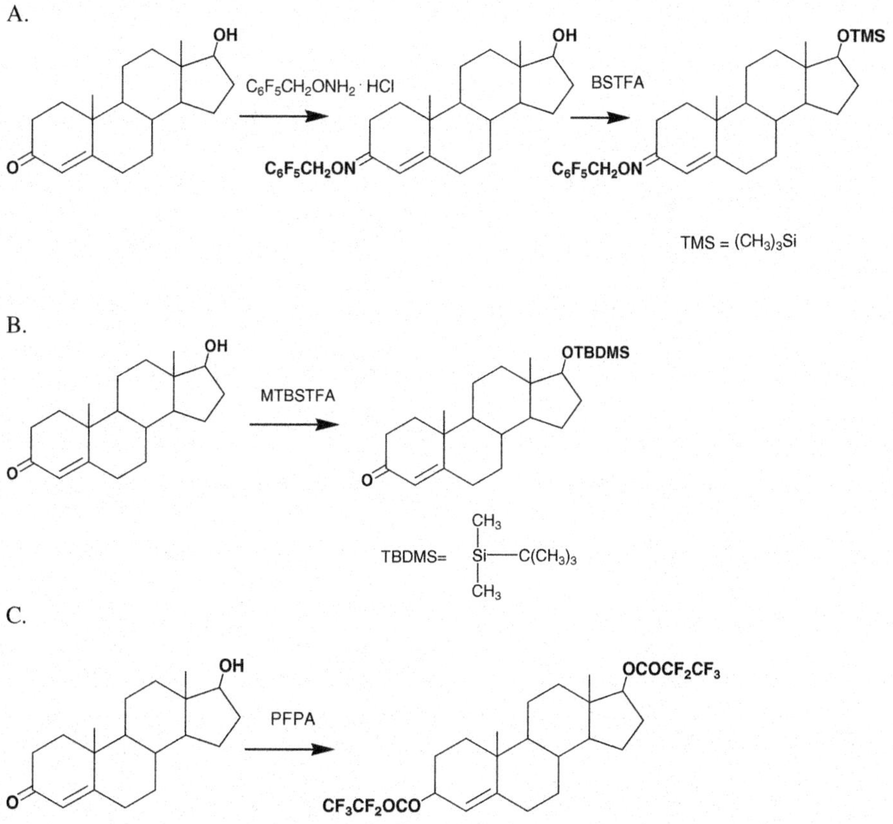

A.

$C_6F_5CH_2ONH_2 \cdot HCl$

$C_6F_5CH_2ON$

BSTFA

$C_6F_5CH_2ON$

$TMS = (CH_3)_3Si$

B.

MTBSTFA

$$TBDMS= \quad \begin{array}{c} CH_3 \\ | \\ Si-C(CH_3)_3 \\ | \\ CH_3 \end{array}$$

C.

PFPA

CF_3CF_2OCO

$OCOCF_2CF_3$

Fig. 48.1. Common derivatives used in GC-MS analysis of testosterone. **(A)** Sequential reaction of testosterone to form the pentafluorobenzyloxime-trimethylsilyl derivative used for negative chemical ionization GC-MS analysis. **(B)** Formation of the t-butyldimethylsilyl (TBDMS) derivative. **(C)** Formation of the pentafluoroproprionic anhydride derivative.

A high purity internal standard is essential for the accurate quantification of testosterone using mass spectrometry. Several internal standards have been utilized, but the more commonly used internal standard testosterone-16,16,17-d_3 is shown in **Fig. 48.2**.

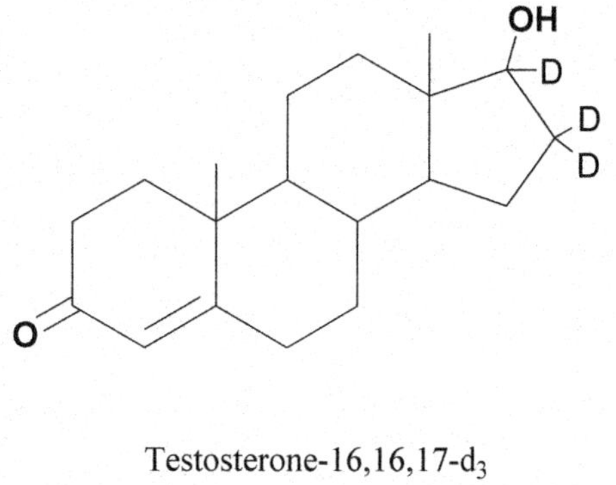

Testosterone-16,16,17-d_3

Fig. 48.2. Deuterium-labeled testosterone.

It should be noted that $(3,4\text{-}^{13}C_2]$ testosterone has also been employed. The advantage of using the d_3 internal standard is that the mass shift is sufficient to eliminate any problems caused by A+2 effects when using silyl derivatives. Generally, it is best to add sufficient internal standard to get good signal-to-noise for accurate peak-area quantification, while limiting the amount added to concentrations similar to those expected in patient specimens. Testosterone-16,16,17-d_3 solutions are stable when stored in HPLC-grade methanol for a period of at least 2 years.

There have been several recent publications using LC-MS-MS for the analysis of testosterone (10, 11). The types of ionization most often used are electrospray (ESI), atmospheric pressure chemical ionization (APCI), and atmospheric pressure photo ionization (APPI). Kushnir et al. compared the overall sensitivity of these three ionization modes for the analysis of underivatized testosterone and concluded that while there was little difference in signal between the three, APPI had the best signal-to-noise ratio. Because of insufficient signal to quantify testosterone in samples from women and children, they derivatized testosterone to form the hydroxylamine in order to improve ionization in the ESI mode (12). We have recently developed a simplified LC-MS-MS analysis of testosterone that eliminates the need for derivatization by taking advantage of the higher sensitivity of an Applied Biosystems API 4000 Q trap mass spectrometer.

Our initial validation studies included experiments with underivatized testosterone as well as the hydroxylamine derivative, using both electrospray ionization (ESI) and atmospheric pressure chemical ionization (APCI). Our conclusion was that ESI was more sensitive than APCI and that the underivatized testosterone gave the best results, although there was not a large difference between the sensitivity of the methods. It should be noted that we did not optimize the LC conditions for the APCI mode and other LC conditions could potentially provide better results for APCI. The LC-MS-MS method is described in **Section 4**.

2. Materials

2.1. Samples

1. Serum or plasma. Store at $-20°C$. (*see* **Note 1**).

2.2. Reagents

1. LC extraction solution ethyl acetate/hexane (3/2, v/v). Add 600 mL of ethyl acetate to 400 mL of hexane and mix by inversion. Stable for 1 year when stored at room temperature.

2. Mobile phase A: To 1 L of deionized water, add 0.63 g of ammonium formate and 0.5 mL of formic acid. Stable for 2 months at room temperature.

3. Mobile phase B: To 1 L of methanol, add 0.63 g of ammonium formate (0.01 mol) and 0.5 mL of formic acid. Stable for 6 months at room temperature.

4. LC-MS-MS reconstitution solvent, 50:50 methanol:water with 0.1% formic acid: Prepare by mixing 50 mL of water, 50 mL of methanol, and 100 uL of concentrated formic acid. Stable for 6 months at room temperature.

5. Pyridine, 99.9+%, HPLC grade.

6. Florox reagent (O-(2,3,4,5,6-pentaflurobenzyl hydroxylamine. HCl), 2.5 g/L in pyridine.

7. Ethyl acetate (HPLC-grade).

8. Derivatizing reagent: N,O-bis(trimethylsilyl)trifluoracetamide (BSTFA) with 10 mL/L trimethylchlorosilane.

2.3. Calibrators

1. Weigh out 10 mg of testosterone into a 10 mL class A volumetric flask, fill to mark with methanol (1 mg/mL or 1000 ng/uL). Make serial dilutions (1:10) to obtain standards at 100, 10, 1, and 0.1 ng/uL in methanol using class A volumetric flasks (*see* **Notes 2** and **3**).

2. Each day of analysis, prepare calibrators at following concentrations:
   a. 1,000 ng/dL: Add 10 uL of 1 ng/uL to a blank extraction tube
   b. 500 ng/dL: Add 50 uL of 0.1 ng/uL to a blank extraction tube
   c. 100 ng/dL: Add 10 uL of 0.1 ng/uL to a blank extraction tube
   d. Add 1,000 uL of water to each standard tube.

2.4. Internal Standards and Quality Control Samples

1. Testosterone-16,16,17-d_3 (MSD Isotopes), dilute with methanol to concentration of 0.1 ng/uL.

2. Four levels of Immunoassay Control Serum (Bio-Rad Lyphochek or equivalent), One targeting ~20 ng/dL should be included in each analytical run (*see* **Note 3**).

2.5. GC-MS Supplies and Equipment

1. GC-MS with negative chemical ionization (methane) capabilities

2. GC-MS column: 15 m × 0.32 mm (i.d.) cross-linked dimethyl polysiloxane (DB-1) with a 0.25 μm film thickness (J&W Scientific, Folsom, CA)

3. Carrier gas: ultrahigh purity helium (99.999%; Airgas, Radnor, PA)

4. Methane (research grade 4.0, 99.99%; Airgas)

5. Tuning solution: Perfluorotributylamine

2.6. LC-MS-MS Supplies and Equipment

1. Applied Biosystem 4000 LC-MS-MS.

2. LC-MS-MS column: Agilent Zorbax Eclipse XDB-C18 (4.6 × 150 mm with a 5 u particle size) and Phenomenex Security Guard Cartridge C18 (4 × 2.0 mm).

3. Multi-tube vortex.

4. Heater/sample concentrator equipped with nitrogen.

5. 100 × 13 mm glass tubes with Teflon lined caps.

3. Methods

3.1. Stepwise Procedure for Negative Chemical Ionization GC-MS Analysis of Testosterone

1. Label a 13 × 100 mm tube for each calibrator, control, and patient sample.

2. Add 50 uL of the 0.1 ng/uL testosterone-16,16,17-d_3 (in methanol) internal standard to each tube.

3. Add 1 mL of each sample, control, or calibrator and vigorously vortex mix for 10 sec.

4. Add 3 mL of ethyl acetate to all specimens. Cap specimens.

5. Vigorously vortex mix for 40 sec.

6. Centrifuge at 1,850 × g for 10 min.

7. Transfer the organic layer to a clean, labeled 100 × 13 mm screw-cap test tube and evaporate to dryness at 40°C under a gentle stream of nitrogen.

8. Add 50 uL of pyridine and 50 uL of Florox reagent. Cap tightly.

9. Heat at 70°C for 1 h.

10. Evaporate to dryness at 40°C under a gentle stream of nitrogen.

11. Add 50 uL of ethylacetate and 50 uL of derivatizing reagent (BSTFA with 10 mL/L trimethylchlorosilane).

12. Cap tightly and heat for 20 min at 70°C.

13. Inject 1 uL into GC-MS (*see* **Section 3.2**).

3.2. GC-MS Conditions

1. This analysis is run on a Thermo TSQ7000 GC-MS-MS instrument using the single quadrupole mode for analysis. The initial GC temperature is 160°C (1 min hold) followed by a temperature program to 280°C at 20°C per min, where it is held for 3 min. The injection port temperature is 260°C and the interface heater is 280°C. The GC is operated in the splitless mode for the first 0.7 min and then switched to a 1:23 split flow. The injection port sweep flow is 1.25 mL/min.

2. Methane (research grade 4.0, 99.99%; Airgas) is used as the reagent gas for ECNCI at a source pressure of 2500 ± 300 mTorr. Perfluorotributylamine is used to optimize electron capture negative chemical ionization (ECNCI) tuning at m/z 414 and 633. The mass spectrometer was operated in the ECNCI mode, monitoring ions at m/z 535.2 ± 0.5 and m/z 538.2 ± 0.5 for testosterone and testosterone16,16,17-d_3, respectively. Dwell times of 0.2 sec were used for each ion. The emission current was 0.3 mA and 70 eV electrons were used. Typical electron multiplier setting was 1500 V.

3. Testosterone concentrations are quantified by comparing peak-area ratios of the 535/538 m/z ion chromatograms. A typical chromatograph from a female specimen containing 6 ng/dL of testosterone is shown in **Fig. 48.3**.

4. Limit of quantification of this method is 5 ng/dL.

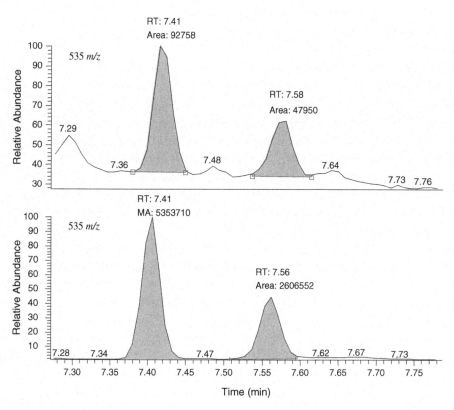

Fig. 48.3. Selected ion monitoring ECNCI chromatographic tracing from a female specimen using the procedure described by Fitzgerald and Herold. This patient specimen had 6 ng/dL of testosterone. The two epimers of the testosterone pentafluorobenzyloxime are shown at 7.41 and 7.58 min. Quantification is performed using the more intense peak at 7.41 min.

3.3. Stepwise Procedure for LC-MS-MS Analysis of Testosterone

1. Label sufficient 13×100 mm tubes for each calibrator, control, and sample. Add 50 uL of the 0.1 ng/uL testosterone-16,16,17-d_3 internal standard to each tube.

2. Add 1 mL of each calibrator, control, and sample to the appropriate tube containing the internal standard.

3. Vigorously vortex mix for 10 sec.

4. Add 3 mL of 3:2 ethyl acetate:hexane to all specimens. Cap specimens.

5. Vortex mix vigorously for 45 sec. Note that it is important that the sample vortex mixes and does not shake as an emulsion will form if the sample is shaken, but rapid vortex mixing yields two well-separated phases after centrifugation. If an emulsion forms, the sample must be prepared again.

6. Centrifuge at $1,850 \times g$ for 10 min.

7. Transfer the organic layer into a 100×13 mm screw-cap test tube and evaporate to dryness at 40°C under a gentle stream of nitrogen.

8. Reconstitute the samples with 100 µL of the water:methanol reconstitution solvent.

9. Inject 2 uL into the LC-MS-MS system.

3.4. LC-MS-MS Conditions and Data Analysis

1. Specimens are analyzed using an isocratic method with 7.5% mobile phase A and 92.5 mobile phase B.

2. The LC run time is 7 min. Full scan MS-MS spectra of underivatized testosterone are shown in **Fig. 48.4**. An example chromatogram from a female specimen near the limit of quantification is shown in **Fig. 48.5** (*see* **Note 4**).

3. The following MS conditions were optimal: curtain gas = 20, collision gas = high, ion-spray voltage = 3,000 V, temperature = 650°C, ion source gas 1 = 50, ion source gas 2 = 50, entrance potential = 10.

4. This analysis is performed using multiple reaction monitoring using the transition of m/z 289.3–109.0 and m/z 289.3–97.1 for testosterone while monitoring m/z 292.2–109.1 and m/z 292.2–97.2 for the internal standard. For each transition we use a dwell time of 200 ms and get about 30 data points across a 15-second peak.

5. The average ratio of the two testosterone peaks resulting from the two MS-MS transitions is determined for the standards and a +/−30% acceptance criteria is used to demonstrate the lack of interfering substances on each patient specimen.

6. Testosterone is quantified using the m/z 289.3–109.0 transition relative to the analogous internal standard transition m/z 292.2–109.1.

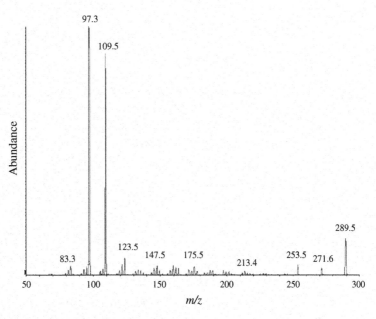

Fig. 48.4. Full scan ESI MS-MS spectra of underivatized testosterone. Prominent ions that are monitored for selected reaction monitoring are 97 and 109 *m/z*.

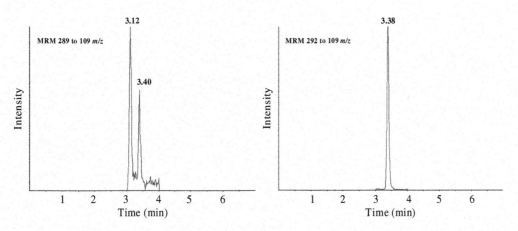

Fig. 48.5. Selected reaction monitoring trace of LC-MS-MS analysis from a female patient. This patient's specimen contained 10 ng/dL testosterone. (**A**) MRM trace of testosterone at 3.40 min (unknown peak at 3.12 min). (**B**) MRM trace of the analogous internal standard ion.

7. This method is linear from 5 ng/dL to 2000 ng/dL of testosterone.

8. Run-to-run precision and accuracy of serum quality control is shown in **Table 48.1**.

9. Interferences (*see* **Notes 5** and **6**).

Table 48.1
Accuracy and precision of run-to-run for LC-MS-MS analysis

Target	Average	Std dev	%CV	%Error	n
32	31	1.3	4.2	1.7	10
167	162	7.7	4.8	3.1	10
470	434	21.5	5.0	7.6	10
898	850	49.2	5.8	5.4	9

4. Notes

1. Sample collection and storage conditions. We typically analyze serum samples that have been obtained from plastic gel serum separator tubes. We have also analyzed plasma (heparin and EDTA) and have found that these methods for MS analysis of testosterone appear to work well with either serum or plasma. When stored at $-20°C$ testosterone is stable in serum specimens at least 10 years (unpublished results in our laboratory).

2. Testosterone stock solutions. Methanolic solutions of testosterone are very stable and can be stored for several years in tightly capped vials at $-20°C$. We use an analytical balance to weigh 10 mg of testosterone and dilute this with 10 mL of methanol in a volumetric flask to prepare a 1 mg/mL solution of testosterone. This standard is then diluted to make 10 ng/uL, 1 ng/uL, and 0.1 ng/uL solutions. We then verify the concentration of the 10 ng/uL solution by UV spectrophotometry. The molar extinction coefficient of testosterone is 15,100 L/mol*cm at 241 nm; thus, a 10 ng/uL standard should have a reading of 0.524 absorbance units (13). In addition to verifying the accuracy of the standard by UV, we also validate the newly prepared standards against the ones in use by mass spectrometry.

3. Testosterone matrix-matched calibrators and controls. There is some disagreement about the best matrix to use when preparing calibrators for testosterone analyses. In our experience it is difficult to obtain serum that does not contain testosterone. Our laboratory routinely evaluates commercially prepared charcoal stripped serum for testosterone and typically finds that it contains in the range of 2–5 ng/dL testosterone. This becomes a significant problem when analyzing specimens from women and children unless the

method of standard addition is used. An alternative is to use an albumin-based matrix since albumin is a major binding protein for testosterone. However, in our experience albumin-based calibrators behave very similar to aqueous-based calibrators; consequently we use aqueous-based calibrators and analyze serum-based controls to monitor the accuracy of the calibrators.

4. Potential Modifications for High Volume or Reference Laboratories: While the simplified extraction LC-MS-MS protocol described above works well, others are using column-switching techniques to quantify testosterone directly from serum specimens (14). Often a turbulent flow column is used as the first column, which is then connected to a reversed-phase column through a switching valve (15). The turbulent flow column separates the larger protein molecules from the rest of the plasma constituents, which are then separated on the reversed-phase column and detected using multiple reaction monitoring. Variations of this method are employed by several commercial laboratories that have dedicated instruments linking a series of up to four LC instruments to a single MS system to directly quantify testosterone from serum specimens. These types of applications are particularly attractive because the only sample preparation involved is the addition of internal standard. By multiplexing several LC systems to a single mass spectrometer the overall throughput of the system and efficiency of the systems are optimized.

5. We evaluated dehydroepiandrosterone (DHEA), dihydrotestosterone, and epi-testosterone as potential interferents. DHEA has a slightly different retention time than testosterone (3.52 and 3.40 min, respectively) but does not ionize well under the conditions employed. When 800 ng of DHEA was injected on column (equivalent to 4,000,000 ng/dL in serum, where the normal range for DHEA is 180–1,250 ng/dL in males (16)), the peak area was equivalent to 8 ng/dL of testosterone; thus, it can be safely concluded that DHEA does not interfere with the analysis of testosterone. We also showed that dihydrotestosterone was well resolved from testosterone and did not interfere at concentrations equivalent to 1,500,000 ng/dL. Epi-testosterone elutes at 3.56 min and is baseline resolved from testosterone, so it does not interfere with this assay.

6. When checking for matrix effects it was noticed that there was a decrease in signal at the retention time of testosterone, but since we are using a deuterium-labeled internal standard that has the same retention time as the analyte of interest; the matrix effect does not cause problems with accurate and precise quantification.

References

1. Fieser LF, Fieser M. (1949) Natural Products Related to Phenanthrene, 3rd edition. Rienhold Publishing Corp, pp. 368–374.

2. Fitzgerald RL, Herold DA. (1996) Serum total testosterone: immunoassay compared with negative chemical ionization gas chromatography-mass spectrometry. *Clin Chem* **42**, 749–755.

3. Taieb J, Mathian B, Millot F, Patricot M, Mathieu E, Queyrel N, Lacroix I, Somma-Delpero C, Boudou P. (2003) Testosterone measured by 10 immunoassays and by isotope-dilution gas chromatography–mass spectrometry in sera from 116 men, women, and children. *Clin Chem* **49**, 1381–1395.

4. Herold DA, Fitzgerald RL. (2004) Letters to the editor: Reliability of Extraction/Chromatography RIAs: response. *Clin Chem* **50**, 778.

5. Herold DA, Fitzgerald RL. (2004) Letters to the editor: Drs. Herold and Fitzgerald respond: *Clin Chem* **50**, 2220–2221.

6. Rosner W, Auchus RJ, Azziz R, Sluss PM, Raff H. (2007) Position statement: Utility, limitations, and pitfalls in measuring testosterone: an endocrinology society position statement. *J Clin Endo Metab* **92**, 405–413.

7. Herold DA, Fitzgerald RL. (2003) Immunoassays for testosterone in women: Better than a guess? (editorial) *Clin Chem* **49**, 1250–1251.

8. Thienpoint LM, Uytfanghe KV Blincko S, Ramsay CS, Xie H, et al. (2008) State of the art of serum testosterone measurement by isotope dilution liquid chromatography tandem mass spectrometry. *Clin Chem* **55**, 1290–1297.

9. Masse R, Wright LA. (1996) Proposed definitive methods for measurement of plasma testosterone and 17α-hydroxyprogesterone. *Clin Biochem* **29**, 321–331.

10. Cawood ML, Field HP, Ford CG, Gillingwater S, Kicman A, Cowan D, Barth JH. (2005) Testosterone measurement by isotope-dilution liquid chromatography-tandem mass spectrometry: Validation of a method for routine clinical practice. *Clin Chem* **51**, 1472–1479.

11. Shiraishi S, Lee PW, Leung A, Goh VH, Swerdloff RS, Wang C. (2008) Simultaneous measurement of serum testosterone and dihydrotestosterone by liquid chromatography-tandem mass spectrometry. *Clin Chem* **54**, 1855–1863.

12. Kushnir MM, Rockwood AL, Roberts WL, Pattison EG, Bunker AM, Fitzgerald RL, Meikle AW. (2006) Performance characteristics of a novel tandem mass spectrometry assay for serum testosterone. *Clin Chem* **52**, 120–128.

13. Yasuda K. (1964) Synthesis of 2-alpha-halo-4-en-3-oxo-steroids. *Chem Pharm Bull* **12**, 1217–24.

14. Morr M, Patel M, Wagner AD, Grant RP. Development and Application of 2D-LC-MS/MS in Clinical Diagnostics. Presented at the 2007 American Society of Mass Spectrometry Annual meeting, Poster WP 167.

15. Ynddal L, Hansen SH. (2003) On-line turbulent-flow chromatography-high-performance liquid chromatography-mass spectrometry for fast sample preparation and quantitation. *J Chromatogr A.* **1020**, 59–67.

16. Burtis CA, Ashwood ER, Bruns DE. (2006) Tietz Textbook of Clinical Chemistry and Molecular Diagnostics, 4th Edition, 2265 pp.

Chapter 49

Determination of Tetrahydrozoline in Urine and Blood Using Gas Chromatography-Mass Spectrometry (GC-MS)

Judy Peat and Uttam Garg

Abstract

Tetrahydrozoline, a derivative of imidazoline, is widely used for the symptomatic relief of conjunctival and nasal congestion; however, intentional or unintentional high doses can result in toxicity manifested by hypotension, tachycardia, and CNS depression. The detection of the drug in blood and urine is helpful in the diagnosis and management of a toxic patient. For the analysis, plasma, serum, or urine is added to a tube containing alkaline buffer and organic extraction solvents, and tetrahydrozoline from the sample is extracted into the organic phase by gentle mixing. After centrifugation, the upper organic solvent layer containing the drug is removed and dried under stream of nitrogen at 40°C. The residue is reconstituted in a hexane–ethanol mixture and analyzed using gas-chromatography-mass spectrometry. Quantitation of the drug is done by comparing responses of unknown sample to the responses of the calibrators using selected ion monitoring. Naphazoline is used as an internal standard.

Key words: Tetrahydrozoline, imidazoline, drug screen, toxicity, mass spectrometry.

1. Introduction

Tetrahydrozoline is a derivative of imidazoline and is widely used in over-the-counter nasal and eye drops for the symptomatic relief of nasal congestion and conjunctival congestion. Other imidazoline derivatives include naphazoline, oxymetazoline, and xylometazoline. These compounds act by stimulating the alpha-adrenergic receptors in the arterioles of the conjunctiva and the nasal mucosa to produce vasoconstriction. Numerous case reports have been published documenting the adverse side effects of these products, which include hypotension, tachycardia, and CNS depression. The drug is so powerful that a very small quantity

U. Garg, C.A. Hammett-Stabler (eds.), *Clinical Applications of Mass Spectrometry*, Methods in Molecular Biology 603, DOI 10.1007/978-1-60761-459-3_49, © Humana Press, a part of Springer Science+Business Media, LLC 2010

administered orally can cause serious clinical effects and may require intubation (1–3). Furthermore, because of the CNS depression, the drug has been used in sexual assault cases (4).

Treatment is largely supportive, but naloxone has been used successfully to treat symptomatic toxicity in a child (5). Thus the ability to detect tetrahydrozoline in biological fluids is desirable and important. Unfortunately, toxicology screening using immunoassays and thin layer chromatography does not detect tetrahydrozoline; however, when the screening is performed using gas chromatography and mass spectrometry (GC-MS), tetrahydrozoline is easily detectable. Although high performance liquid chromatography (HPLC) methods for estimation of tetrahydrozoline are described (6–8), GC-MS is preferred as it identifies the compound with higher certainty, making it more applicable in forensic situations. Quantitation of tetrahydrozoline may be helpful for studying the pharmacokinetics of the drug.

2. Materials

2.1. Sample

Randomly collected urine, serum, or plasma (collected in heparin or EDTA tubes) is an acceptable sample for this procedure. Refrigerated samples are stable for 1 week.

2.2. Reagents

1. Tetrahydrozoline hydrochloride and naphazoline hydrochloride (Sigma-Aldrich, Louis, MO).
2. Buffer salts mixture: Sodium carbonate:sodium bicarbonate:sodium chloride (1:1:12). The mixture is stable at room temperature for 1 year.
3. Extraction solvent: Methylene chloride:cyclohexanes:isopropanol (9:9:2). The extraction solvent is stable at room temperature for 1 year.
4. Human drug-free urine and serum (UTAK Laboratories, Inc., Valencia, CA).

2.3. Standards and Calibrators

1. Primary tetrahydrozoline standard (1 mg/mL) in methanol was prepared using tetrahydrozoline hydrochloride.
2. Methanolic standard solutions of 100, 10, 1, and 0.1 µg/mL were prepared by serial dilutions of the primary standard.
3. Working calibrators for urine and serum were made according to **Table 49.1** (*see* **Note 1**).
4. Standards and calibrators are stable for 6 months when stored at −20°C.

Table 49.1

Preparation of calibrators. Final volume of each calibrator is 1 mL

Calibrator	Drug-free serum(s) or urine(u) (µL)	100 µg/mL standard (µL)	10 µg/mL standard (µL)	1 µg/mL standard (µL)	0.1 µg/mL standard (µL)	Final concentration (µg/mL)
0	1000	0	0	0	0	0
1	900 (s)	0	0	0	100	0.010
2	975 (s)	0	0	25	0	0.025
3	950 (s)	0	0	50	0	0.050
4	900 (s)	0	0	100	0	0.100
5	900 (u)	0	100	0	0	1.0
6	950 (u)	50	0	0	0	5.0
7	900 (u)	100	0	0	0	10.0
8	800 (u)	200	0	0	0	20.0

2.4. Internal Standard and Quality Controls

1. Primary naphazoline internal standard (1 mg/mL) was prepared in methanol using naphazoline HCl (Sigma-Aldrich, Louis, MO).

2. Working methanolic internal standard (IS) of 100 µg/mL concentration was prepared from the primary internal standard.

3. Quality control samples were prepared according to **Table 49.2** (*see* **Note 2**).

4. The internal standards and quality control samples are stable for 6 months when stored at −20°C.

Table 49.2

Preparation of quality controls samples. Final volume of each control is 10 mL

Controls	Drug-free serum (s) or urine (u)(mL)	1 mg/mL (µL)	10 µg/mL (µL)	Final concentration (µg/mL)
Low serum	9.98 (s)	0	20	0.020
High serum	9.90 (s)	0	100	0.100
Low urine	9.98 (u)	20	0	2.0
High urine	9.90 (u)	100	0	10.0

2.5. Supplies	1. 13 × 100 mm test tubes with Teflon-lined screw-caps (Fisher Scientific, Fair Lawn, NJ). These tubes are used both for drug extraction and concentration of the extracts.
	2. Autosampler vials (12 × 32 mm with crimp caps) with 0.3 mL limited volume glass inserts (P.J. Cobert Associates, Inc., St. Louis, MO).
	3. Gas chromatography column: Zebron ZB-1, 15 m × 0.25 mm × 0.5 μm (Phenomenex, Torrance, California).
2.6. Equipment	1. TurboVap®1 V Evaporator (Zymark Corporation, Hopkinton, MA, USA).
	2. A gas chromatograph/mass spectrometer (GC-MS), model 6890 N/5973 operated in electron impact mode (Agilent Technologies, Wilmington, DE).

3. Methods

3.1. Stepwise Procedure	1. Prepare an unextracted standard by adding 100 μL of the 10 μg/mL standard and 100 μL of working internal standard solution to a 13 × 100 mm concentration tube. Dry down the contents under nitrogen, at 40°C, and reconstitute in 100 μL of hexanes:ethanol (1:1).
	2. Label a 13 × 100 mm tube for each calibrator, control, and sample to be analyzed. Add 1 g of buffer salt and 3 mL of extraction solvent to each extraction tube.
	3. Add 1 mL of sample, calibrator, or control.
	4. Add 100 μL of working IS solution to each tube.
	5. Cap and rock the tubes for 5 min.
	6. Centrifuge tubes for 5 min at ∼1,600 × g.
	7. Transfer the upper organic layer to appropriately labeled 13 × 100 mm test tube. Discard the bottom aqueous phase (*see* **Note 3**).
	8. Concentrate organic layer to dryness using TurboVap1V Evaporator at 40°C (*see* **Note 4**).
	9. Reconstitute the residue in 100 μL hexanes:ethanol (1:1). Transfer the contents to autosampler vials.
3.2. Instrument Operating Conditions	Inject 1 μL onto the GC-MS for analysis. The instrument's operating conditions are given in **Table 49.3**.
3.3. Data Analysis	1. Representative GC-MS chromatogram of tetrahydrozoline and naphazoline is shown in **Fig. 49.1**. GC-MS selected ion chromatograms are shown in **Fig. 49.2**. Electron impact

Table 49.3
GC-MS operating conditions

Initial oven temp	90°C
Initial time	1.0 min
Ramp 1	32°C/min
Temp. 2	170°C
Time	2.0 min
Ramp 2	20°C/min
Final temp	270°C
Time	5.0 min
Mode	Splitless
Injector temp	250 °C
Column pressure	7.5 psi
Purge time	0.4 min
MS source temp	230°C
MS mode	Electron impact at 70 eV
MS tune	Autotune

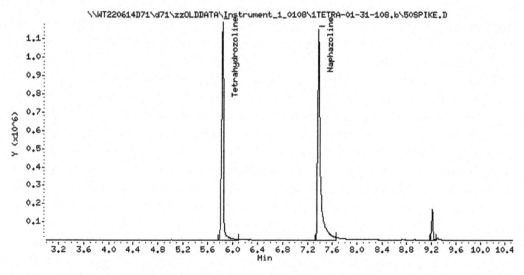

Fig. 49.1. GC-MS chromatogram of tetrahydrozoline and naphazoline (internal standard) for 10 µg/mL calibrator.

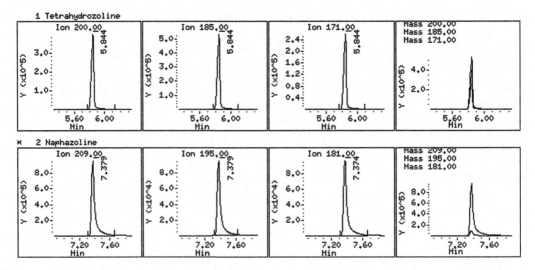

Fig. 49.2. Selected ion chromatograms of tetrahydrozoline and naphazoline (internal standard).

ionization mass spectra of these compounds are shown in the **Figs. 49.3** and **49.4**, respectively (*see* **Note 5**). Ions used for identification and quantification is listed in **Table 49.4**.

2. Analyze data using Target Software (Thru-Put Systems, Orlando, FL) or similar software. The quantifying ions (**Table 49.4**) are used to construct standard curves of the peak-area ratios (calibrator/internal standard pair) vs. concentration. These curves are then used to determine the concentrations of the controls and unknown samples.

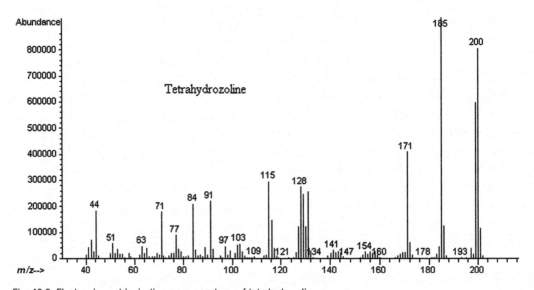

Fig. 49.3. Electron impact ionization mass spectrum of tetrahydrozoline.

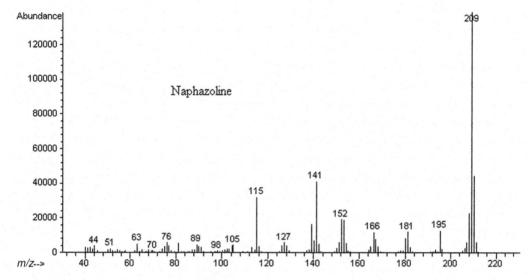

Fig. 49.4. Electron impact ionization mass spectrum naphazoline.

Table 49.4
Quantitation and qualifying ions for tetrahydrozoline

	Quantitation ion	Qualifier ions
Tetrahydrozoline	200	171,185
Naphazoline	209	181,195

3. The linearity/limit of quantitation of the method is 10–20 µg/mL. Samples in which the drug concentrations exceed the upper limit of quantitation should be diluted with appropriate negative urine, serum, or plasma and retested.

4. Standard curves should have a correlation coefficient (r^2) >0.99.

5. Typical intra- and inter-assay imprecision is <10%.

6. Quality control: The run is considered acceptable if calculated concentrations of drugs in the controls are within +/− 20% of target values. Quantifying ion in the sample is considered acceptable if the ratios of qualifier ions to quantifying ion are within +/− 20% of the ion ratios for the calibrators.

4. Notes

1. Separate calibrator curves are run for serum and urine as urine generally has much higher concentration of tetrahydrozoline as compared to serum.

2. When possible, calibrators and controls should be made from different batches of drugs and at different times by different analysts.

3. Remove the upper organic phase as much as possible without disturbing interface and making sure that no aqueous layer is transferred.

4. Do not over-dry the extract. This will result in poor recovery and failed run.

5. Electron impact ionization spectra are needed in the initial stages of method setup to establish retention times and later on, if there is a need for change in quantifying or qualifying ions. They are not needed for routine quantitaion.

Acknowledgement

We acknowledge the help of David Scott in preparing the figures.

References

1. Spiller, H. and Griffith, J. (2008) Prolonged cardiovascular effects after unintentional ingestion of tetrahydrozoline. *Clin Toxicol (Phila)*, **46**, 171–2.

2. Daggy, A., Kaplan, R., Roberge, R. and Akhtar, J. (2003) Pediatric Visine (tetrahydrozoline) ingestion: case report and review of imidazoline toxicity. *Vet Hum Toxicol*, **45**, 210–2.

3. Osterhoudt, K.C. and Henretig, F.M. (2004) Sinoatrial node arrest following tetrahydrozoline ingestion. *J Emerg Med*, **27**, 313–4.

4. Spiller, H.A., Rogers, J. and Sawyer, T.S. (2007) Drug facilitated sexual assault using an over-the-counter ocular solution containing tetrahydrozoline (Visine). *Leg Med (Tokyo)*, **9**, 192–5.

5. Holmes, J.F. and Berman, D.A. (1999) Use of naloxone to reverse symptomatic tetrahydrozoline overdose in a child. *Pediatr Emerg Care*, **15**, 193–4.

6. Ali, M.S., Ghori, M. and Saeed, A. (2002) Simultaneous determination of ofloxacin, tetrahydrozoline hydrochloride, and prednisolone acetate by high-performance liquid chromatography. *J Chromatogr Sci*, **40**, 429–33.

7. Rojsitthisak, P., Wichitnithad, W., Pipitharome, O., Sanphanya, K. and Thanawattanawanich, P. (2005) Simple HPLC determination of benzalkonium chloride in ophthalmic formulations containing antazoline and tetrahydrozoline. *PDA J Pharm Sci Technol*, **59**, 332–7.

8. Santoni, G., Medica, A., Gratteri, P., Furlanetto, S. and Pinzauti, S. (1994) High-performance liquid chromatographic determination of benzalkonium and naphazoline or tetrahydrozoline in nasal and ophthalmic solutions. *Farmaco*, **40**, 751–4.

Chapter 50

Quantitation of 25-OH-Vitamin D (25OHD) Using Liquid Tandem Mass Spectrometry (LC-MS-MS)

Ravinder J. Singh

Abstract

25-OH-vitamin D (25OHD) is ordered and interpreted clinically in light of calcium balance and home-ostasis. Various methodologies including immunoassay and chromatography are described for the measurement of 25OHD in biological fluids. It is well reported that all existing methods are challenging and require a high level of technical expertise. While this is also true of the methods using liquid chromatography-tandem mass spectrometry (LC-MS-MS), the technique is gaining favor in various laboratories because it offers advantages of greatly improved specificity and sensitivity.

Key words: Tandem mass spectrometry, endocrine, vitamin D, 25-OH-vitamin D, bone.

1. Introduction

Vitamin D has gained great interest largely due to its role in bone health and numerous reports of deficiencies within the general population. In the circulation, the majority of vitamin D circulates bound to a specific transport protein, vitamin D-binding protein (DBP). Vitamin D is hydroxylated in the liver at the carbon 25-position giving rise to 25-OH vitamin D3, the most abundant circulating, but biologically inactive form of the vitamin. Biological activity is conferred by the final hydroxylation step catalyzed in the kidney by a 1-α-hydroxylase enzyme, resulting in the production of 1,25-(OH)2-vitamin D. The activity of the renal α-hydroxylase is under tight control so that the production of 1,25-(OH)2-vitamin D remains constant over a wide range of substrate (25-OH-vitamin D) concentrations.

U. Garg, C.A. Hammett-Stabler (eds.), *Clinical Applications of Mass Spectrometry*, Methods in Molecular Biology 603,
DOI 10.1007/978-1-60761-459-3_50, © Humana Press, a part of Springer Science+Business Media, LLC 2010

The main regulators of 1-α-hydroxylase activity are calcium, parathyroid hormone, and phosphate. Low serum calcium stimulates the secretion of parathyroid hormone, which acts to increase the conversion of 25-OH-vitamin D to 1,25-(OH)2-vitamin D. Of the circulating vitamin D metabolites, 25-OH-vitamin D is the most abundant form and has the longest half-life (approximately 1–2 weeks). Its concentration serves as the best index of skin synthesis and dietary intake of vitamin D. Nutritional vitamin D deficiency, rickets, or osteomalacia, is associated with chronic low 25-OH-vitamin D levels, and subjects who are intoxicated with vitamin D can have concentrations above 80 ng/mL.

Characteristic seasonal fluctuations are seen for serum 25-OH-vitamin D concentration, which is a reflection of the amount of sunlight to which a person is exposed. Concentrations of 25-OH-vitamin D are highest in late summer and lowest in spring. Most evidence indicates that both natural vitamin D3 and synthetic vitamin D2 are metabolized by the same enzyme systems so that ingested vitamin D2 is also converted first to 25-OH-vitamin D2 and then to 1, 25-(OH)2-vitamin D2. From the standpoint of assay development, it is critical that any vitamin D2 metabolites present in serum be included in the total assayed fraction.

A number of methods are being used by clinical laboratories for the measurement of the vitamin for patient care, including HPLC-UV, HPLC-EC, radioimmunoassay (with low throughput), automated chemiluminescence immunoassays (high throughput), and liquid chromatography-tandem mass spectrometry (LC-MS-MS) (1–3). It is well reported that there are challenges in all of these methods and high-level technical expertise is required for each. One challenge that has both analytical and clinical implications is the variability encountered with some of the methods, a problem reflected in proficiency surveys. When the results of a recent College of American Pathologists (CAP) survey are compared, the impact of methodology and lack of standardization is dramatic. For one sample a mean of 75 ng/mL was obtained using LC-MS-MS with <15% variability between the reporting laboratories. For this same sample, laboratories using chemiluminescence immunoassays reported a range of results from 41 to 96 ng/mL (4, 5). There could be many reasons for these variations, including drifts in the reagents being manufactured, but there is a clear and urgent need for harmonization and standardization.

To address the issue of standardization, NIST is developing quality control materials (human serum, SRM 972) that will contain 25-OH-D2, 25-OH-D3, and the metabolite 3-epi-25-OH-D at four different concentrations as characterized by LC-MS-MS. SRM is especially important for assays for which the cross-reactivity with these metabolites is not well defined (4). LC-MS-MS is becoming the technique of choice for various reference laboratories and a robust method is presented in this chapter.

2. Materials

2.1. Samples

Serum. Collect in phlebotomy tube without a gel-barrier, i.e., plain red. Stable up to 7 days at 4–8°C. Freeze the sample if assay cannot be performed within 7 days.

2.2. Reagents and Buffers

1. 50% Methanol: Combine 2 L of methanol and 2 L of deionized H_2O. Mix. Store at room temperature. Reagent is stable for 6 months (*see* **Note 1**).

2. Mobile phase A: Add 50 μL of formic acid to 1 L of HPCLgrade water. Mix and degas. Store at ambient temperature. Reagent is stable for 6 months.

3. Mobile phase B: Add 50 μL of formic acid to 1 L of methanol. Mix and degas. Store at ambient temperature. Reagent is stable for 6 months.

4. Stock Estriol: Dissolve 25 mg of estriol in 25 mL of methanol. Store at −20°C. Stable for 2 years.

5. Reconstitution Solvent: Dilute 700 mL of methanol to 1 L with deionized H_2O. Add 1 mL of 1 mg/mL stock estriol for a final concentration of 1 μg/mL. Store at −20°C. Reagent is stable for 6 months.

6. Bovine Serum Albumin Standard Buffer, 0.1 M PBS, 1.0% BSA, 0.9% NaCl, pH 7.4: Dissolve 2.14 g $Na_2HPO_4 \cdot 7\ H_2O$, 0.268 g $NaH_2PO_4 \cdot H_2O$, 0.9 g NaCl, 1.0 g bovine serum albumin in 75 ml of deionized H_2O. Adjust pH to 7.4 (with dilute NaOH or HCl) and bring to 100 mL volume with deionized H_2O. Use this buffer to prepare standards described in the calibration section of this procedure. Store at 4°C. Calibrators are stable for 10 years.

7. 25OHD2 and 25OHD3 powders - minimum 98% purity (Sigma Chemicals).

8. Charcoal Stripped Human serum.

2.3. Standards and Calibrators

1. Stock I, 30 μg/mL: Dissolve separately 10 mg 25OHD3 and 25OHD2 powder in 100 mL of ethanol, which will provide stock concentration of approximately 100 μg/ml. Make 1:5, 1:10, and 1:20 dilutions for each 25OHD2 and 25OHD3 in pure ethanol and read each of the dilutions on a UV spectrophotometer at 264 nm against an ethanol blank. The molar extinction coefficient of 18,300 should be used to assign the final concentrations for the above dilutions. Based on the absorbance assign the final concentration to the stock. Dilute the stock solutions of 25OHD3 and 25OHD2 to a final concentration of 50 μg/mL. Store in 5 mL aliquots at −70°C in crimp-cap vials. Reagent is stable for 10 years.

Table 50.1
Preparation of calibrators

Standard	BSA standard buffer	Amount of 2.5 µg/mL 25OH vitamin D stock 2 standard (mL)	25OH vit D std concentration (ng/mL)
Std 6	Fill to 100 mL	4.0	100
Std 5	Fill to 100 mL	2.0	50
Std 4	Fill to 100 mL	1	25
Std 3	Fill to 100 mL	0.4	10
Std 2	Fill to 100 mL	0.2	5
Std 1	Fill to 100 mL	0	0

2. Stock II, 2.5 µg/mL: Dilute 10 mL of 25OHD3 and 25OHD2 Stock I to 200 mL with reconstitution solvent. Store in 10 mL aliquots at −70°C in crimp-cap vials.

3. Calibrators: Prepare the calibrators according to **Table 50.1**.

4. LC-MS-MS performance Check Sample (75 ng/mL): 30 ug/mL 25OH D2/D3 Stock Standard 1 volumetric. Dilute 1.25 mL of 30 ug/mL 25OH D2/D3 Stock Standard 1 with 500 mL of diluting solvent. Store at −20°C. Stable for 10 years.

2.4. Internal Standard

1. Stock internal standard, 1 mg/mL: Weigh 10 mg of 26, 26, 26, 27, 27, 27-d6 25-hydroxy vitamin-D3. (>98% purity) into a 10 ml volumetric flask and bring to volume with ethanol. Store at −20°C. Reagent is stable for 10 years.

2. Working internal standard, 200 ng/mL: Add 20 µL of stock internal standard to 0.1 L of reconstitution solvent. Stable for 2 years when stored at −20°C.

2.5. Quality Controls

1. Three controls low, medium, and high are prepared by spiking charcoal stripped human serum (SeraCare Life Sciences) (*see* **Note 2**).

2. Pour approximately 500 ml of the charcoal stripped human serum into three 1 L volumetric flasks. Label the three flasks as low, medium, and high and add 2, 8, and 40 mL of 25OHD stock II (2.5 µg/mL), respectively.

3. Place a magnetic stir bar in each flask and mix the contents on a stir-plate for 5 min.

4. After testing each pool in an assay, aliquot them into 20 mL plastic scintillation vials. Store at −70°C. Controls are stable for 10 years.

5. Run quality controls with each run (*see* **Note 3**).

2.6. Supplies

1. Autosampler vials with caps.

2. Disposable culture tubes 13 × 100 mm.

3. Disposable culture tubes 12 × 75 mm.

4. Extraction Cartridges: Strata-X Reverse Phase- (please apply the same change to OHPG methods as well) 30 mg/mL (Phenomenex).

5. Supelcosil LC-18 Analytical Column. 3.3 cm × 4.6 mm, 3 µm (Supelco).

6. The pre-column filter, C12 (Max RP) 4 mm L × 2.0 mm ID (Phenomenex).

2.7. Equipment

1. Tandem Mass Spectrometer with an ESI or APCI.

2. HPLC with autosampler.

3. Evaporator.

3. Methods

3.1. Sample Preparation

1. Pipet 0.1 mL of each standard, control, and sample to the appropriately labeled 13 × 100 mm glass tubes.

2. Add 25 µL of working internal standard to each tube and vortex. Incubate at least 10 min.

3. Add 1 ml of deionized H_2O to each tube and vortex.

3.2. Solid Phase Extraction

1. Condition extraction cartridges by applying 2.0 mL of acetonitrile to each cartridge at a rate of <2.0 mL/min. Discard eluates.

2. Condition extraction cartridges by applying 2.0 mL of 35:65 CH_3CN–water to each cartridge at a rate of <2.0 mL/min. Discard eluates.

3. Apply 1.0 mL of deionized H_2O to each extraction cartridge at a rate of <2.0 mL/min and discard eluates.

4. Apply each prepared standard, control, and sample to the appropriate conditioned cartridge at a rate of <2.0 mL/min. Discard eluates.

5. Apply 2.0 mL of 35:65 CH_3CN–water to each cartridge at a rate of <2.0 mL/min. Discard eluates. Wipe any hanging drops of fluid from cartridge nipples.

6. Place 12 × 75 mm glass tubes under each cartridge and apply 2.0 mL of CH3CN to each at a rate of <2.0 mL/min. Collect eluates.

7. Place the tubes containing the eluates into an evaporator and dry under nitrogen.

8. Add 150 μL of reconstitution solvent to each dried eluate tube and gently mix.

9. Transfer the reconstituted samples to the autosampler vials and inject 30 μL into LC-MS-MS.

3.3. Instrument Operating Conditions

LC-MS-MS is used with an ESI or APCI source to monitor ion pairs in multiple reaction monitoring mode (**Table 50.2**) (*see* **Notes 4** and **5**). The flow rate of the mobile phase is 1 mL/min with no split of the post-column flow to the mass spectrometer (*see* **Note 6**). The solvent gradient is as follows: equilibrate of the HPLC column with 20% mobile phase B for 1 min which is then raised to 100% over 2 min for elution of analytes. Return to 20% mobile phase B for 1 min to recondition before the next injection. To assure that system is optimized and is working well to run patient samples, the signal-to-noise for LC-MS-MS performance check sample should at least be 100 but not less than 50.

Table 50.2
MRM transitions

Analyte	Q1	Q3	Qualifier ion
25OHD3	401	383	365
25OHD2	413	395	377
25OH D3-d6	407	389	n/a

3.4. Data Analysis

The data are collected and analyzed using mass spectrometer vendor specific software. The quantitation of 25OHD2 and 25OHD3 are made from their respective standard curves using the peak-area ratio of analyte to that of the internal standard. Total 25OHD is reported as the sum of 25OHD2 and 25OHD3. The calibration curve is constructed using weighting of $1/x$ and linear regression. The limit of quantification of the method is 5 ng/mL. The linearity is up to 200 ng/mL with very high correlation coefficient (>0.999) using linear regression. The inter-day precision should be less than 10% for a control with a 25OHD mean value of ~50 ng/mL. During method validation it is critical to make sure that relative recovery of 25OHD2 and 25OHD3 from human serum/plasma matrix is 100 ±10% and carry over is less than .01%.

4. Notes

1. The organic solvents used in the extraction and mobile phases are flammable. They should be handled with care and used in an exhaust hood. Use eye protection when working with acids or bases. The preparation of the reagents required for in-house LC-MS-MS assays should be conducted by individual laboratories under their institutionally regulated standard procedures.

2. Assay 0.1 ml of the charcoal stripped serum prior to use to make sure that it is negative for 25OHD.

3. Mean and acceptable limits are established by assaying each pool level 20 times over multiple days. The means, standard deviations, and %CVs are calculated for each level of control. An established %CV, based on assay historical performance and medical relevance, is used to calculate the ± 2 SD range. All QC values are graphed on paper QC charts and entered into lab information system for each assay; and monitored for being inbounds, trends, and shifts. QC charts are reviewed according to laboratory quality rules.

4. 25OHD2 and 25OHD3 signal-to-noise is optimized by infusing 25OHD2 and 25OHD3 solutions at a concentration of 10 µg/ml (**Figs. 50.1** and **50.2**). The mass of the ion should be optimized down to the tenth. This method is based on off-line SPE cleanup of the serum/plasma sample, to avoid suppression, before injection into the LC-MS-MS.

5. Laboratories, which use in-house LC-MS-MS have responsibility for many steps of the assay. The LC-MS-MS technology for testing of human samples is not approved by the FDA, and manufacturers of LC-MS-MS instrumentation are not responsible for troubleshooting the assays. Laboratories performing 25-OH-D testing by LC-MS-MS technology have differences in their standard operating procedures, and thus inter-laboratory CVs in the range of 20% are observed in proficiency testing surveys.

6. Check with the manufacturer if the source is compatible with high flow rates.

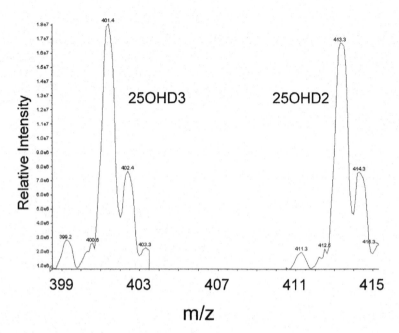

Fig. 50.1. Q1 Scan of 25OH D2 and 25OH D3 during infusion of solution into LC-MS-MS.

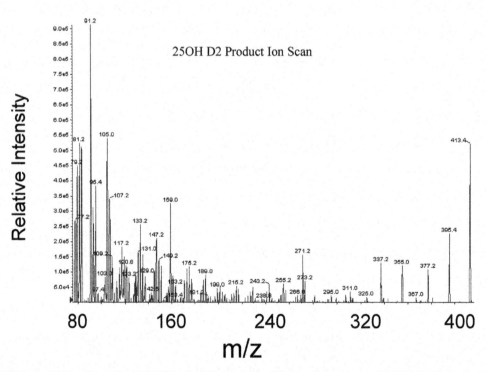

Fig. 50.2. Product Ion Scans for 25OH D2 and 25OH D3 during infusion of solution into LC-MS-MS. (**A**) Product ion spectrum of 401. (**B**) Product ion spectrum of 413.

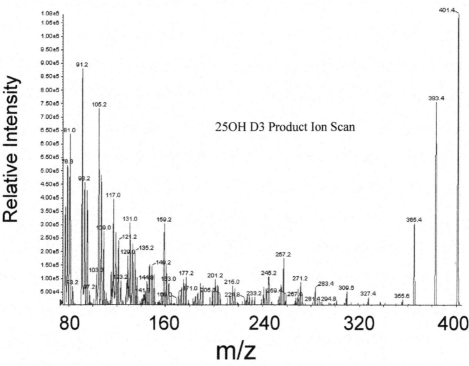

Fig. 50.2. (continued)

References

1. Jones G. (1978) Assay of vitamin D2 and D3 and 25-hydroxyvitamins D2 and D3 in human plasma by high-performance liquid chromatography. *Clin Chem* **24**:287–298.

2. Higashi T, Awada D, Shimada K. (2001) Simultaneous determination of 25-hydroxyvitamin D-2 and 25-hydroxyvitamin D-3 in human plasma by liquid chromatography-tandem mass spectrometry employing derivatization with a Cookson-type reagent. *Biol Pharm Bull* **24**:738–743.

3. Singh RJ, Taylor RL, Reddy GS, Grebe SK. (2006) C-3 epimers can account for a significant proportion of total circulating 25-hydroxyvitamin D in infants, complicating accurate measurement and interpretation of vitamin D status. *J Clin Endocrinol Metab* **91**:3055–3061.

4. Binkley N, Krueger D, Cowgill CS, Plum L, Lake E, Hansen KE, et al. (2004) Assay variation confounds the diagnosis of hypovitaminosis D: a call for standardization. *J Clin Endocrinol Metab* **89**:3152–3157.

5. Singh RJ. (2008) Are clinical laboratories prepared for accurate testing of 25-hydroxy vitamin D? *Clin Chem* **54**:221–223.

Chapter 51

Identification and Quantitation of Zolpidem in Biological Matrices Using Gas Chromatography-Mass Spectrometry (GC-MS)

Timothy P. Rohrig, Lydia A. Harryman, and Melissa C. Norton

Abstract

Zolpidem is a short-acting nonbenzodiazepine hypnotic used for the treatment of insomnia. It works quickly and has a short half-life of 2–3 h. Significant side effects are reported including dizziness, amnesia, and even hallucinations. In this method, zolpidem is extracted from blood or other biological fluids/tissues using solid phase extraction technology for analysis. The extract is then assayed using selected ion monitoring gas chromatography/mass spectrometry for absolute confirmation of the compound. The quantitation of this drug is made from standard curves constructed using selected ion monitoring and peak-area ratios of analyte vs. internal standard of quantifying ions.

Key words: Zolpidem, mass spectrometry, gas chromatography, Solid Phase Extraction (SPE).

1. Introduction

Zolpidem is used for the short-term treatment of insomnia, defined as difficulty falling and staying asleep. An imidazopyridine derivative, the drug was first introduced in Europe in 1986 and in the United States in 1993 (1). Zolpidem usually acts within 15 min following oral administration and has a short half-life of 2–3 h. It is converted to pharmacologically inactive metabolites (**Fig. 51.1**) that are excreted in urine (1).

Side effects of zolpidem usage include daytime drowsiness, dizziness, amnesia, headache, and nausea (1). Hallucinations have been reported, especially by patients receiving concurrent antidepressant medications (2). Driving while under the drug's influence is considered dangerous due to the variability and severity of the side effects that may be encountered.

U. Garg, C.A. Hammett-Stabler (eds.), *Clinical Applications of Mass Spectrometry*, Methods in Molecular Biology 603, DOI 10.1007/978-1-60761-459-3_51, © Humana Press, a part of Springer Science+Business Media, LLC 2010

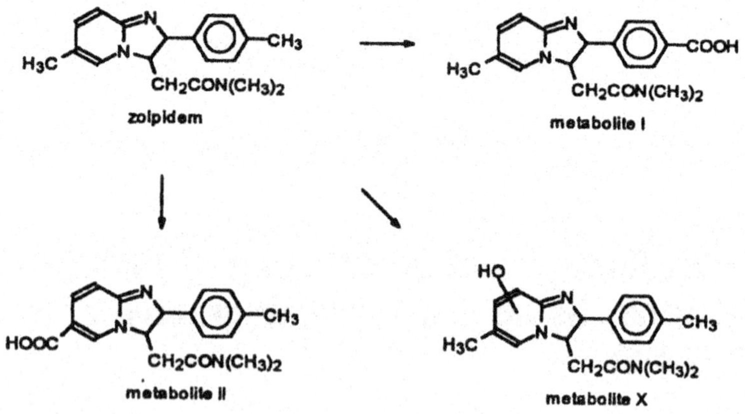

Fig. 51.1. Metabolism of Zolpidem.

This hypnotic drug is measured in the biological specimens for toxicological and forensic purposes. Samples are generally screened by immunoassay for the presence of Zolpidem. Zolpidem is extracted from the biological specimens using solid phase extraction and confirmed by gas chromatography mass spectrometry.

2. Materials

2.1. Sample

Preferred samples include postmortem tissues, urine, and fluoridated blood from the cardiac and femoral area collected during autopsies, and antemortem urine, and fluoridated blood collected from outside agencies intended for human performance testing. Aliquots should be stored in glass tubes or plastic jars at a temperature range of 2–8°C until the time of analysis.

2.2. Reagents and Buffers

1. Monobasic sodium phosphate, acetic acid, methanol, hexane, Isopropanol, methylene chloride, ammonium hydroxide, and ethyl acetate were purchased from Fisher Scientific (Fair Lawn, NJ) or equivalent, and were of analytical grade.

2. 100 mM Sodium phosphate buffer: pH 6.0. Dissolve 13.84 g of sodium phosphate (monobasic) in 500 mL of DI water; dilute to 1,000 mL. Buffer is stable at room temperature for 6 months.

3. 1.0 N Acetic Acid. Stable at room temperature for 6 months.

4. 50/50 Hexane/Ethyl Acetate. Stable at room temperature for 1 year.

5. Elution Solvent: methylene chloride:Isopropanol:ammo-nium hydroxide (78:20:2) is stored in an amber bottle at room temperature and is prepared daily (*see* **Note 1**).

6. Bovine blood (Lampire Biological Laboratories, Pipersville, PA).

7. Urine (In house, Wichita, KS).

2.3. Calibrators/ Calibration

1. Primary standard: Zolpidem, 1 mg/mL (Cerilliant Corporation, Round Rock, TX).

2. Secondary standard (100 μg/mL): Prepared by transferring 1.0 mL of the primary standard into a 10 mL volumetric flask and diluting with methanol. Secondary standard is stable for 1 year at room temperature (*see* **Note 2**).

3. Tertiary standard (1 μg /mL): Prepared by transferring 0.1 mL of secondary standard to a 10 mL volumetric flask and diluting with methanol. Tertiary standard is stable for 1 year at room temperature.

4. Working calibrators: Prepared according to **Table 51.1** for each analytical run.

Table 51.1
Preparation of calibrators

Calibrator	Negative blood or urine (mL)	Tertiary Standard (mL)	Concentration (ng/mL)
1	1.95	0.050	25
2	1.90	0.100	50
3	1.80	0.200	100
4	1.60	0.40	200
5	1.00	1.0	500

2.4. Internal Standard and Quality Controls

1. The primary internal standard: Cyproheptadine HCl (Sigma-Aldrich, St. Louis, MO).

2. Secondary internal standard (1 mg/mL): Prepared by transferring 112.7 mg of primary internal standard into a 100 mL volumetric flask and diluting with methanol (*see* **Note 2**).

3. Tertiary internal standard (10 μg /mL): Prepared by transferring 100 μL of secondary internal standard to a 10 mL volumetric flask and diluting with methanol. Tertiary internal standard is stable for 1 year at room temperature.

4. Working internal standard (1 µg /mL): Prepared by transferring 1.0 mL of tertiary internal standard to a 10 mL volumetric flask and diluting with methanol.

5. Quality control samples: An in-house control is used to validate sample analysis from a second set of prepared standards. The in-house control is made according to **Table 51.2**. The control is prepared fresh for each batch. The target value for the quality control sample is established in-house.

Table 51.2
Preparation of in-house controls

Control	Negative blood or urine (mL)	Tertiary Standard (mL)	Concentration (ng/mL)
Positive	1.80	0.200	100

2.5. Supplies

1. UCT Clean Screen solid phase extraction (SPE) columns (3) (CSDAU206 or equivalent) were purchased from United Chemical Technologies, INC. (Bristol, PA)

2. Auto sampler vials (12 × 32 mm; clear crimp), 0.300 mL conical with spring inserts, and 11 mm crimp seals with FEP/NatRubber were purchased from P.J. MicroLiter Analytical Supplies, INC. (Suwanne, GA)

3. 16 × 100 mm glass culture tubes were purchased from Fisher Scientific (Fair Lawn, NJ)

4. GC column used: RTX-5MS 15 m × 0.25 mm × 0.25 µm (Restek International, Bellefonte, PA)

2.6. Equipment

1. Cerex System 48 Positive Pressure Manifold for solid phase extraction from SPEware Corporation (Baldwin Park, CA).

2. A gas chromatograph/mass spectrometer system (GC-MS; 6890/5973, Agilent Technologies (Wilmington, DE).

3. Pierce, 18835 Reacti-Therm III[TM] Heating Module with Reacti-Vap Evaporator, (Rockford, IL).

4. Fisher Vortex was purchased from Fisher Scientific (Fair Lawn, NJ).

3. Methods

3.1. Preparation of Tissue Homogenates

1. Portions of tissue samples are weighed (approximately 20 g, if available) to generate one part tissue to three parts DI water, creating a 1:4 tissue homogenate.

3.2. Stepwise
Procedure

1. Pipette 200 µL of working internal standard (1 µg /mL) to appropriately labeled 16 × 100 mm glass culture tubes. Batches should include calibrators, negative control, samples, and positive control.

2. Add 2 mL of blood, urine, or 4 g of a 1:4 tissue homogenate.

3. Add 2 mL of DI water. Mix/vortex.

4. Add 2 mL of 100 mM sodium phosphate buffer, pH 6.0. Mix/vortex (*see* **Note 3**).

5. Centrifuge samples at ~~1,300 × g force for ~10 min.

6. Place appropriately labeled 16 × 100 mm glass culture tubes in the collection rack of the positive pressure manifold.

7. Install labeled SPE columns (CSDAU206 or equivalent) on the positive pressure manifold in their respective slots.

8. Add in order the following reagents and aspirate each, under low pressure (<3″ Hg) to prevent sorbent from drying: 3 mL of methanol, 3 mL of water, 3 mL of sodium phosphate buffer 100 mM pH 6.0.

9. Load samples at ~1–2 mL/min (~2–3″ Hg).

10. Wash the columns with the following reagents, in order, and aspirate each under low pressure: 3 mL of deionized water, 1 mL of 1.0 N acetic acid, and 3 mL of methanol.

11. Dry columns for approximately 5 min under full pressure.

12. Wash the columns with the following reagents, in order, and aspirate each under low pressure (<3″ Hg) to prevent sorbent from drying: 2 mL of hexane, 3 mL of hexane/ethyl acetate [50:50], and 3 mL of methanol.

Table 51.3
GC operating conditions

Initial oven temp.	100°C
Initial time	1.0 min
Ramp 1	30°C/min
Temp. 2	300°C
Time	1.33 min
Injector temp	250°C
Detector temp	300°C
Purge time on	0.5 min.
Column pressure	11.7 psi

Table 51.4
Quantitation and qualifying ions for Zolpidem

	Quantitation ion	Qualifier ions
Zolpidem	235	219, 307
Cyproheptadine	287	215

```
QUANTITATION REPORT FOR Zolpidem ON : Milhous

Data File          : C:\MSDChem\1\DATA\0126\26Jan09A.02p\zol03.D
Tune File Name     : C:\MSDChem\1\5973N\atune.u
Tune Date          : 26 Jan 2009   2:36 pm                     Mult : 0
Acq Method Name    : ZOL.M Calib date : 27 Jan 2009 8:51 am
Sample Name        : Std 3
Acquisition date   : 26 Jan 2009   3:14 pm

Retention Time    7.51 Zolpidem         +/- 2.00% =    7.35 -   7.66 min
Retention Time    6.33 Cyproheptadine   +/- 2.00% =    6.20 -   6.46 min
R.R.T. =   1.186  Unknown target ion / ISTD target ion =    1.40

Zolpidem        => 235.0 =   7225329  307.0 =    952078  219.0 =    661696
Cyproheptadine  => 287.0 =   5149010  215.0 =   2899033

Zolpidem        => 307.0/235.0 =   13.2 +/-  20.0% rel =   10.6 -   15.8
Zolpidem        => 219.0/235.0 =    9.2 +/-  20.0% rel =    7.4 -   11.0
Cyproheptadine  => 215.0/287.0 =   56.3 +/-  20.0% rel =   45.0 -   67.6


Concentration =   236.00 ** CONCENTRATION > 0.00 ng LIMIT **<==
```

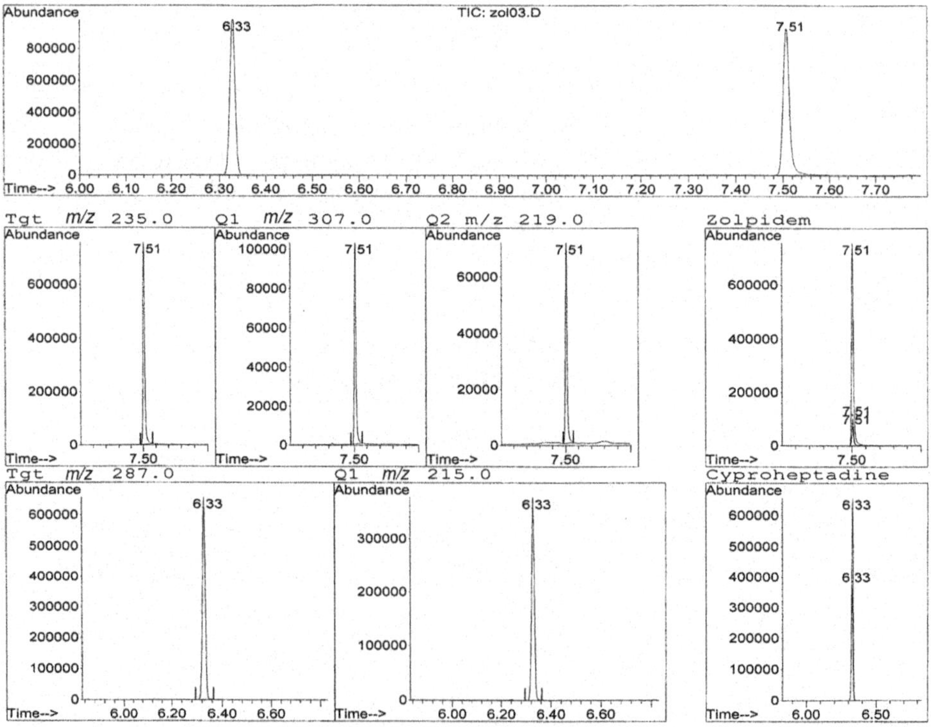

```
Zolpidem        : RT extraction window from   7.00 to   8.01 min
Cyproheptadine  : RT extraction window from   5.83 to   6.83 min

Page 132                              Printed on : 10 Mar 2009 2:18 pm
```

Fig. 51.2. GC-MS selected ion chromatograms for Zolpidem.

13. Dry columns for approximately 5 min under full pressure.

14. Add 3 mL of elution solvent (MeCl, IPA, NH$_4$OH).

15. Collect elute slowly, ~1–2 mL/min.

16. Dry eluates under a gentle stream of air/nitrogen at ~40°C.

17. Add 0.100 mL of ethyl acetate to the residue. Vortex well.

18. Transfer to appropriately labeled autosampler vials for GC-MS analysis.

3.3. Instrument Operating Conditions

The instrument's operating conditions are given in **Table 51.3**.

3.4. Data Analysis

1. The data was analyzed using ChemStation (Agilent Technologies). The quantitation of drugs was made from standard curves using selected ion monitoring and peak-area ratios of analyte vs. internal standard of quantifying ions (**Table 51.4**). GC-MS selected ion chromatogram is shown in the **Fig. 51.2** and mass spectra is shown in **Fig. 51.3**. The run is considered acceptable if calculated concentrations of drugs in the controls are within 20% of target values. Quantifying ion in the sample is considered acceptable if the ratios of qualifier ions are within 20% of the ion ratios for the calibrators. The linearity/limits of quantitation of the method are 25–500 ng/mL (*see* **Note 4**).

2. The retention time of the case analyte matches that of the calibrator retention time ±2%, based on the quantitation ion.

3. Typical coefficient of correlation of the standard curve is >0.99.

4. Typical intra- and inter-assay variations are <10%.

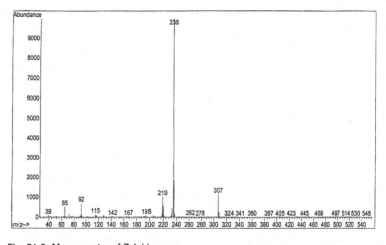

Fig. 51.3. Mass spectra of Zolpidem.

5. All ion chromatograms for the analyte, calibrator, and control exhibit acceptable chromatography.

6. The quantitative value for the positive control is within acceptable range.

7. The signal-to-noise is $> \sim 3:1$.

---

4. Notes

1. Ammonium hydroxide in the elution solvent is highly volatile. It is recommended the solution be prepared fresh daily.

2. Concentration verification by UV-visible spectrophotometry.

3. Check sample pH, pH should be 6.0 ± 0.5. Adjust pH according with the appropriate weak acid or base.

4. Samples in which the drug concentrations exceed the upper limit of quantitation can be diluted with appropriate negative urine or blood.

References

1. R.C. Baselt 5th Edition (2000). "Disposition of Toxic Drugs and Chemicals in Man."

2. C.J. Elko, J.L. Burgess, W.O. Robertson (1998). Zolpidem associated hallucinations and serotonin reuptake inhibition: a possible interaction. *Clin. Tox.* 36: 195–203.

3. Application Manual (1991). CLEAN SCREEN Extraction Column. Worldwide Monitoring Corp., Horsham, PA.

Chapter 52

Identification and Quantitation of Zopiclone in Biological Matrices Using Gas Chromatography-Mass Spectrometry (GC-MS)

Timothy P. Rohrig, Melissa C. Norton, and Lydia A. Harryman

Abstract

Zopiclone is a nonbenzodiazapine hypnotic used for the treatment of insomnia. Significant side effects include daytime drowsiness, dizziness, lightheadedness, bitter taste, dry mouth, headache, and upset stomach. A single method for confirmation and quantitation of zopiclone was developed for biological specimens and tissues. Zopiclone is extracted from the biological matrix using solid phase extraction technology. The drug is confirmed using gas chromatography mass spectrometry for toxicological and forensic purposes.

Key words: Zopiclone, mass spectrometry, gas chromatography, Solid Phase Extraction (SPE).

1. Introduction

Zopiclone is a nonbenzodiazapine hypnotic agent in the cyclopyrrolone class used for the treatment of insomnia. Zopiclone is rapidly absorbed with an estimated half-life of 6 h and reaches a peak concentration within 1 h from administration (1). It is extensively metabolized in the liver by oxidation, demethylation, and decarboxylation. Around half of the dose is converted to inactive metabolites by decarboxylation (2).

Zopiclone is extracted from the biological specimens using solid phase extraction, and is screened and confirmed by gas chromatography mass spectrometry.

U. Garg, C.A. Hammett-Stabler (eds.), *Clinical Applications of Mass Spectrometry*, Methods in Molecular Biology 603, DOI 10.1007/978-1-60761-459-3_52, © Humana Press, a part of Springer Science+Business Media, LLC 2010

2. Materials

2.1. Sample

Preferred samples include postmortem tissues, urine, and fluoridated blood from the cardiac and femoral area collected during autopsies, and antemortem urine, and fluoridated blood collected from outside agencies intended for human performance testing. Aliquots should be stored in glass tubes or plastic jars at a temperature range of 2–8°C until the time of analysis.

2.2. Reagents and Buffers

1. Monobasic sodium phosphate, glacial acetic acid, hexane, methanol, methylene chloride, Isopropanol, ammonium hydroxide, and ethyl acetate were purchased from Fisher Scientific (Fair Lawn, NJ) or Sigma-Aldrich (St. Louis, MO), and were of analytical grade.

2. Sodium phosphate buffer: 100 mM, pH 6.0. Dissolve 13.84 g of sodium phosphate (monobasic) in 500 mL of DI water, dilute to 1,000 mL. Buffer is stable at room temperature for 6 months.

3. 1 N Acetic acid. Stable at room temperature for 6 months.

4. Elution solvent: Methylene chloride:Isopropanol:ammonium hydroxide (78:20:2) is stored in an amber bottle at room temperature and is prepared fresh daily (*see* **Note 1**).

5. Bovine drug-free blood (Lampire Biological Laboratories, Pipersville, PA) and human drug-free urine was collected in-house (Wichita, KS).

2.3. Calibrators/ Calibration

1. Primary standard: 1 mg/mL zopiclone (Lipomed, INC., Cambridge, MA).

2. Secondary standard (100 μg/mL): Prepared by transferring 1.0 mL of each primary standard into a 10 mL volumetric flask and diluting with acetonitrile. Secondary standard is stable for 1 year at room temperature (*see* **Note 2**).

3. Tertiary standard (1 μg /mL): Prepared by transferring 100 μL of secondary standard to a 10 mL volumetric flask and diluting with acetonitrile. Tertiary standard is stable for 1 year at room temperature.

4. Working calibrators are made according to **Table 52.1**. The calibrators are prepared with each batch.

2.4. Internal Standard and Quality Controls

1. The primary internal standard: Cyproheptadine HCl (Sigma-Aldrich, St. Louis, MO).

2. Secondary internal standard (1 mg/mL): Prepared by transferring 112.7 mg of primary internal standard into a 100 mL volumetric flask and diluting with methanol. Secondary internal standard is stable for 1 year at room temperature.

Table 52.1
Preparation of calibrators

Calibrator	Negative blood or urine (mL)	Tertiary standard (mL)	Concentration (ng/mL)
1	3.96	0.040	10
2	3.9	0.100	25
3	3.8	0.200	50
4	3.5	0.500	125

3. Tertiary internal standard (10 µg/mL): Prepared by transferring 100 µL of secondary internal standard to a 10 mL volumetric flask and diluting with methanol. Tertiary internal standard is stable for 1 year at room temperature.

4. Working internal standard (1 µg/mL): Prepared by transferring 1.0 mL of tertiary internal standard to a 10 mL volumetric flask and diluting with methanol. Working internal standard is stable for one year at room temperature.

5. Quality control samples: An in-house control is used to validate sample analysis from a second set of prepared standards. The in-house control is made according to **Table 52.2.** The control is prepared for each batch. The target value for the quality control sample is established in-house.

Table 52.2
Preparation of in-house control

Control	Negative blood or urine (mL)	Tertiary standard (mL)	Concentration (ng/mL)
Positive	3.8	0.200	50

2.5. Supplies

1. UCT Clean Screen solid phase extraction (SPE) columns (CSDAU206 or equivalent) (United Chemical Technologies, INC., Bristol, PA).

2. Autosampler vials (12 × 32 mm; clear crimp), 0.300 mL conical with spring inserts, and 11 mm crimp seals with FEP/NatRubber (MicroLiter Analytical Supplies, INC., Suwanee, GA)

3. 16 × 100 mm glass culture tubes (Fisher Scientific, Fair Lawn, NJ)

4. GC column used: RTX-5MS 15 m × 0.25 mm × 0.25 µm (Restek International, Bellefonte, PA).

2.6. Equipment

1. Cerex System 48 Positive Pressure Manifold for solid phase extraction from SPEware Corporation (Baldwin Park, CA)

2. A gas chromatograph/mass spectrometer system (GC-MS; 6890/5973), Agilent Technologies (Wilmington, DE)

3. Pierce, 18835 Reacti-Therm III™ Heating Module with Reacti-Vap Evaporator, (Rockford, IL)

4. Fisher Vortex was purchased from Fisher Scientific (Fair Lawn, NJ)

3. Methods (3, 4, 5)

3.1. Preparation of Tissue Homogenates

Portions of tissue samples are weighed (approximately 20 g, if available) to generate one part tissue to three parts deionzed water, creating a 1:4 tissue homogenate.

3.2. Stepwise Procedure

1. Pipette 25 μL of working internal standard (1 μg/ml) to appropriately labeled 16 × 100 mm glass culture tubes. Batches should contain calibrators, negative control, samples, and positive control.

2. Add 4 mL of whole blood, urine, or 4 g of a 1:4 tissue homogenate.

3. Add 2 mL of DI water.

4. Add 2 mL of 100 mM sodium phosphate buffer, pH 6.0. Mix/vortex (*see* **Note 3**).

5. Centrifuge for ∼10 min at ∼2000 rpm (∼800 ×g).

6. Place appropriately labeled 16 × 100 mm glass culture tubes in the collection rack of the positive pressure manifold.

7. Install labeled SPE columns (CSDAU206 or equivalent) on the positive pressure manifold in their respective slots.

8. Add in order the following reagents and aspirate each, under low pressure (<3″ Hg) to prevent sorbent from drying: 3 mL of methanol, 3 mL of DI water, 3 mL of sodium phosphate buffer: 100 mM, pH 6.0.

9. Load samples at ∼1–2 mL/min (∼2–3″ Hg).

10. Wash the columns with the following reagent, in order, and aspirate each under low pressure: 3 mL of DI water, 1 mL of 1.0 N acetic acid.

11. Dry columns for approximately 5 min under full pressure.

12. Wash the columns with the following reagent, in order, and aspirate each under low pressure (<3″ Hg) to prevent sorbent from drying: 1 mL of hexane, 3 mL of methanol.

13. Dry columns for approximately 5 min under full pressure.

14. Add 3 mL of elution solvent (MeCl/IPA/NH$_4$OH).

15. Collect elute slowly, ~1–2 mL/min.

16. Dry eluates under a gentle stream of air/nitrogen at ~40°C.

17. Add 50 μL of ethyl acetate to the residue. Vortex well.

18. Transfer to appropriately labeled autosampler vials for GC-MS analysis.

3.3. Instrument Operating Conditions

The instrument's operating conditions are given in **Table 52.3**.

Table 52.3
GC operating conditions

Initial oven temp.	100°C
Initial time	1.0 min
Ramp 1	30°C/min
Temp. 2	300°C
Time	1.33 min
Injector temp.	250°C
Detector temp.	300°C
Purge time on	0.5 min.
Column pressure	11.7 psi

3.4. Data Analysis

1. The data was analyzed using ChemStation (Agilent Technologies). The quantitation of drugs was made from standard curves using selected ion monitoring and peak-area ratios of analyte vs. internal standard of quantifying ions. The run is considered acceptable if calculated concentrations of drugs in the controls are within 20% of target values. Quantifying ion in the sample is considered acceptable if the ratios of qualifier ions are within 20% of the ion ratios for the calibrators. *See* **Table 52.4** and **Fig. 52.2** for ions. GC-MS selected ion chromatogram is shown in the **Fig. 52.1**.

2. The linearity/limits of quantitation of the method is 10–125 ng/mL (*see* **Note 4**).

3. The retention time of the case analyte matches that of the calibrator retention time ±2%, based on the quantitation ion.

4. Typical coefficient of correlation of the standard curve is >0.99.

Table 52.4
Quantitation and qualifying ions for zopiclone and cyproheptadine

	Quantitation ion	Qualifier ions
Zopiclone	143	245, 217
Cyproheptadine	287	215

```
QUANTITATION REPORT FOR Zopiclone ON : Milhous
Data File              : C:\MSDCHEM\1\DATA\0211\11FEB08A.04P\ZOPC01.D
Tune File Name         : C:\MSDchem\1\5973N\atune.u
Tune Date              : 11 Feb 2008    4:00 pm              Mult : 1
Acq Method Name        : ZOPICL.M Calib date : 12 Feb 2008 7:36 am
Sample Name            : Cal
Acquisition date       : 11 Feb 2008    5:57 pm

Retention Time    7.97 Zopiclone         +/- 2.00% =  7.81 -   8.13 min
Retention Time    6.08 Cyproheptadine    +/- 2.00% =  5.95 -   6.20 min
R.R.T. =  1.312   Unknown target ion / ISTD target ion =    0.19

Zopiclone       => 143.0 =    252542  245.0 =    152642  217.0 =     67479
Cyproheptadine  => 287.0 =   1295852  215.0 =   1243726

Zopiclone       => 245.0/143.0 =     60.4 +/-   20.0% rel =   48.3 -    72.5
Zopiclone       => 217.0/143.0 =     26.7 +/-   20.0% rel =   21.4 -    32.0
Cyproheptadine  => 215.0/287.0 =     96.0 +/-   20.0% rel =   76.8 -   115.2

Concentration =    25.00 ** CONCENTRATION > 0.00 ng/mL LIMIT **<==
```

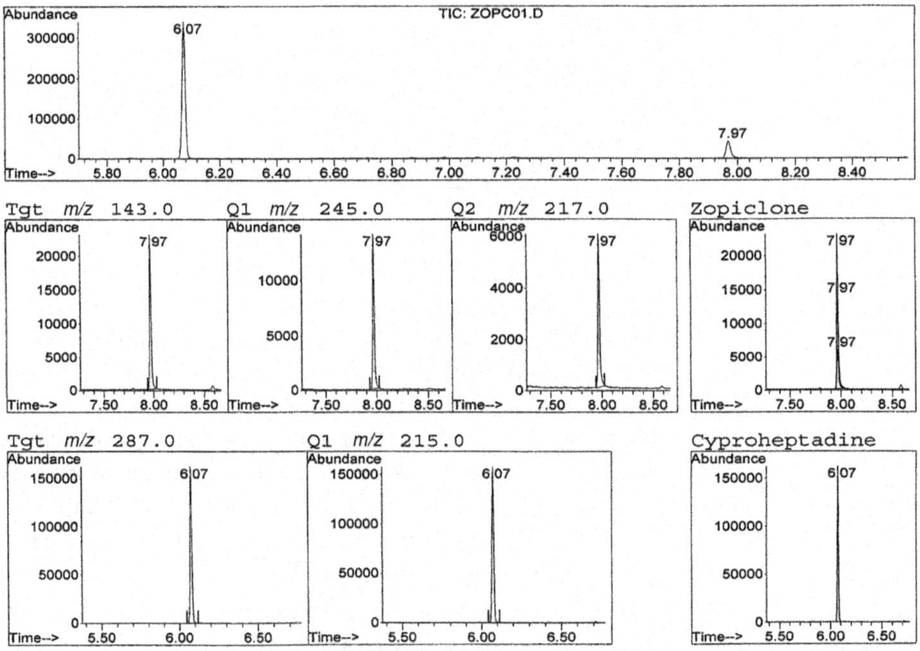

```
Zopiclone        : RT extraction window from   7.77 to   8.17 min
Cyproheptadine   : RT extraction window from   5.87 to   6.27 min

Page 25543                        Printed on : 12 Feb 2008 7:36 am
```

Fig. 52.1. GC-MS selected ion chromatograms of zopiclone.

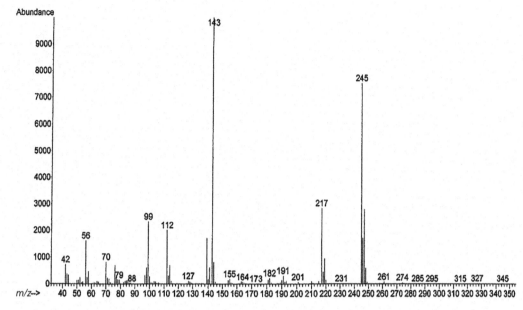

Fig. 52.2. Mass spectra of zopiclone.

5. Typical intra- and inter-assay variations are <10%.

6. All ion chromatograms for the analyte, calibrator, and control exhibit acceptable chromatography.

7. The quantitative value for the positive control is within acceptable range.

8. The signal-to-noise is > ~ 3:1.

4. Notes

1. Ammonium hydroxide in the elution solvent is highly volatile. It is recommended the solution be prepared fresh daily.

2. Concentration verification by UV-visible spectrophotometry.

3. Check sample pH, pH should be 6.0 ± 0.5. Adjust pH according with the appropriate weak acid or base.

4. Samples in which the drug concentrations exceed the upper limit of quantitation can be diluted with appropriate negative urine or blood.

References

1. Baselt, R.C. 5th Edition (2000). "Disposition of Toxic Drugs and Chemicals in Man."

2. *Clarke's Analysis of Drugs and Poisons* (2004), pp. 1717–1719; Pharmaceutical Press

3. Application Manual (1991). *CLEAN SCREEN* Extraction Column. Worldwide Monitoring Corp., Horsham, PA.

4. Boniface, P. and Russell S. (1996). Two Cases of Fatal Zopiclone Overdose. *J. Anal. Tox.* 20: 131–133.

5. Bocxlaer, J., Meyer, E., Clauwaert, K., Lambet, W., Piette, M., and De Leenheer, A. (1996). Analysis of Zopiclone (Imovane®) in Postmortem Specimens by GC-MS and HPLC with Diode-Array Detection. *J. Anal. Tox.* 20: 52–54.

SUBJECT INDEX

Note: The letters 'f' and 't' following the locators refer to figures and tables respectively

U. Garg, C.A. Hammett-Stabler (eds.), *Clinical Applications of Mass Spectrometry*, Methods in Molecular Biology 603, DOI 10.1007/978-1-60761-459-3, © Humana Press, a part of Springer Science+Business Media, LLC 2010

535

CPSIA information can be obtained
at www.ICGtesting.com
Printed in the USA
BVOW04*2345090617
486520BV00004B/1/P

9 781607 614586